ACS SYMPOSIUM SERIES **564**

Sulfur Compounds in Foods

Cynthia J. Mussinan, EDITOR
International Flavors & Fragrances

Mary E. Keelan, EDITOR
International Flavors & Fragrances

Developed from a symposium sponsored
by the Division of Agricultural and Food Chemistry
at the 206th National Meeting
of the American Chemical Society,
Chicago, Illinois
August 22-27, 1993

American Chemical Society, Washington, DC 1994

Library of Congress Cataloging-in-Publication Data

Sulfur compounds in foods / Cynthia J. Mussinan, editor, Mary E. Keelan, editor.

p. cm.—(ACS symposium series, ISSN 0097–6156; 564)

"Developed from a symposium sponsored by the Division of Agricultural and Food Chemistry at the 206th National Meeting of the American Chemical Society, Chicago, Illinois, August 22–27, 1993."

Includes bibliographical references and indexes.

ISBN 0–8412–2943–0

1. Food—Composition—Congresses. 2. Sulphur compounds—Congresses. 3. Flavor—Congresses.

I. Mussinan, Cynthia J., 1946- . II. Keelan, Mary E., 1965- . III. American Chemical Society. Division of Agricultural and Food Chemistry. IV. American Chemical Society. Meeting (206th: 1993: Chicago, Ill.) V. Series.

TX511.S85 1994
664—dc20   94–20826
                                                            CIP

The paper used in this publication meets the minimum requirements of American National Standard for Information Sciences—Permanence of Paper for Printed Library Materials, ANSI Z39.48–1984.

Copyright © 1994

American Chemical Society

All Rights Reserved. The appearance of the code at the bottom of the first page of each chapter in this volume indicates the copyright owner's consent that reprographic copies of the chapter may be made for personal or internal use or for the personal or internal use of specific clients. This consent is given on the condition, however, that the copier pay the stated per-copy fee through the Copyright Clearance Center, Inc., 27 Congress Street, Salem, MA 01970, for copying beyond that permitted by Sections 107 or 108 of the U.S. Copyright Law. This consent does not extend to copying or transmission by any means—graphic or electronic—for any other purpose, such as for general distribution, for advertising or promotional purposes, for creating a new collective work, for resale, or for information storage and retrieval systems. The copying fee for each chapter is indicated in the code at the bottom of the first page of the chapter.

The citation of trade names and/or names of manufacturers in this publication is not to be construed as an endorsement or as approval by ACS of the commercial products or services referenced herein; nor should the mere reference herein to any drawing, specification, chemical process, or other data be regarded as a license or as a conveyance of any right or permission to the holder, reader, or any other person or corporation, to manufacture, reproduce, use, or sell any patented invention or copyrighted work that may in any way be related thereto. Registered names, trademarks, etc., used in this publication, even without specific indication thereof, are not to be considered unprotected by law.

PRINTED IN THE UNITED STATES OF AMERICA

1994 Advisory Board

ACS Symposium Series

M. Joan Comstock, *Series Editor*

Robert J. Alaimo
Procter & Gamble Pharmaceuticals

Mark Arnold
University of Iowa

David Baker
University of Tennessee

Arindam Bose
Pfizer Central Research

Robert F. Brady, Jr.
Naval Research Laboratory

Margaret A. Cavanaugh
National Science Foundation

Arthur B. Ellis
University of Wisconsin at Madison

Dennis W. Hess
Lehigh University

Hiroshi Ito
IBM Almaden Research Center

Madeleine M. Joullie
University of Pennsylvania

Lawrence P. Klemann
Nabisco Foods Group

Gretchen S. Kohl
Dow-Corning Corporation

Bonnie Lawlor
Institute for Scientific Information

Douglas R. Lloyd
The University of Texas at Austin

Cynthia A. Maryanoff
R. W. Johnson Pharmaceutical
   Research Institute

Julius J. Menn
Western Cotton Research Laboratory,
   U.S. Department of Agriculture

Roger A. Minear
University of Illinois
   at Urbana–Champaign

Vincent Pecoraro
University of Michigan

Marshall Phillips
Delmont Laboratories

George W. Roberts
North Carolina State University

A. Truman Schwartz
Macalaster College

John R. Shapley
University of Illinois
   at Urbana–Champaign

L. Somasundaram
DuPont

Michael D. Taylor
Parke-Davis Pharmaceutical Research

Peter Willett
University of Sheffield (England)

Foreword

THE ACS SYMPOSIUM SERIES was first published in 1974 to provide a mechanism for publishing symposia quickly in book form. The purpose of this series is to publish comprehensive books developed from symposia, which are usually "snapshots in time" of the current research being done on a topic, plus some review material on the topic. For this reason, it is necessary that the papers be published as quickly as possible.

Before a symposium-based book is put under contract, the proposed table of contents is reviewed for appropriateness to the topic and for comprehensiveness of the collection. Some papers are excluded at this point, and others are added to round out the scope of the volume. In addition, a draft of each paper is peer-reviewed prior to final acceptance or rejection. This anonymous review process is supervised by the organizer(s) of the symposium, who become the editor(s) of the book. The authors then revise their papers according to the recommendations of both the reviewers and the editors, prepare camera-ready copy, and submit the final papers to the editors, who check that all necessary revisions have been made.

As a rule, only original research papers and original review papers are included in the volumes. Verbatim reproductions of previously published papers are not accepted.

M. Joan Comstock
Series Editor

Contents

Preface ix

1. Sulfur Compounds in Foods: An Overview 1
   Cynthia J. Mussinan and Mary E. Keelan

   ANALYSIS AND FLAVOR CHARACTERISTICS

2. Comparison of Gas Chromatographic Detectors for the Analysis
   of Volatile Sulfur Compounds in Foods 8
   B. S. Mistry, G. A. Reineccius, and B. L. Jasper

3. Chemiluminescence Detection of Sulfur Compounds
   in Cooked Milk 22
   Jeffrey S. Steely

4. Sulfur Volatiles in *Cucumis melo* cv. Makdimon (Muskmelon)
   Aroma: Sensory Evaluation by Gas Chromatography–
   Olfactometry 36
   S. Grant Wyllie, David N. Leach, Youming Wang,
   and Robert L. Shewfelt

5. Sulfur-Containing Flavor Compounds in Beef: Are They Really
   Present or Are They Artifacts? 49
   Arthur M. Spanier, Casey C. Grimm, and James A. Miller

6. Facts and Artifacts in *Allium* Chemistry 63
   Eric Block and Elizabeth M. Calvey

7. Effect of L-Cysteine and N-Acetyl-L-cysteine on Off-Flavor
   Formation in Stored Citrus Products 80
   M. Naim, U. Zehavi, I. Zuker, R. L. Rouseff, and S. Nagy

8. Modulation of Volatile Sulfur Compounds in Cruciferous
   Vegetables 90
   Hsi-Wen Chin and Robert C. Lindsay

FORMATION

9. **Thioglucosides of *Brassica* Oilseeds and Their Process-Induced Chemical Transformations** 106
   Fereidoon Shahidi

10. **Kinetics of the Formation of Methional, Dimethyl Disulfide, and 2-Acetylthiophene via the Maillard Reaction** 127
    F. Chan and G. A. Reineccius

11. **Kinetics of the Release of Hydrogen Sulfide from Cysteine and Glutathione During Thermal Treatment** 138
    Yan Zheng and Chi-Tang Ho

12. **Volatile Sulfur Compounds in Yeast Extracts** 147
    Jennifer M. Ames

13. **Generation of Furfuryl Mercaptan in Cysteine–Pentose Model Systems in Relation to Roasted Coffee** 160
    Thomas H. Parliment and Howard D. Stahl

14. **Heat-Induced Changes of Sulfhydryl Groups of Muscle Foods** 171
    Fereidoon Shahidi, Akhile C. Onodenalore, and Jozef Synowiecki

15. **Important Sulfur-Containing Aroma Volatiles in Meat** 180
    Donald S. Mottram and Marta S. Madruga

16. **Volatile Compounds Generated from Thermal Interactions of Inosine-5′-monophosphate and Alliin or Deoxyalliin** 188
    Tung-Hsi Yu, Chung-May Wu, and Chi-Tang Ho

17. **Thermal Degradation of Thiamin (Vitamin B_1): A Comprehensive Survey of the Latest Studies** 199
    Matthias Güntert, H.-J. Bertram, R. Emberger, R. Hopp, H. Sommer, and P. Werkhoff

18. **Formation of Sulfur-Containing Flavor Compounds from [^{13}C]-Labeled Sugars, Cysteine, and Methionine** 224
    R. Tressl, E. Kersten, C. Nittka, and D. Rewicki

FUNCTIONAL PROPERTIES

19. **Sulfur Compounds in Wood Garlic (*Scorodocarpus borneensis* Becc.) as Versatile Food Components** 238
    Kikue Kubota and Akio Kobayashi

20. **Sulfur-Containing Heterocyclic Compounds with Antioxidative Activity Formed in Maillard Reaction Model Systems** 247
    Jason P. Eiserich and Takayuki Shibamoto

21. **Mechanisms of Beneficial Effects of Sulfur Amino Acids** 258
    Mendel Friedman

22. **Inhibition of Chemically Induced Carcinogenesis by 2-*n*-Heptylfuran and 2-*n*-Butylthiophene from Roast Beef Aroma** 278
    Luke K. T. Lam, Jilun Zhang, Fuguo Zhang, and Boling Zhang

Author Index 292

Affiliation Index 292

Subject Index 293

Preface

THE CONTRIBUTION OF SULFUR COMPOUNDS to the flavor and off-flavor characteristics of foods has been studied for many years. Advances in analytical techniques and instrumentation and mechanistic studies have led to a better understanding of the formation and mode of action of these compounds. Recent interest, however, has focused on their importance in areas other than flavor. The role of dietary sulfur compounds as anticarcinogens, antimicrobials, and antioxidants has been and continues to be extensively studied.

Although each of the afore-mentioned areas could itself be the subject of a book, the purpose of this book is to provide an overview of the entire subject. Various chapters discuss the organoleptic characteristics, analysis, formation, and functional properties of this important class of compounds.

We would like to thank everyone who contributed to the success of the symposium upon which this book is based. We would also like to thank our chapter authors and reviewers. Finally, we gratefully acknowledge the financial support of the following companies: Dragoco, Elan, Givaudan-Roure, Hershey, International Flavors & Fragrances, Kraft, and McCormick.

CYNTHIA J. MUSSINAN
International Flavors & Fragrances
Research & Development
1515 Highway 36
Union Beach, NJ 07735

MARY E. KEELAN
International Flavors & Fragrances
800 Rose Lane
Union Beach, NJ 07735

March 23, 1994

Chapter 1

Sulfur Compounds in Foods
An Overview

[1]Cynthia J. Mussinan and [2]Mary E. Keelan

[1]International Flavors and Fragrances, Research and Development, 1515 Highway 36, Union Beach, NJ 07735
[2]International Flavors and Fragrances, 800 Rose Lane, Union Beach, NJ 07735

The importance of sulfur compounds to the flavor and off-flavor characteristics of foods is well known and continues to be studied. Mechanistic studies have led to a greater understanding of the formation and mode of action of these compounds. Recently, however, more attention has been given to some of their non-sensory characteristics, particularly to their antimicrobial and anticarcinogenic effects.

For years researchers have investigated the sulfur compounds present in various foods. Cooked foods typically contain numerous sulfur compounds, especially heterocyclic compounds like thiazoles, thiophenes, thiazolines, etc. In 1986, Shahidi et al. (1) reported that 144 sulfur compounds had been identified in beef. Other heated food systems like bread, potato products, nuts, popcorn, and coffee also contain many sulfur compounds. Aliphatic thiols have been found in fruits, vegetables, dairy products etc., as well as in heated foods. No discussion of the occurrence of sulfur compounds in foods would be complete without mention of their major role in the various allium species. Indeed, more than half of the volatile compounds reported in garlic, onion, leek, and chive contain sulfur (2). Comprehensive reviews of the literature concerning the role of thiazoles, thiophenes, and thiols in food flavor through 1975 can be found in Maga's series of review articles (3-5).

Sensory Properties

While the occurrence of sulfur compounds is of interest, their sensory properties can be critical to the flavor of a food. According to Boelens and van Gemert (2) "most of the volatile sulfur compounds are essential constituents of the material. That means they are necessary for the sensory quality, but that they are not characteristic. A small number of the sulfur compounds, however, are characteristic compounds..." These compounds, by themselves, can be recognized as having the same basic organoleptic character as the material. They are the so-called character-impact compounds. Examples are shown in Table I. Since some of these compounds have extremely low threshold values, they can be important to the flavor of a food at the ppb level. For example, one of the compounds reported in onion, propyl propane thiosulfonate, has a powerful and distinct odor of freshly cut onion (*14*). Since the threshold of this compound (1.5ppb) is several hundred times lower than its concentration (0.5ppm), it undoubtedly makes a significant contribution to the overall onion flavor (*15*).

Table I. Character-Impact Sulfur Compounds

Food	Compound	Reference
Bread	2-acetyl thiazoline	6
Cabbage	dimethyl sulfide	7
Coffee	furfuryl mercaptan	8
Grapefruit	1-p-menthene-8-thiol	9
Passionfruit	2-methyl-4-propyl-1,3-oxathiane	10
Potato	methional	11
Tomato	2-isobutyl thiazole	12
Truffle	bis(methylthio)methane	13

Although sulfur compounds are important contributors to the flavor of many foods, they have also been implicated in off-flavor development (Chin and Lindsay, Chapter 8, Naim et al., Chapter 7). Because they have such low thresholds, only small quantities can have a deleterious effect on the organoleptic character of a food. Even in meat where sulfur compounds are very important to the organoleptic character, high levels of volatile sulfides can form undesirable flavors (*16*). Although hydrogen sulfide may be important to the aroma of various fresh citrus juices, dimethyl sulfide has been implicated in off-flavors in canned orange and grapefruit juices. The latter compound probably comes from heat degradation during processing (*17*). Hydrogen sulfide, dimethyl sulfide, and dimethyl disulfide have been associated with the cooked, cabbage-like, sulfurous odor and flavor that occurs when milk is heated. This undesirable character develops during ultra-high-temperature (UHT) sterilization, but typically disappears after several days of storage (*18*).

Analysis

The minute quantities of sulfur compounds found in many foods makes their analysis and quantitation a challenging problem. Extraction without further fractionation, will, in many cases, not result in a high enough concentration of these trace sulfur constituents to permit their identification by gas chromatography/mass spectrometry (GC/MS). Element-selective detectors are useful tools for quickly locating sulfur compounds in complex mixtures and for monitoring their presence through subsequent isolation and fractionation steps (*19*, Mistry et al., Chapter 2). The most commonly used sulfur detector is the flame photometric detector (FPD) (*20*). More recently, a new chemiluminescence detector (SCD) was developed for GC (*21*, Steely, Chapter 3). The SCD is reported to be sensitive, selective and linear. In a 1988 applications note (*22*), dimethyl sulfide was determined in beer by headspace analysis without any preconcentration. Levels of this compound in beer are typically 1-1000ppb. Atomic emission detection (AED) has recently been commercially developed for gas chromatography. The AED can be tuned to selectively detect any element in any compound that can be eluted from a GC (*23*). Using these detectors as guides, the analyst can target those fractions or areas of a chromatogram containing sulfur for careful analysis by GC/MS or other hyphenated techniques like supercritical fluid chromatography/mass spectrometry (SFC/MS).

Artifacts

When attempting to analyze for sulfur constituents, the possibility of artifact formation must be carefully considered (Block and Calvey, Chapter 6; Spanier et al., Chapter 5). Many sulfur compounds are thermally unstable. Therefore, it's not surprising that the high temperatures typically encountered in the inlet of a gas chromatograph can lead to artifact formation. As Block (*24*) pointed out, many of the thiosulfinates and related sulfoxides found in garlic and onion are thermally unstable, in some cases even at room temperature. He suggests that "many of the so-called 'new compounds' being reported are simple artifacts resulting from decomposition of the fragile primary flavorants in the injection port, column, and heated transfer line connecting the GC to the MS." In order to guard against this misinterpretation, researchers need to test their analytical methods using standard compounds and conditions identical to those under which the analysis is run to verify their stability. In cases where instability is evident, alternative, low-temperature methods such as liquid chromatography or NMR need to be used.

Formation

The mechanism of formation of sulfur compounds has also been studied in depth, either directly, or using model systems (Tressl et al. Chapter 18). Nonvolatile precursors to volatile sulfur compounds in foods are the sulfur containing amino acids (cysteine/cystine, and methionine), reducing sugars, and thiamin (vitamin B1). The amino acids and reducing sugars react in the so-called Maillard reaction first described by L.C. Maillard in 1912 (*25*) and since extensively studied (*26*). The dicarbonyls which form during this reaction catalyze the Strecker degradation of cysteine to mercaptoacetaldehyde, acetaldehyde, hydrogen sulfide and other compounds (*27*). Similarly, the Strecker degradation of methionine produces methional which further hydrolyzes to methyl mercaptan and hydroxypropionaldehyde. The volatile sulfur compounds, hydrogen sulfide and methyl mercaptan, are highly odored and quite reactive. Both will readily react with carbonyl compounds and carbon-carbon double bonds to yield many potent flavor compounds (*28*). In addition to the Maillard reaction which involves a thermal degradation, methionine may degrade by enzymatic means to form dimethyl sulfide, another very powerful flavor compound (*28*). These small, reactive molecules may serve as precursors for other sulfur containing flavor compounds. For example, hydrogen sulfide may react with furanones leading to the formation of sulfur containing heterocycles (*29*). Methanethiol may also further react forming mono-, di- and trisulfides (*29*). The thermal degradation of thiamin (vitamin B1) provides another source of sulfur containing flavor compounds, many of them heterocyclic (*30*, Guntert et al., Chapter 17).

Functional Properties

While the sensory properties of sulfur compounds have been of interest for many years, recent attention has focused more heavily on some of the other functional properties of these compounds. Certain sulfur compounds have been found to have antioxidative, antimicrobial, and human medicinal properties.

Antioxidative Properties. When cooked meat is refrigerated, a rancid or stale flavor usually develops within 48 hrs. This character has been termed "warmed-over flavor"(WOF) and is generally attributed to the oxidation of lipids. Various synthetic and natural antioxidants have been used to reduce the development of WOF. Among the natural antioxidants used are the sulfur containing amino acid cysteine, and various Maillard reaction products. Eiserich and Shibamoto (Chapter 20) found that certain volatile sulfur heterocycles derived from Maillard reaction systems can function as antioxidants.

Natural phenols of white grapes undergo browning reactions during oxidation resulting in the formation of an undesirable brown color. Using caffeic acid as a model for fruit juice phenols, Cilliers and Singleton (*31*) studied the effect of thiols on browning. The presence of cysteine or glutathione was found to reduce the oxidation of caffeic acid, thus protecting it from browning.

Medicinal Properties. According to Lawson (*32*), garlic is one of the most researched medicinal plants with about 940 research papers published between 1960 and 1992. The biological studies have been primarily concerned with its cardiovascular, antimicrobial and anticancer effects as well as its hypoglycemic, heavy-metal poisoning antidote, and liver-protective effects (*32*). Most of the studies on garlic have focused on its organosulfur compounds because studies have shown that removal of the thiosulfinates eliminates many of these effects. Almost all human studies on the lipid-lowering effects of garlic and garlic products showed significant decreases in serum cholesterol and serum triglyceride. Reports that onion and garlic contain hypoglycemic (blood sugar reducing) agents go back about 50 years. In fact, the ancient literature reveals their use in treating diabetes (*33*).

Garlic, and to a lesser extent onion, have been found to possess antimicrobial properties. They are extensively used in the Oriental diet and have been used to treat various ailments, especially those associated with bacterial infections (*34*). The antibacterial effect of garlic has been attributed to allicin, S-2-propenyl 2-propenethiosulfinate. Garlic extract has been found effective against various strains of influenza virus in mice (*33*). Kubota and Kobayashi (Chapter 19) isolated two sulfur compounds from the fruit of *Scorodocarpus borneensis* Becc.("wood garlic") which were found to possess antibacterial and/or antifungal properties. Neither of these compounds had been previously found in the allium genus.

Various organic sulfides present in Allium have been found to have anticarcinogenic activity. For example, allyl sulfide, a constituent of garlic oil, inhibited colon cancer in mice exposed to 1,2-dimethylhydrazine, and allyl methyl trisulfide, allyl methyl disulfide, allyl trisulfide, and allyl sulfide all inhibited benzo[a]pyrene-induced neoplasma of the forestomach and lung in female mice (35). Lam et al. (Chapter 22) investigated the ability of 2-n-butyl thiophene, a constituent of roast beef aroma, to inhibit chemically induced carcinogenesis in three different tumor systems. This compound was found to be effective in the forestomach, lung, and colon models.

Conclusion

The chemistry of sulfur compounds in foods is very complex and continues to be extensively studied. Both cooked and uncooked foods contain organoleptically important sulfur-containing compounds. Conversely, the off-odors of numerous foods have been attributed to sulfur compounds. In addition to their sensory properties, recent work has been increasingly geared towards other functional properties of these compounds, especially antioxidant, antimicrobial, and anticarcinogenic effects. These areas will, no doubt, continue to be the subject of research for years to come.

Literature Cited

1. Shahidi, F.; Rubin, L.J.; D'Souza, L.A. *CRC Crit. Rev. Food Sci. Nutr.* **1986**, *24*, 141.
2. Boelens, M.H.; van Gemert, L.J. *Perf. & Flav.* **1993**, *18*, 29.
3. Maga. J.A.; *CRC Crit. Rev. Food Sci. Nutr.* **1975**, *6(2)*, 153.
4. Maga. J.A.; *CRC Crit. Rev. Food Sci. Nutr.* **1975**, *6(3)*, 241.
5. Maga. J.A.; *CRC Crit. Rev. Food Sci. Nutr.* **1976**, *7(2)*, 147.
6. Teranishi, R.; Buttery, R.G.; Guadagni, D.G. In *Geruch-und Geschmackstoffe*; Drawert, F., Ed.; Nurnberg: Verlag Hans Carl, 1975; 177-186.
7. Buttery, R.G. In *Flavor Science: Sensible Principles and Techniques*; Acree, T.E.; Teranishi, R., Eds.; ACS Professional Reference Book; American Chemical Society: Washington, DC, 1993; 259-286.
8. Tressl, R.; Silwar, R. *J.Agric. Food Chem.* **1981**, *29*, 1078.
9. Demole, E.; Enggist, P.; Ohloff, G. *Helv. Chim Acta* **1982**, *65*, 1785.
10. Winter, M.; Furrer, A.; Willhalm, B.; Thommen, W. *Helv. Chim Acta* **1976**, *59(5)*, 1613.
11. Buttery, R.G.; Seifert, R.M.; Guadagni, D.G.; Ling, L.C. *J. Agric. Food Chem.* **1971**, *19*, 969.
12. Buttery, R.G.; Teranishi, R.; Ling, L.C.; *J. Agric. Food Chem.* **1987**, *35*, 540-544.
13. Sloot, D.; Harkes, P. *J. Agric. Food Chem.* **1975**, *23*, 356.
14. Boelens, M.; de Valois, P.J.; Wobben, H.J.; van der Gen, A. *J. Agric. Food Chem.* **1971**, *19*, 984.

15. Mussinan, C.J. *Chemtech* **1980**, *10(10)*, 618.
16. Bailey M.E.; Rourke, T.J.; Gutheil, R.A.; Wang, C.Y-J. In *Off-Flavors in Foods and Beverages*; Charalambous, G., Ed.; Elsevier: New York, 1992; 127-169.
17. Shaw, P.E.; Wilson III, C.W.; *J. Agric. Food Chem.* **1982**, *30(4)*, 685.
18. Azzara, C.D.; Campbell, L.B. In *Off-Flavors in Foods and Beverages*; Charalambous, G., Ed.; Elsevier: New York, 1992; 329-373.
19. Mussinan, C.J. In *Flavor Science: Sensible Principles and Techniques*; Acree, T.E.; Teranishi, R., Eds.; ACS Professional Reference Book; American Chemical Society: Washington, DC, 1993; 169-224.
20. Brody, S.S.; Chaney, J.E. *J. Gas Chromatogr.* **1966**, *4*, 42.
21. Shearer, R.L.; O'Neal, D.L.; Rios, R.; Baker, M.D. *J. Chromatogr. Sci.* **1990**, *28(1)*, 24.
22. Jones, L.; Application Note No. 003; *Sievers Research, Inc.*: Boulder, CO, 1988.
23. Buffington, R. *GC-Atomic Emission Spectroscopy Using Microwave Plasmas*; Hewlett-Packard Company: Avondale, PA, 1988.
24. Block, E.J. *J. Agric. Food Chem.* **1993**, *41*, 692.
25. Maillard, L.C. *Compt. Rend. Acad. Sci.* Paris **1912**, *66*, 154.
26. Waller, G.R.; Feather, M.S., Eds.; *The Maillard Reaction in Foods and Nutrition*; ACS Symposium Series 215; American Chemical Society: Washington, DC, 1983.
27. de Roos, K.B. In *Flavor Precursors: Thermal and Enzymatic Conversions*; Teranishi, R.; Takeoka, G.R.; Guntert, M., Eds.; ACS Symposium Series 490; American Chemical Society: Washington, DC, 1992; 203-216.
28. Scarpellino, R.; Soukup, R.J. In *Flavor Science: Sensible Principles and Techniques*; Acree, T.E.; Teranishi, R., Eds.; ACS Professional Reference Book; American Chemical Society: Washington, DC, 1993; 309-335.
29. Schutte, L. In *Phenolic, Sulfur, and Nitrogen Compounds in Food Flavors*; Charalambous, G.; Katz, I., Eds: ACS Symposium Series 26; American Chemical Society: Washington, DC, 1976; 96-113.
30. Guntert, M.; Bruning, J.; Emberger, R.; Hopp, R.; Kopsel, M.; Surbury, H.; Werkhoff, P. In *Flavor Precursors: Thermal and Enzymatic Conversions*; Teranishi, R.; Takeoka, G.R.; Guntert, M., Eds.; ACS Symposium Series 490; American Chemical Society: Washington, DC, 1992; 140-163.
31. Cilliers, J.J.L.; Singleton, V.L. *J. Agric. Food Chem.* **1990**, *38*, 1789.
32. Lawson, L.D. In *Human Medicinal Agents from Plants*; Kinghorn, A.D.; Balandrin, M.F., Eds.; ACS Symposium Series 534; American Chemical Society: Washington, DC, 1993; 306-330.
33. Fenwick, G.R.; Hanley, A.B. *CRC Crit. Rev. Food Sci. Nutr.* **1985**, *23(1)*, 1.
34. Elnima, E.I.; Ahmed, S.A.; Mekkawi, A.G.; Mossa, J.S. *Pharmazie* **1983**, *38(11)*, 747.
35. Weinberg, D.S.; Manier, M.L.; Richardson, M.D.; Haibach, F.G.; Rogers, T.S. *J. High Res. Chromatogr.* **1992**, *15(10)*, 641.

RECEIVED April 5, 1994

ANALYSIS AND FLAVOR CHARACTERISTICS

Chapter 2

Comparison of Gas Chromatographic Detectors for the Analysis of Volatile Sulfur Compounds in Foods

B. S. Mistry[1], G. A. Reineccius[1], and B. L. Jasper[2]

[1]Department of Food Science and Nutrition, University of Minnesota, 1334 Eckles Avenue, St. Paul, MN 55108
[2]Coca-Cola Foods, 2651 Orange Avenue, Plymouth, FL 32768

> Sulfur compounds play a major role in determining the flavor and odor characteristics of many food substances. Often sulfur compounds are present in trace levels in foods making their isolation and quantification very difficult for chromatographers. This study compares three gas chromatographic detectors; the flame photometric detector, sulfur chemiluminescence detector and the atomic emission detector, for the analysis of volatile sulfur compounds in foods. The atomic emission detector showed the most linearity in its response to sulfur; the upper limit of the linear dynamic range for the atomic emission detector was 6 to 8 times greater than the other two detectors. The atomic emission detector had the greatest sensitivity to the sulfur compounds with minimum detectable levels as low as 1 pg.

Sulfur compounds are found in a variety of food substances. Organic sulfur compounds, especially those with low boiling points, have been found to be major contributors to the odor and flavor of many foods (*1*). Although the concentration of these sulfur compounds in foods is low, their contribution to the overall flavor profile may be important due to their low flavor thresholds (*2-4*). Selective and accurate detection of trace levels of sulfur compounds in complex food matrices is a challenging task for chromatographers. The analysis of trace levels of sulfur compounds is complicated due to two factors: the sorption and loss of sulfur compounds in the chromatographic system and problems with isolation and accurate detection of these compounds in complex food systems (*5*). Although there are several methods for the analysis of volatile sulfur compounds, gas chromatography is widely used since it has sulfur-specific detectors that can greatly increase sensitivity and allow for quantification of sulfur compounds (*6*).

An evaluation of gas chromatographic detectors typically involves a study of dynamic range, mimimum detectable level and selectivity. Dynamic range is one of the major response characteristics of detectors. It is the range of sample concentrations for which the detector can provide accurate quantitation. The dynamic range varies for different detectors. The minimum detectable level is defined as the amount of sample where the sample peak height is two or three times the noise height (ie, the signal-to-noise ratio, S/N, is 2 or 3). Selectivity indicates which categories of compounds will give a detector response (7). This study compares the response characteristics for the three gas chromatographic detectors most commonly used in the analysis of sulfur compounds; flame photometric detector (FPD), sulfur chemiluminescence detector (SCD) and atomic emission detector (AED).

Materials and Methods

A stock solution (0.1%) of seven sulfur compounds (Aldrich Chemical Co., Milwaukee, WI) in Table I was prepared in dichloromethane. Dilutions for the standard curve were then prepared in dichloromethane.

Table I. Formulas and Molecular Weights of Sulfur Compounds

Compound	Formula	Molecular Weight
Butyl sulfide	$[CH_3(CH_2)_3]_2S$	146.3
Butyl disulfide	$[CH_3(CH_2)_3]_2S_2$	178.4
Butyl mercaptan	$[CH_3(CH_2)_3]SH$	90.2
1,2-ethane dithiol	$HSCH_2CH_2SH$	94.2
Thiophenol	C_6H_5SH	110.2
Phenyl sulfone	$(C_6H_5)_2SO_2$	218.3
Ethyl-p-toluene sulfonate	$CH_3C_6H_4SO_3C_2H_5$	200.2

For the determination of FPD response characteristics, a 5880A Hewlett Packard (Kenneth Square, PA) gas chromatograph equipped with a FPD was used. A Sievers SCD Model 350 B (Sievers Research, Inc., Boulder, CO) was used for SCD response characteristics. A Hewlett Packard 5890A gas chromatograph configured to a 5921A AED was used for response characteristics of the AED.

Identical gas chromatographic conditions and column were used in the evaluation of all three detectors. A DB-5 (J&W Scientific, Folsom, CA) column, 30 m length x 0.32 mm i.d.,1 mm film thickness, was used. The column head pressure was 15 psi helium. Splitless injections with a 2 min. valve delay were performed; injection volume, 2 ml. Gas chromatographic operating parameters were as follows: injection port temperature, 200°C; detector temperature, 225°C. The oven was temperature programmed from an initial temperature of 35°C for 1 min. to 210°C at a program rate of 7°C/min. The oven was maintained at 210°C for 15 min.

The following parameters were used for the atomic emission detector: spectrometer purge flow, nitrogen @ 2l/min. There was solvent back flush. Transfer line temperature was 250°C; cavity temperature was 250°C; water temperature was 65°C. Element wavelengths for sulfur, carbon and nitrogen were 181.4, 193.0 and 174.3 nm, respectively.

Analyses were conducted in duplicate.

Flame Photometric Detector (FPD)

The principle of operation of a flame photometric detector (FPD) is that sulfur- or phosphorus-containing compounds produce chemiluminescent species when burned in a flame ionization detector-type flame. The chemiluminescent species emit light at characteristic wavelengths. Light of the desired wavelength (393 nm for sulfur compounds) enters the photomultiplier through an optical filter and a signal is produced (Figure 1). Fused silica capillary columns can be run directly in to the base of the FPD flame. This minimizes sample loss or tailing due to adsorption. The detection zone (chemiluminescent zone) is above the flame. Since the flame is shielded by a burner cap, the photomultiplier tube can only detect the area above the flame. Hence, the flame conditions are critical to successful operation. The chemical reactions that occur in the flame and detection zone depend on the gas flow and temperature which must, therefore, be optimized to allow components burned in the flame to emit in the detection zone. The ratios of hydrogen/oxygen and the total gas flow are very critical to optimized selectivity and sensitivity of the FPD detector. FPD temperature is very important; sulfur response decreases with increased detector temperature (7). Sulfur is detected as an S=S molecule; the molecule is created in a metastable state and upon decay, it releases energy in the form of a photon of specific wavelength.

The FPD suffers from several drawbacks (7,8). The light emitted is not linear with concentration of sulfur; it is approximately proportional to the square of the sulfur atom concentration. Also, undesired light absorption by way of hydrogen quenching or self-quenching can occur in the flame. Hydrogen quenching can occur from collisions when there is a high concentration of carbon dioxide in the flame due to a hydrocarbon peak eluting at the same time as the sulfur compound. Self-quenching can occur at high concentrations of the heteroatom species whereby the photon is prevented from reaching the photomultiplier tube. Condensation of water, especially with halogenated solvents or samples, will cause corrosion in the detection zone or fogging of the window to the photomultiplier tube (7,9).

Dynamic Range of the FPD. Calibration curves of log peak area vs log square of sulfur atom concentration injected were plotted for each sulfur compound. A calibration curve for butyl mercaptan is shown in Figure 2. The other six sulfur compounds showed similar calibration curves. Figure 2 shows that the dynamic range is linear up to a value of 3.7 on the x axis; this value corresponds to 74 ng sulfur injected or 208 ng or 104 ppm of butyl mercaptan injected. All 7 sulfur compounds showed a linear dynamic range up to ca. 200 ng of the compounds injected.

Minimum Detectable Level of the FPD. The minimum detectable levels of the sulfur compounds on the FPD were calculated for a signal/noise ratio (peak heights) of 2 and are listed in Table II.

Selectivity of the FPD. The minimum detectable levels of four non-sulfur compounds, methyl anthranilate, pyridine, heptane and caproic acid ethyl ester, on the FPD at a signal/noise ratio of 2 were 368, 784.5, 759.8 and 760 ng, respectively.

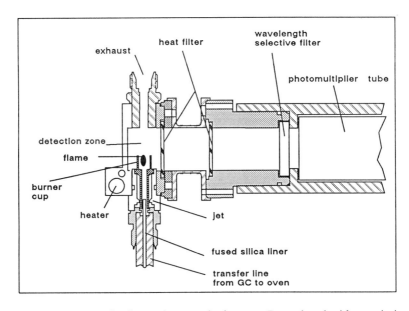

Figure 1. Diagram of a flame photometric detector (Reproduced with permission from ref. 7. Copyright 1987 Hewlett-Packard.).

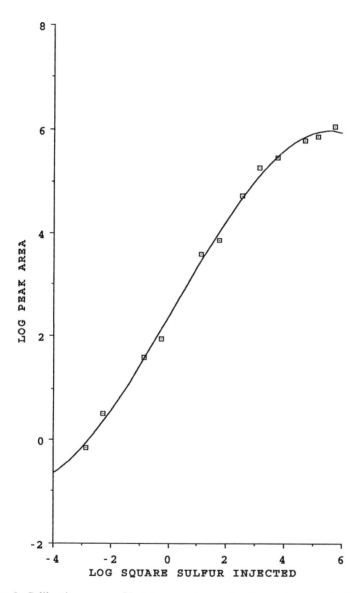

Figure 2. Calibration curve of butyl mercaptan on the flame photometric detector.

These data indicate that non-sulfur compounds containing nitrogen, hydrogen, carbon or oxygen at high enough concentrations will give a potentially misleading response on the FPD.

Table II. Minimum Detectable Levels of Sulfur Compounds (S/N = 2)

Compound Injected	Minimum Detectable Level (pg)		
	FPD	SCD	AED
Butyl sulfide	198.56	20	1.99
Butyl disulfide	99.12	10	0.99
Butyl mercaptan	208.32	40	10.42
1,2-ethane dithiol	100.92	200	10.09
Thiophenol	205.04	2000	102.52
Phenyl sulfone	200.32	100	1.00
Ethyl-p-toluene sulfonate	210.88	200	1.05

Sulfur Chemiluminescence Detector (SCD)

The operation of the sulfur chemiluminescence detector (SCD) is based on the combustion of sulfur-containing compounds in a hydrogen-rich/air flame of a flame ionization detector to form sulfur monoxide (SO) (equation 1). Sulfur monoxide is then detected based on an ozone-induced, highly exothermic chemiluminescent reaction to form electronically excited sulfur dioxide (SO_2^*) (equation 2). The excited sulfur dioxide, upon collapse to the ground state, emits light with a maximum intensity around 350 nm that is detected in a manner similar to that of the FPD (equation 3) (2,5).

Sulfur compounds + H_2/air = SO + other products   (1)
$SO + O_3 = SO_2^* + O_2$ (2)
$SO_2^* = SO_2 + h\upsilon$ (3)

A schematic of the Sievers Model 350 B SCD used in this study is shown in Figure 3 (10). The SCD 350 B produces ozone by corona discharge using compressed air or oxygen. Sulfur compounds are combusted in a flame ionization detector; the combustion gases are collected by a ceramic probe located ca. 0.4 cm above the flame jet. The probe then transfers the gases to the chemiluminescence reaction cell through black PFA tubing in less than 1 sec. The SO and ozone are mixed in the reaction cell in front of the photomultiplier tube. A UV band pass filter (300-450 nm) located between the reaction window and the photomultiplier tube selectively transmits the radiation generated from the SO and O_3 reaction. A pressure transducer measures the pressure in the reaction cell. Under typical operating conditions, the reaction cell operates at 8-10 torr which is accomplished using a vacuum pump. The vacuum pump also samples and transfers combustion gases to the reaction cell, transfers ozone from the ozone generator to the reaction cell, reduces non-radiative

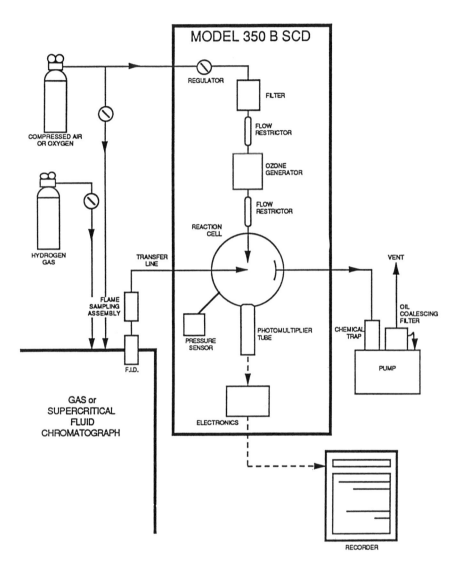

Figure 3. Schematic of the Sievers 350 B sulfur chemiluminescence detector (Reproduced with permission from ref. 10. Copyright 1990 Sievers.).

quenching by collision of SO_2^* in the cell and prevents condensation or transfer of water produced in the flame. The vacuum pump is protected by a Hopcalite™ trap which removes unreacted ozone, oxides of nitrogen and other corrosive gases. An oil coalescing filter is placed on the vacuum pump exhaust to trap vaporized oil and return it to the pump oil reservoir (*10*).

Dynamic Range of the SCD. Calibration curves for the sulfur compounds were plotted as peak areas vs ng compound injected. The upper limit of the dynamic linear range for all sulfur compounds with the exception of ethyl-p-toluene sulfonate, was ca. 200 ng of the compound injected. The upper limit for ethyl-p-toluene sulfonate was ca. 1000 ng.

Minimum Detectable Level of the SCD. The minimum detectable levels (S/N = 2) of the sulfur compounds are shown in Table II.

Selectivity of the SCD. The selectivity of the SCD with respect to non-sulfur compounds is shown in Table III. All the compounds in Table III were injected neat (2 ml) into the gas chromatograph except methyl anthranilate which was injected at a 1000 ppm level in dichloromethane.

Table III. Selectivity of the SCD for Non-Sulfur Compounds

Compound	*SCD Response*
Methylene chloride	none
Cyclopentane	slight
Pyridine	small unsymmetrical peak
Methyl anthranilate	none
Hexane	slight
Pentane	none

Atomic Emission Detector (AED)

A schematic diagram of the GC-AED is shown in Figure 4 (*7*). The carrier gas moves the vaporized sample from the injection port through the column to the detector. The detector consists of a helium discharge chamber which is powered by a microwave generator. The light source may be a microwave induced plasma (MIP) which is a helium or other inert gas discharge in a fused silica tube. The microwave generator emits energy (magnetron) to power the plasma. The sample molecules are cleaved into atoms, which in turn are raised to an excited state and emit light. This light is dispersed by the diffraction grating of a spectrometer into different wavelengths and measured by a light sensor (photomultiplier tube or photodiode array). The photodiode array can monitor a portion of spectrum at a time; elements which emit

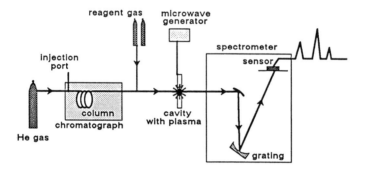

Figure 4. Schematic of the GC-AED (Reproduced with permission from ref. 7. Copyright 1987 Hewlett-Packard.).

light within a 25 to 40 nm range can generally be run simultaneously. For example, it is possible to obtain simultaneous chromatograms for N, S and C or C, H, Cl and Br.

The AED has a few requirements for smooth and efficient running of the system. Helium has become the popular choice for plasma gas for MIPs since it can provide more energy for sample excitation. Ultrahigh purity helium gas is required since spectra of impurities can interfere with the sample. The discharge tube should be cooled to prolong its life and prevent interaction of samples with the tube walls. The solvent component of a sample should be vented; if not, it may either extinguish the plasma or form deposits on the discharge tube causing noise or peak tailing. Solvent venting may be done by reversing the flow of helium in the discharge tube and venting the solvent through the column/discharge tube interface. A reagent or scavenger gas needs to be added to the eluent to maintain peak shape and discourage carbon deposition. Further, the system should be kept leak-free since leaks cause errors, especially in the measurement of oxygen or nitrogen (7, 11).

Dynamic Range of the AED. The calibration curves for all sulfur compounds on the AED were obtained as plots of peak area vs ng of compound injected. The calibration curves for all sulfur compounds (except ethyl-p-toluene sulfonate and phenyl sulfone) were linear to an upper limit of 1200 to 1550 ng of compound injected. Calibration curves for the latter two compounds were linear even at 2000 ng of compound injected (the maximum limit used in this study). Figure 5 shows the dynamic ranges for both butyl mercaptan and phenyl sulfone on the AED.

Minimum Detectable Level of the AED. The minimum detectable levels (S/N = 2) of the sulfur compounds on the AED are shown in Table II.

Selectivity of the AED. Selectivity for an element on the AED can be optimized by suppressing the response of undesired elements on the desired channel by a process called background correction. Also, the elemental identity can be confirmed using the "snapshot" option. A snapshot is a selected segment of the emission spectrum showing specific elemental emission wavelengths (12). A snapshot of a sulfur peak is shown in Figure 6. The presence of sulfur is confirmed by the sulfur lines at wavelengths 180.7, 182.0 and 182.6 nm (triplet). Non-sulfur compounds such as methyl anthranilate, pyridine, heptane and caproic acid ethyl ester showed peaks on the sulfur channel at 3168, 785, 1520 and 189.6 ng injected, respectively. Figure 7 shows the presence of a pyridine peak (785 ng injected) on the nitrogen, carbon and sulfur channels. These non-sulfur peaks were not suppressed with background correction at these concentrations. At lower concentrations, background correction suppressed the non-sulfur peaks which were then absent on the sulfur channel. To confirm elemental identity, snapshots of the peaks of the four non-sulfur compounds were taken. There were no sulfur triplet lines at wavelengths 180.7, 182 and 182.6 for the non-sulfur compounds.

Comparison of the FPD, SCD and AED

The AED exhibited best detector response in terms of dynamic range, sensitivity and

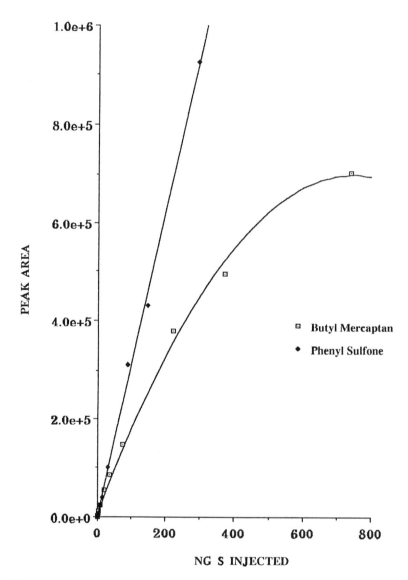

Figure 5. Dynamic ranges of butyl mercaptan and phenyl sulfone on the atomic emission detector.

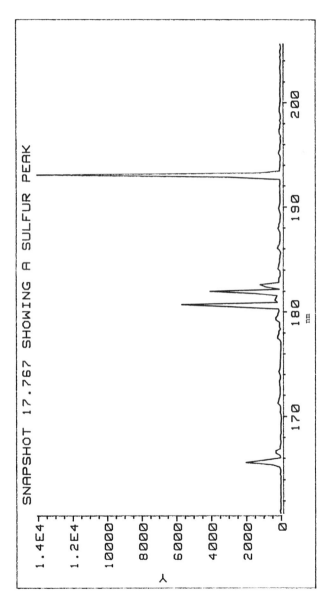

Figure 6. Snapshot of a sulfur-containing peak.

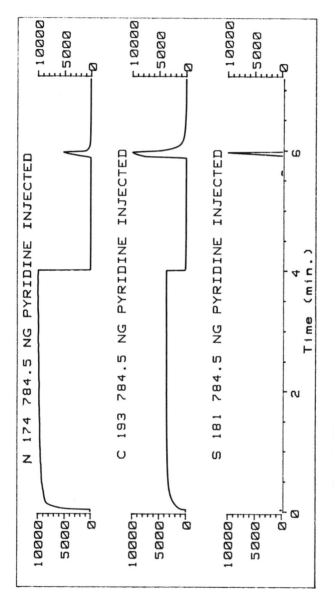

Figure 7. Chromatogram of pyridine on the C, N, S channels on the AED.

selectivity; the FPD and SCD demonstrated comparable detector response to each other for the sulfur compounds used in this study.

The upper limit of the linear dynamic range for the AED was 6 to 8 times greater than for the FPD or the SCD (i.e. 1200-1550 ng of the compounds injected for the AED vs 200 ng for the FPD and SCD). Two sulfur compounds, ethyl-p-toluene sulfonate and phenyl sulfone, exhibited linearity even at concentrations of 2000 ng for the AED. Thus, the AED was more linear in response to sulfur than the FPD or SCD. Minimum detectable levels (MDL) of the sulfur compounds on the AED were as low as 1 pg; thiophenol had the highest MDL at 103 pg. On the FPD, the MDL ranged from 100 to 200 pg; on the SCD, from 10 to 200 pg with MDL for thiophenol at 2000 pg. The SCD did not show any detector response to the non-sulfur compounds; these compounds showed peaks on the FPD and AED at sufficiently high levels of injection. However, the AED has a clear advantage over the FPD in that when the selectivity of the AED is in question, "snapshots" of the element can be used to prove the absence or presence of the element. Also, the AED is capable of accumulating more than one element-specific chromatogram; these may be placed in a single data file for analysis. For example, with one injection of a given compound/s, the AED can accumulate data on carbon, nitrogen and sulfur at the same time; this reduces analysis time and the need for element-specific detectors.

Literature Cited

1. *Aroma Research;* Maarse, H. and Goenens, P., Ed.; Proc. Internat. Symp. on Aroma Research; Zeit, The Netherlands, 1975.
2. Burmeister, M.; Drummond, C.; Pfisterer, E.; Hysert, D. *Amer. Soc. Brewing Chemists J.* **1992**, *50*, pp 53.
3. Takken, H.; Linde, L.; De Valois, P.; Dort, H.; Boelens, M. In *Phenolic, Sulfur and Nitrogen Compounds in Foods*; Charalambous, G. and Katz, I., Ed.; ACS Symp. Series 26; ACS: Washington, D.C., 1976.
4. Shaw, P; Ammons, J.; Bramman, R. *J. Agric. Food Chem.* **1980**, *28*, pp. 778.
5. Hutte, R.; Johansen, N.; Legier, M. *J. High Resol. Chromat.* **1990**, *13*, pp. 421.
6. Supelco Technical Bulletin # 876, Supelco, Inc., 1991.
7. Buffington, R.; Wilson, M. *Detectors for Gas Chromatography*; Hewlett-Packard, 1987.
8. Farwell, S.; Brinaga, C. *J. Chromat. Sci.* **1986**, *24*, pp. 483.
9. Benner, R.; Stedman, D. *Anal. Chem.* **1989**, *61*, pp. 1266.
10. Sievers. *SCD 350B Operation and Service Manual;* Sievers Research, Inc.: Boulder, CO, 1990.
11. Buffington, R. *GC-Atomic Emission Spectroscopy using Microwave Plasmas*, Hewlett-Packard, 1988.
12. David, F.; Sandra, P. *HP Applic. Note.* 1991, *228-136*.

RECEIVED April 19, 1994

Chapter 3

Chemiluminescence Detection of Sulfur Compounds in Cooked Milk

Jeffrey S. Steely

Hershey Foods Corporation, 1025 Reese Avenue, Hershey, PA 17033

> Many sulfur flavor volatiles have low threshold values making them a challenge to detect and identify in food matrices. The sensitivity, selectivity, and lack of quenching has established the Sulfur Chemiluminescence Detector (SCD) as a valuable gas chromatographic detector for studying sulfur flavor chemistry. The bland, pleasant flavor of fluid milk can change dramatically as it is processed, especially with the application of heat. High temperature processing can lead to the generation of a sulfurous off-flavor. It is widely held that heat denaturation of sulfur bearing whey and fat-globule membrane proteins liberate "free sulfhydryl" groups which are responsible for this cooked flavor. Several sulfides and thiols have been found to increase in concentration as a result of ultra-high temperature (UHT) processing. Incorporation of oxygen into milk post-process has been found to diminish this flavor defect. This research employs gas chromatography coupled with the SCD to characterize volatile sulfur compounds found in UHT processed milk.

Milk is a complex and versatile food. Consumed in its native state, it provides sustenance for mammalian species in the suckling stage of development by supplying essential nutrition. It can also serve as an important constituent of a well balanced diet. In addition to this simplest form of consumption, milk is used as a key ingredient in many types of processed foods. It not only is transformed into such basic foodstuffs as butter, cheese, yogurt, and ice cream, but utilized in a number of products for its unique structural and flavor attributes.

The flavor of milk is an important factor affecting its consumption. Ideally, fresh milk should possess a clean, bland, pleasantly sweet taste (1). Off-flavors can arise from the way in which it is handled, processed, and stored. These off-flavors have been categorized based on the conditions responsible for their creation (Table I) (2).

Table I. Categories of Off-Flavors in Milk

Causes	Descriptors
1. HEATED	Cooked, caramelized, scorched
2. LIGHT-INDUCED	Light, sunlight, activated
3. LIPOLYZED	Rancid, butyric, bitter, goaty
4. MICROBIAL	Acid, bitter, fruity, malty, putrid, unclean
5. OXIDIZED	Papery, cardboard, metallic, oily, fishy
6. TRANSMITTED	Feed, weed, cowy, barny
7. MISCELLANEOUS	Absorbed, astringent, bitter, chalky, chemical, flat, foreign, lacks freshness, salty

SOURCE: Reproduced with permission from reference 2. Copyright 1978 American Dairy Science Association.

Raw fluid milk is routinely processed by exposure to a thermal regimen in order to render it safe for human consumption. The extent to which the milk is exposed to heat generates a range of recognized off-flavors. Duration of the treatment varies according to the intended use of the processed product. These treatments range in intensity from simple pasteurization (13 seconds at 75 °C) to retort sterilization (30 minutes at 110 °C) *(3)*.

Heat-induced flavor generation has been described as cooked or sulfurous, heated or rich, caramelized, and scorched *(2)*. The kind of off-flavor developed depends on the severity of the heat treatment. Milder treatments such as pasteurization impart a slightly cooked flavor. Ultra-high temperature (UHT) processing yields a strong sulfurous note with an accompanying rich flavor impact. Sterilized milk typically exhibits a caramelized flavor quality. Scorched flavor can be induced by prolonged exposure to high temperature contact surfaces such as in heat exchangers. This is sometimes seen in dried milk powders.

The basic constituents of milk - protein, lipid, carbohydrate - can serve as precursor substrates for the formation of a wide variety of flavor compounds. Nearly two hundred volatiles have reportedly been found in fresh and processed milk *(4)*. Numerous research efforts have focused on the conditions and mechanisms of off-flavor development in milk. The chemical compounds responsible for these off-flavors have been characterized *(3,5-7)*. Most fluid milk processing is carefully controlled so that the appearance of caramelized and scorched flavor notes rarely occurs. The rich flavor associated with the thermal formation of diacetyl and various lactones is not objectionable to most consumers and, therefore, not a serious concern. Conversely, the sulfurous off-flavor in cooked milk is of concern and is especially prevalent in freshly processed UHT milk.

As previously mentioned, cooked milk flavor is described as being sulfurous or cabbage-like. The compounds responsible for this defect have been traditionally referred to as "free sulfhydryl groups" and specific spectrophotometric tests have been developed for measuring these compounds *(8,9)*. They are thought to arise from the thermal denaturation of whey proteins, mainly ß-lactoglobulin, as well as proteins associated with the fat-globule membrane. Sulfur containing amino acids liberated from the protein react to form an assortment of sulfides, thiols, and other minor sulfur bearing compounds. Another proposed mechanism is the formation of sulfur volatiles from heat labile precursors introduced via feed from plants *(10)*.

It is generally agreed in the literature that the compounds responsible for the cooked flavor defect are hydrogen sulfide, methanethiol, dimethyl sulfide, and dimethyl disulfide *(3,7,11-13)*. Of these four compounds, however, not all investigators agree on the single most important contributor. In addition, several other sulfur compounds have reportedly been found in heated milk. These are: iso-butanethiol, dimethyl trisulfide, methyl isothiocyanate, ethyl isothiocyanate, isothiocyanate, 2,4-dithiapentane, 2,3,4-trithiapentane, and benzothiazole *(13)*. Due to the significant impact sulfur compounds have on flavor, it is to be expected that all of the aforementioned perform in concert to express the sulfurous defect.

UHT processing of milk is an especially attractive method for extending the shelf-life of dairy products. When applied to a product that is subsequently aseptically packaged, the possibilities for innovative product applications become numerous. UHT processing does give rise to some unique considerations and difficulties. Incomplete deactivation of lipase and losses of important nutrients and vitamins have been observed post-process in UHT milk *(14-16)*. By far, however, the greatest amount of attention has been focused on the off-flavor issue, ie., how much off-flavor is there, how quickly it diminishes, and how can it be minimized *(11,17-20)*.

The cabbage-like flavor associated with freshly processed UHT milk has been observed to dissipate over time. The mechanism for this flavor fade is thought to be an oxidation of the sulfur moieties formed during thermal processing. Shibamoto et al. *(5)* postulated that the dimethyl sulfoxide and dimethyl sulfone in raw milk could be metabolic products of dimethyl sulfide oxidation within the cow. Ferretti *(11)* was able to inhibit cooked flavor formation with the addition of organic thiolsulfonates and thiosulfates prior to processing. Fink and Kessler *(18,19)* concluded that oxidation of free SH-groups in UHT milk was responsible for the complete disappearance of cooked flavor. Recently, Andersson and Öste *(20)* attempted, with mixed results, to accelerate the disappearance of cooked flavor by adding oxygen to totally degassed milk and the headspace of the storage container.

Sulfur Compound Detection

Selective GC detectors aid in the detection and identification of compounds containing specific elements: halogens with electron capture detector (ECD) or electrolytic conductivity detector (ELCD); nitrogen and phosphorus with nitrogen-phosphorus detector (NPD); sulfur and phosphorus with flame photometric detector (FPD); and sulfur with sulfur chemiluminescence detector (SCD). The development of the atomic emission detector (AED) has enabled investigators to utilize a single detector for any element of interest.

A flavor isolate can contain hundreds of compounds and achieving baseline separation of all components in one GC run is not always possible. With a selective detector, compounds bearing the element of interest can be located and quantified even in the presence of co-eluting compounds. Ideally, the detector should only respond to the analyte containing the species of interest. In practice, however, this is not always the case. Compounds bearing the same specific element of interest can produce uneven detector response depending upon molecular characteristics such as size, additional functionalities, oxidative state, unsaturation, or other elemental interference.

Selective detection of sulfur compounds is especially important in foods and beverages. The unique chemistry of the sulfur heteroatom and its powerful interaction with the olfactory system presents a particularly challenging task for the analytical chemist. Of the thousands of volatile compounds detected and identified in foods and beverages, about 10% of these are sulfur compounds (21). With the recent improvements in detection methods and the large number of realistically conceivable sulfur compounds not yet found in foods, it is probable that several hundred new sulfur flavor volatiles will be detected and identified.

Flame Photometric Detector. To date, the most utilized detector for sulfur bearing species has been the FPD. Based on a technique divulged in a West German Patent (22), Brody and Chaney (23) extended the concept to gas chromatography and introduced this specific detector for phosphorus or sulfur containing compounds. In the FPD, column effluent is incinerated in a hydrogen/oxygen flame producing an interrelated product mix of sulfur species from the original sulfur compound. The fuel mix is hydrogen rich and leads to the formation of H_2S, HS, S, S_2, SO and SO_2 when sulfur compounds are added (24). The FPD is configured to photometrically detect the chemiluminescence of the excited S_2^* species, with the emission of photons exhibiting a maximum at 394 nm (23). When the sulfur compound contains carbon, as in most GC flavor effluents, the complexity of the flame chemistry increases due to CS, CS_2, and OCS species.

Difficulties in its use and non-uniformity of results has led to frustration in the application of the FPD to sulfur compound detection (24). Most often cited problems with the FPD (23-25) are its non-linear response, reduced response with co-eluting hydrocarbons (quenching), and inconsistent selectivity and sensitivity.

Other schemes to selectively detect sulfur compounds from GC effluents have been devised (26-29). Generally, these detection techniques suffer from one or all of a number of disadvantages including varying sensitivities, operational complexity, large size, or prohibitive cost. Detection based on chemiluminescence continues to receive a great amount of attention due to its inherent selectivity and sensitivity.

Sulfur Chemiluminescence Detector. Another sulfur selective detector based on photometric sensing of a chemiluminescent reaction has become commercially available (Sievers Research, Inc., Boulder, CO). Similar to the FPD, the detection scheme of the SCD incorporates combustion of the GC column effluent in a hydrogen rich flame. However, the sulfur species detected is an excited state of sulfur dioxide (SO_2^*), not

S_2^*. The SO_2^* is produced by the reaction of SO with ozone (O_3) in a separate reaction cell.

This reaction mechanism was first proposed by Halsted and Thrush *(30)* when studying the kinetics of elementary reactions involving the oxides of sulfur. Visible and UV spectroscopic studies *(31)* confirmed that the chemiluminescent emission was from SO_2^*. Recently, it has also been confirmed that the sulfur-analyte molecule from the GC effluent is converted to SO in the flame of the SCD *(32)*. Even though SO is a free radical, it can be sufficiently stabilized in a flow system under reduced pressure *(33,34)* to be sampled and transferred to a vessel to react with introduced O_3. Based on these operational principles, Benner and Stedman *(35)* concluded that SO produced in a flame could be easily detected. They modified a redox chemiluminescence detector *(36)* to produce what was termed a Universal Sulfur Detector (USD). A linear response between 0.4 ppb and 1.5 ppm (roughly equal to 3 to 13,000 pg of S/sec) was demonstrated with equal response to the five sulfur compounds tested. This detection scheme has been utilized as the basis for the commercially available GC detector.

Evaluation of the commercially available GC detector, now known as the sulfur chemiluminescence detector, was detailed by Shearer et al. *(34)*. Emphasis was focused on chemical and petrochemical applications, but the fundamental characteristics are relevant to other fields as well. The detector can be interfaced with any GC having a flame ionization detector. A ceramic sampling probe is positioned above the tip of the FID flame. Approximately 90-95% of the flame combustion products *(37)* are drawn into the ceramic probe via a transfer line, under reduced pressure, to the reaction chamber. The formation of SO in the flame is optimized by adjusting the air and hydrogen flows to the FID. Oxygen or compressed air is passed through an ozone generator situated adjacent to the reaction cell. The SO from the flame combustion of sulfur compounds reacts with O_3 under reduced pressure to produce SO_2^*, which then emits photons as it returns to its ground state. The light is detected by a photomultiplier tube (PMT) with a UV band pass filter (225 - 450 nm) located between the reaction cell and the PMT. The combination of the flame and UV band pass filter virtually eliminates interferences from non-sulfur containing analytes. The reduced pressure is provided by a rotary vane pump which also serves to sweep the cell of product and unreacted effluent. A Hopcalite trap is used for removal of unreacted ozone and oxides of nitrogen prior to the trap. The signal from the PMT is processed by the electronics of the SCD for output to a recording device. The FID signal and the SCD response can be recorded simultaneously due to the mounting configuration of the SCD probe. However, the flow rate of hydrogen relative to air in the flame for optimum SO production results in a reduction in FID sensitivity by about two orders of magnitude *(34)*.

The reported advantages of the SCD *(38)* over other existing sulfur selective detection systems include linear dynamic range between three and five orders of magnitude, high selectivity for sulfur (10^5 for compounds containing heteroatoms; 10^6 for hydrocarbons *(34)*), absence of quenching due to co-eluting compounds, lower detection limits, and a nearly equimolar response for all sulfur compounds.

Not all GC capillary columns are appropriate for use with the SCD. Optimal dimensions and stationary phases have been investigated *(37)*. A range of column lengths (15 - 100 meters), interior diameters (0.25 - 0.53 mm i.d.), and film thicknesses (0.5 μm - 5.0 μm df.) were explored. Less stable stationary phases tend to foul the ceramic probe and build up inside the reaction cell of the SCD. Fairly thick film coatings of methyl silicone tend to work out best for retaining and separating sulfur compounds with minimal bleed problems.

The SCD has been optimized for use mainly as a GC detector. Since only about one quarter of all organic compounds can be analyzed by GC, the SCD has also been interfaced with supercritical fluid chromatography (SFC) *(39-42)* and high performance liquid chromatography (HPLC) *(43)*. Performance characteristics of the SCD interfaced with the SFC are similar to the GC/SCD coupling except for decreased sensitivity in some cases. This is a manifestation of the gas flow balance required for proper flame chemistry in the FID. While technically feasible, utilizing the SCD as an HPLC detector is limited to microbore columns and many special considerations since a liquid eluent must be efficiently vaporized and combusted.

The weakest aspect of the SCD is the probe interface to the FID. Proper probe positioning is critical to adequately sample the combustion effluent for optimal performance. Probe positioning is accomplished mostly by trial and error; ie., the probe is manually adjusted until maximum response is achieved. Failure to position the probe at the same location prior to each analysis causes variations in sensitivity. Flame gas flow optimization can also be tedious, especially if the probe is in a slightly changed position from the previous analyses. A recent modification *(44,45)* to the GC effluent interface eliminates these drawbacks by replacing the ceramic sampling probe with a separate burner assembly containing a high temperature furnace. The GC column exit is plumbed directly into the assembly and flows of both hydrogen and oxygen are introduced into the furnace to produce a flameless combustion of analytes. Several advantages are realized with this modification. The total effluent from the GC column is sampled rather than the reduced amount sampled by the ceramic probe. The closed burner assembly produces lower operating pressure in the reaction cell which increases the intensity of the chemiluminescence. Concerns about probe fouling are eliminated and GC column phase selection is greatly expanded. Finally, the ratio of hydrogen to oxygen in the burner can be increased substantially, which in turn increases the amount of SO produced during combustion.

The aim of this research is to utilize the SCD to detect the sulfur compounds found in UHT processed milk. An integrated analytical approach to sampling, detection, and identification is used which includes purge and trap volatile concentration followed by thermal desorption, GC separation, SCD detection, and mass spectral identification. The sulfur compounds identified should be chiefly responsible for the cooked flavor in UHT milk. In addition, the mechanism for the disappearance of this sulfurous flavor will be explored.

Analyses

Instrumentation and Methods. A Sievers Sulfur Chemiluminescence Detector Model 350 was utilized for the selective detection of sulfur compounds. The electronics were set to integrate photon counts for 0.12 seconds. Supply air to the ozone generator was set at 8 psi., and the reaction vessel pressure was measured at 12-14 torr.

Both a Varian 3400 GC/FID and a Hewlett-Packard 5890A GC/FID were employed during these studies. Analog outputs from the SCD and FID were recorded simultaneously on a dual channel flatbed recorder (Kipp & Zonen BD110). FID and SCD chromatograms were also acquired into a PE Nelson data system during the HP GC runs. Sulfur signals from both GC's were similar, but the Varian FID trace was extremely noisy when operated with the SCD probe inserted. The GC columns used for the studies were 30m x 0.32mm id. fused silica capillaries with 4μm film stationary phase of methyl polysiloxane (Supelco SPB-1). The ovens were temperature programmed as follows: initial temperature, 35 °C; initial time, 1 min; 35 °C to 250 °C at 4°C/min; final time, 30 min. Helium carrier gas at 1 ml/min. FID detectors were operated at 250 °C. Optimal flame gas flows for the SCD on the Varian 3400 were: 392 ml/min air, 175 ml/min H_2, 20 ml/min N_2 makeup gas; and with the HP 5890A they were: 346 ml/min air, 174 ml/min H_2, 30 ml/min N_2.

Volatile compounds were sparged from UHT milk samples in a Dynatherm Dynamic Thermal Stripper (DTS) 1000 and trapped onto multi-bed adsorbent traps. Sorbent traps were packed with a layered progression of Tenax TA/Ambersorb XE-340/Charcoal to effectively trap all purged volatiles. The oven, block, and tube heaters were set at 50 °C. Times and flow rates were: 5 min oven pre-heat, 45 min. purge, 20 minute dry; and 50 ml/min N_2 purge, 30 ml/min N_2 dry flow.

Trapped headspace volatiles were thermally desorbed with a Dynatherm Thermal Desorption Unit (TDU) ACEM 900. The narrow bore focusing trap was packed with a two-part bed consisting of Tenax TA and Carboxen 1000. Times and temperatures for desorption were: 10 min. dry, 3 min. sorbent tube heat at 235 °C, 3 min. focusing trap heat at 235 °C, valve box and transfer line temperatures at 235 °C.

For identifications, the same conditions for the thermal desorption, column, and oven program were used with a Hewlett-Packard 5890 Series II GC interfaced with a HP 5971A MSD.

All experiments were conducted with pasteurized 2% fat milk acquired from a local dairy. The milk was homogenized in two stages at 2500/500 psi. UHT processing was accomplished with a plate heat exchanger unit at 144.4 °C for 4 seconds. Flow rate was 2000 ml/min. Samples were collected in sterile stainless steel vessels. Aliquots were aseptically transferred to 250 ml amber bottles and sealed with teflon lined lids. For volatile content analysis, 10 ml of UHT milk was transferred to a 100 ml sparging vessel and treated as described previously.

Samples held at ambient temperature were analyzed at time 0, 1 day, 3 day, and 5 day. Three month old samples stored at 4 °C were also analyzed. Additionally, fresh samples which were sparged with a 10% O_2/90% N_2 gas blend were analyzed directly after sparging and at 1 and 2 days.

Results

The selective detection of volatile sulfur compounds from freshly processed UHT milk is shown in Figure 1. The top chromatogram is the FID response while the bottom trace is from the SCD. The signals were recorded simultaneously by the PE Nelson data system and are displayed on the same time axis. The y-axis scales are adjusted

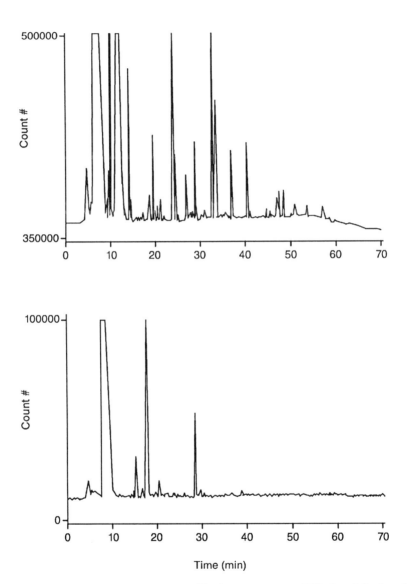

Figure 1. Freshly processed UHT milk chromatograms; FID signal (top) and SCD signal (bottom).

for display purposes. The FID baseline signal is normally quite high with the flame gases adjusted to favor optimal conditions for sulfur detection. The most abundant peaks seen in the SCD trace are barely visible in the FID chromatogram. The largest peak appearing around 10 minutes on the SCD chromatogram is actually buried on the trailing shoulder of the largest peak in the FID trace. Positive identifications of the sulfur compounds seen in the SCD chromatogram are listed in Table II.

Table II. Sulfur Peak Identifications

Peak Number	Compound
1	Hydrogen Sulfide
2	Sulfur Dioxide
3	Methanethiol
4	Dimethyl Sulfide
5	Carbon Disulfide
6	Butanethiol
7	Dimethyl Disulfide
8	Dimethyl Sulfoxide
9	Dimethyl Sulfone
10	Dimethyl Trisulfide

The identifications are referenced to the expanded view of the SCD chromatogram seen in Figure 2. It should be noted that while ten identifications have been made, a number of unidentified compounds remain.

During the initial phases of these investigations, the temperatures used to sparge volatile compounds from the milk were seen to have a significant effect on the sulfur compounds. Temperatures in excess of 70 °C caused some peaks to decrease while others increased. Figure 3 is a comparison between UHT milk stripped at lower (50 °C) and higher temperatures (90 °C). The effect of this treatment is demonstrated here by the disappearance of the dimethyl sulfide peak. An increase in dimethyl sulfone is seen at the same time.

Three approaches were used to explore cooked flavor fade. The first was to introduce oxygen into freshly processed UHT milk. The other two were merely to store the product at two different temperatures to determine what changes, if any, occur over time. All three showed no dramatic quantitative changes in sulfide or thiol concentration. Small decreases, however, were observed with noticeable increases of dimethyl sulfone.

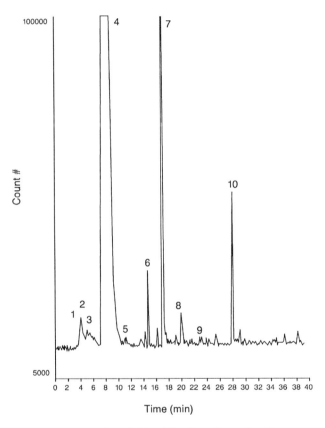

Figure 2. Sulfur compound peak identifications from freshly processed UHT milk (SCD trace).

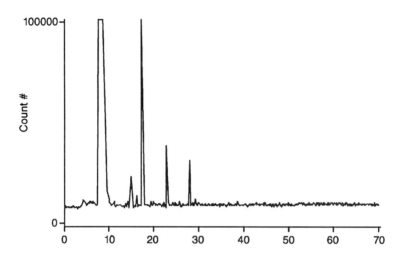
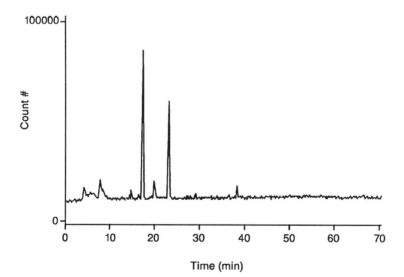

Figure 3. Effect of stripping temperature on sulfur compounds in UHT milk; 50 °C (top) vs. 90 °C (bottom).

Discussion

The SCD has proven to be a useful tool in the detection of volatile sulfur compounds. The list of compounds identified in these studies agrees closely qualitatively with what has been reported in the literature. Further work is required to identify the smaller peaks seen in the SCD chromatograms. Some of them are presumably already on the list of reported compounds. While quantitative numbers have not been generated, it appears very clear that the major contributor to the cooked flavor in UHT milk is dimethyl sulfide. Other investigators alternately credit the "major contributor" title to either hydrogen sulfide or methanethiol. The differences in conclusions drawn could very well be attributed to the method of analysis. Conclusions based on a photometric method for quantifying the "free sulfhydryl" content of UHT milk are general at best. The reactivity of sulfur compounds, especially along its oxidative pathway demand careful consideration of analytical conditions.

The temperature to which thiols and sulfides are exposed is more critical than originally thought. Thiols are converted to disulfides without much oxidative pressure. It appears that if enough energy is pumped into the system, sulfides can be oxidized to the sulfoxide and eventually the sulfone. From the results of sparging UHT milk at 90 °C, it is clear that this level of thermal input is adequate to cause at least partial conversion of the dimethyl sulfide to form dimethyl sulfone. The presence of dimethyl sulfoxide, which is the intermediate product between the sulfide and sulfone, establishes the link in this reaction chain in UHT milk. The effect of stripping temperature on dimethyl sulfide obliges one to examine all aspects of sample treatment during analysis. The high temperatures used in the thermal desorption of volatiles from the packed sorbent traps remains a concern. While standards run under the same desorption conditions show no evidence of thermally induced degradation, additional experiments need to be conducted to confirm these findings with actual samples.

The cooked flavor of UHT milk is observed to fade with time. The storage samples analyzed for this study indicate only a slight change in sulfur volatile content. The changes observed, however, do suggest a similar mechanism as seen with thermally induced oxidation. The organoleptic character of the oxidative series methanethiol-dimethyl sulfide-dimethyl sulfoxide-dimethyl sulfone demonstrates the feasibility of this as the active mechanism for flavor fade. Methanethiol is the most powerful of the group with a decrease in olfactory activity until arriving at the flavorless dimethyl sulfone. While this scheme is overly simplistic it can serve as a template for further investigations into the complete mechanism of UHT cooked flavor fade.

Conclusions

Sulfur compound investigation can be significantly aided by the use of the SCD. It has been proven to be a sensitive, selective GC detector for sulfur volatiles in food matrixes. In order to achieve accurate results, sample handling and analytical conditions need to be controlled carefully to minimize any reactions involving sulfur flavor volatiles. Thiols and sulfides have been shown to be susceptible to oxidation when exposed to moderate amounts of heat. Chemical evidence for the proposed mechanism of cooked flavor fade has been presented.

Acknowledgments

The author extends thanks to Dr. Jorge O. Bouzas for supplying the UHT milk. Thanks also to Dr. Jennifer L. Weist-Schwartz for analytical instrumental assistance.

Literature Cited

1. Shipe, W.F. In *The Analysis and Control of Less Desirable Flavors in Foods and Beverages;* Charalambous, G., Ed.; Academic Press Inc.: Orlando, FL, 1980; p 201.
2. Shipe, W.F.; Bassette, R.; Deane, D.D.; Dunkley, W.L.; Hammond, E.G.; Harper, W.J.; Kleyn, D.H.; Morgan, M.E.; Nelson, J.H.; Scanlan, R.A. *J. Dairy Sci.* **1978**, *61*, 855.
3. Badings, H.T.; van der Pol, J.J.G.; Neeter, R. In *Flavour '81*; Charalambous, G., Ed.; Walter de Gruyter & Co: Berlin, New York, 1981; p 683.
4. Maarse, H.; Visscher, C.A., Eds. *Volatile Compounds in Food: Qualitative and Quantitative Data.* Sixth edition, Volume I. TNO-CIVO Food Analysis Institute: The Netherlands, 1989; p 387.
5. Shibamoto, T.; Mihara, S.; Nishimura, O.; Kamiya, Y.; Aitoku, A.; Hayashi, J. In *The Analysis and Control of Less Desirable Flavors in Foods and Beverages;* Charalambous, G., Ed.; Academic Press Inc.: Orlando, FL, 1980; p 241.
6. Azzara, C.D.; Campbell, L.B.; In *Off-Flavors in Foods and Beverages;* Charalambous, G., Ed.; Elsevier Science Publishers B.V., 1992; p 329.
7. Jaddou, H.A.; Pavey, J.A.; Manning, D.J. *J. Dairy Res.* **1978**, *45*, 391.
8. Koka, M.; Mikolajcik, E.M.; Gould, I.A. *J. Dairy Sci.* **1968**, *51*, 217.
9. Patrick, P.S; Swaisgood, H.E. *J. Dairy Sci.* **1976**, *59*, 594.
10. Keenan, T.W.; Lindsay, R.C. *J. Dairy Sci.* **1968**, *51*, 112.
11. Ferretti, A. *J. Agric. Food Chem.* **1973**, *21*, 939.
12. Christensen, K.R.; Reineccius, G.A. *J. Dairy Sci.* **1992**, *75*, 2098.
13. Badings, H.T.; Neeter, R. *Neth. Inst. Dairy Res.* **1980**, *34*, 9.
14. Andersson, I.; Öste, R. *Milchwissenschaft.* **1992**, *47*, 223.
15. Andersson, I.; Öste, R. *Milchwissenschaft.* **1992**, *47*, 299.
16. Choi, I.W.; Jeon, I.J. *J. Dairy Sci.* **1993**, *76*, 78.
17. Zadow, J.G.; Birtwistle, R. *J. Dairy Res.* **1973**, *40*, 169.
18. Fink, R.; Kessler, H.G. *Milchwissenschaft.* **1986**, *41*, 90.
19. Fink, R.; Kessler, H.G. *Milchwissenschaft.* **1986**, *41*, 152.
20. Andersson, I.; Öste, R. *Milchwissenschaft.* **1992**, *47*, 438.
21. Boelens, M.H.; van Gemert, L.J. *Perf. Flav.* **1993**, *18*, 29.
22. Draegerwerk, H.; Draeger, W. West German Patent 1,133,918, 1962.
23. Brody, S.S.; Chaney, J.E. *J. Gas Chrom.* **1966**, *4*, 42.
24. Farwell, S.O.; Barinaga, C.J. *J. Chrom. Sci.* **1986**, *24*, 483.
25. Olesik, S.V.; Pekay, L.A.; Paliwoda, E.A. *Anal. Chem.* **1989**, *61*, 58.

26. Spurlin, S.R.; Yeung, E.S. *Anal. Chem.* **1982**, *54*, 318.
27. Nelson, J.K.; Getty, R.H.; Birks, J.W. *Anal. Chem.* **1983**, *55*, 1767.
28. Johnson, J.E.; Lovelock, J.E. *Anal. Chem.* **1988**, *60*, 812.
29. Chasteen, T.G.; Silver, G.M.; Birks, J.W.; Fall, R. *Chromatographia.* **1990**, *30*, 181.
30. Halstead, C.J.; Thrush, B.A. *Proc. R. Soc. London.* **1966**, *295*, 363.
31. Akimoto, H.; Finlayson, B.J.; Pitts Jr., J.N. *Chem. Phys. Lett.* **1971**, *12*, 199.
32. Martin, H.R.; Glinski, R.J. *Appl. Spectrosc.* **1992**, *46*, 948.
33. Halstead, C.J.; Thrush, B.A. *Proc. R. Soc. London.* **1966**, *295*, 380.
34. Shearer, R.L.; O'Neal, D.L.; Rios, R.; Baker, M.D. *J. Chrom. Sci.* **1990**, *28*, 24.
35. Benner, R.L.; Stedman, D.H. *Anal. Chem.* **1989**, *61*, 1268.
36. Nyardy, S.A.; Barkley, R.M.; Sievers, R.E. *Anal. Chem.* **1985**, *57*, 2074.
37. Hutte, R.S.; Johansen, N.G.; Legier, M.F. *J. High Res. Chrom.* **1990**, *13*, 421.
38. Gaines, K.K.; Chatham, W.H.; Farwell, S.O. *J. High Res. Chrom.* **1990**, *13*, 489.
39. Chang, H.-C.K.; Taylor, L.T. *J. Chrom.* **1990**, *517*, 491.
40. Pekay, L.A.; Olesik, S.V. *J. Microcol. Sep.* **1990**, *2*, 270.
41. Sye, W.F.; Zhao, Z.X.; Lee, M.L. *Chromatographia,* **1992**, *33*, 507.
42. Howard, A.L.; Taylor, L.T. *Anal. Chem.* **1993**, *65*, 724.
43. Chang, H.-C.K.; Taylor, L.T. *Anal. Chem.* **1991**, *63*, 486.
44. Shearer, R.L. *Anal. Chem.* **1992**, *64*, 2192.
45. Shearer, R.L.; Poole, E.B.; Nowalk, J.B. *J. Chrom. Sci.* **1993**, *31*, 82.

RECEIVED March 23, 1994

Chapter 4

Sulfur Volatiles in *Cucumis melo* cv. Makdimon (Muskmelon) Aroma

Sensory Evaluation by Gas Chromatography−Olfactometry

S. Grant Wyllie[1], David N. Leach[1], Youming Wang[1], and Robert L. Shewfelt[2]

[1]Centre for Biostructural and Biomolecular Research, School of Science, University of Western Sydney, Hawkesbury, Richmond, New South Wales, 2753 Australia
[2]Department of Food Science and Technology, University of Georgia, Georgia Station, Griffin, GA 30323

> Simultaneous distillation-extraction (SDE) of muskmelon (*Cucumis melo* cv. Makdimon) and subsequent gas chromatographic analysis using an atomic emission detector identified twenty sulfur volatiles, including S-methyl esters of ethanethioic, butanethioic and pentanethioic acids, and dimethyltrisulfide. From aroma extraction dilution analysis (AEDA) seven key compounds were identified, four of which contained sulfur; S-methyl thiobutanoate, 3-(methlythio)-propyl acetate, 3-(methylthio)propanal and possibly dimethyltetrasulfide. Of these seven key sensory responses only two are described as fruity/floral (ethyl 2-methylpropanoate and 3-(methylthio)propyl acetate), the other five components being responsible for the earthy, stale musk aroma of this melon. However, comparison of odor units suggests a higher weighting than the AEDA results for the common fruity ester notes from methyl and ethyl 2-methylbutanoate and 2-methylbutyl acetate. Preliminary tests indicate that blending the flesh of this melon prior to SDE activates lipoxygenase and hydrolase activity and alters the aroma profile appreciably.

The aroma profiles of a range of melon varieties and cultivars have been the subject of a number of investigations as summarized by Maarse (*1*). The results show that the extracts obtained using a variety of techniques were qualitatively and quantitatively dominated by a number of typical fruit esters. Some samples also contained significant amounts of unsaturated C_6 and C_9 aldehydes and alcohols exhibiting green and melon-like aromas, presumably derived by lipoxygenase activity (*2*). The sulfur compounds dimethyldisulfide and ethyl 2-(methylthio)acetate detected in this early work on melons would also contribute to the overall sensory profile provided they are present above their odor threshold. A series of sulfur volatiles, methyl (methylthio)acetate, ethyl (methylthio)acetate, 2-(methylthio)ethyl acetate, 3-(methylthio)propyl acetate, methyl 3-(methylthio)propanoate, ethyl 3-(methylthio)propanoate, 3-(methylthio)propanitrile, 3-(methylthio)propan-1-ol and 3-(methylthio)ethanol have now been identified in melons (*3,4,5,6*). The methylthioether esters have been identified in melons subjected to various extraction techniques including cold solvent extraction with Freon (*5,6*) and simultaneous distillation extraction at

normal (*3*) and reduced pressure (*7,8*). However, the alcohols and the nitrile compound have only been detected when cold solvent extraction was used.
The presence of sulfur volatiles has now been reported in a wide range of fruits and has been found to correlate well with quality in strawberries (*9*) and black currant buds (*10*). Detailed studies of the volatiles in pineapple (*11,12,13*) also reported the presence of several thioether esters including methyl (methylthio)acetate, ethyl (methylthio)acetate, 3-(methylthio)propyl acetate, methyl 3-(methylthio)propanoate, ethyl 3-(methylthio)propanoate and the alcohol, 3-(methylthio)propan-1-ol. The sulfur esters 3-(methylthio)propyl acetate and ethyl 3-(methylthio)propanoate were also detected in Asian pears (*14*), and fifteen sulfur volatiles, including 3-(methylthio)hexyl acetate, have recently been reported in yellow passionfruit (*15*).

Sensory aspects of muskmelon aroma using aroma extract dilution analysis have been determined by Schieberle *et al.* (*16*). They concluded that the most significant contributors to the aroma of the melon examined were methyl 2-methylbutanoate, (Z)-3-hexenal, (E)-2-hexenal, and ethyl 2-methylpropanoate. (E,Z)-2,6-nonadienal and (E)-2-nonenal were also contributers. No sulfur compounds were identified by this group although two compounds, described as potato-like and fatty, were not identified. It is our experience over several years that extracts prepared by simultaneous distillation-extraction from fully ripened melons of the reticulatus type show no discernable levels of unsaturated C_6 or C_9 aldehydes or alcohols. Further, the aroma of the intact fruit does not have the green notes associated with C_6 aldehydes and alcohols. These results are consistent with the finding of Lester (*17*) that there is no lipoxygenase activity in the middle-mesocarp tissue from the netted muskmelon 'Perlita'. Lipoxygenase activity however, was detected in the hypodermal tissue of fruit that was at least 30 days postanthesis. The major characteristic of the aroma of intact muskmelon fruit seems to be a blend of the fruity ester-like aroma with a musk note. We consider that the latter is provided, at least in part, by a range of sulfur compounds.

This work reports additional sulfur volatiles in *Cucumis melo* cv. Makdimon and investigates the sensory significance of sulfur volatiles in melon aroma as determined by gas chromatography-olfactometry and aroma extraction dilution analysis (AEDA). The results of blending the melon flesh prior to extraction are also presented.

Experimental Procedures

Melons used in this study were from authenticated seed obtained from commercial seed producers and were grown in the open in a fertilized plot at UWS, Hawkesbury or using an NFT hydroponic system with a controlled nutrient mix at Conisseur's Choice, Llandilo, NSW. Melons were tagged at anthesis, harvested at full slip, stored at 4°C and extracted within 48 h. Melon samples were taken from the whole fruit by cutting it into longitudinal sections, the edible portion (middle-mesocarp) removed and cut into small pieces (5x5 cm).

This sample was then subjected to simultaneous distillation-extraction (SDE) for 1.5 h using pentane as the extracting solvent, or blended using a Waring blender for 1 min prior to SDE. A known mass of butyl hexanoate, as internal standard, was added to the flesh prior to distillation. The extract was concentrated (1 mL) in a Kuderna-Danish flask attached to a Snyder column using a bath temperature of 45°C. The concentrated extracts were then chromatographed using a Hewlett-Packard 5890 Series II gas chromatograph fitted with an SGE BP1, 25 m x 0.22 mm i.d. capillary column of 1.0 µm film thickness. Data was acquired under the following conditions; initial temperature:45°C, 5.0 min, program rate: 5°C/min, final temperature: 240°C, final time: 10.0 min, injector temperature:250°C, transfer line temperature: 280°C, carrier gas: He at 22 cm/s, split ratio:1:40. The column was terminated at a

Hewlett-Packard Mass Selective Detector (MSD)(HP 5971A). The ion source was run in the EI mode at 170 °C using an ionisation energy of 70 eV. The scan rate was 0.9 scans/sec. Data from the MSD was stored and processed using a Hewlett-Packard Vectra QS20 computer installed with Mustang software and the Wiley Mass Spectral Library. Kovats indices were calculated against external hydrocarbon standards. Concentrations were determined from the internal standard, butyl hexanoate, and are not corrected for detector response.

The melon extract was analyzed for sulfur volatiles using a Hewlett-Packard 5890 Series II gas chromatograph coupled to a Hewlett-Packard Atomic Emission Detector (HP5921A). The gas chromatograph was fitted with an OV1 column, 25 m x 0.25 mm i.d. and a film thickness of 0.5 µm. The carrier gas was helium, set at a head pressure of 15 psi. The oven was temperature programed as follows; initial temperature: 40°C held for 5 min; program rate: 5°C/min; final temperature: 220°C.

For the GC effluent sniffing experiments the concentrated extracts were chromatographed using a Pye Unicam GCV. The outlet from a nonpolar column (J&W DB1, 30 m x 0.32 mm i.d., 0.32 µm film thickness) was divided 1:1 using an outlet splitter (SGE,Australia) with one arm connected to an FID detector and the other to a sniffing port (SGE, Australia) flushed with humidified air at 500 mL/min. Chromatographic conditions were: initial temperature: 60°C; initial time: 2 min; program rate: 4°C/min; final temperature: 200°C; injector temperature: 220°C; detector temperature: 220°C; carrier gas: N_2 at 10 psi. The sensory response to the column effluent was recorded as outlined by Miranda-Lopez et al. (18) except that the response was recorded in parallel with the FID using the second channel of a computing integrator (DAPA Scientific Pty. Ltd., Perth, Australia). The resultant aromagram gives a series of peaks which record the intensity and time of the response. Descriptors of each response were recorded during each run on a tape recorder that was subsequently synchronised with the aromagram. The aroma extract dilution analysis was performed by injecting decreasing amounts of the extract into the gas chromatograph. The volumes injected were 1.0, 0.33, 0.1 and 0.05 µL giving effective dilutions of 3, 10 and 20 respectively.

Results and Discussion

Yabumoto, Jennings and Yamaguchi (19) in a prescient comment in 1977 pointed out that many believed sulfur compounds were important in fruit and vegetable quality, although in melon aroma up to that time only dimethyldisulfide had been identified. The evaluation of the contribution of sulfur volatiles to melon aroma required a detailed analysis of all sulfur components present in the extract. Analysis of a Makdimon aroma extract recorded using flame photometric detection (4) indicated at least five sulfur components were present. Responses obtained from the sniffing port, however, suggested additional sulfurous notes at very low concentrations and often obscured in the FID trace. Subsequent analysis using the more sensitive atomic emisson detector revealed additional sulfur containing peaks. The GC-AED traces for carbon and sulfur from a Makdimon melon SDE extract are shown in Figure 1. Twenty sulfur volatiles were detected using the sulfur selective mode, and this work focused on the contribution of these components to the aroma profile. Peak identification and the approximate concentration of each component are shown in Table I. The series of S-methyl acid esters, dimethyl trisulfide and the tentatively identified dimethyltetrasulfide, have not been previously reported in melon aroma.

To assess the relative sensory significance of the esters and the sulfur volatiles to the overall melon aroma, the SDE extract was analyzed by aroma extraction dilution analysis (AEDA) using gas chromatography-olfactometry (GCO). The FID and sensory responses to the most dilute extract are shown in Figure 2, and the complete AEDA data is summarized in Table II. Of the seven peaks identified as significant odorants at the highest dilution, four contain sulfur, S-methyl thiobutanoate, 3-(methylthio)propanal, 3-(methylthio)propyl acetate and,

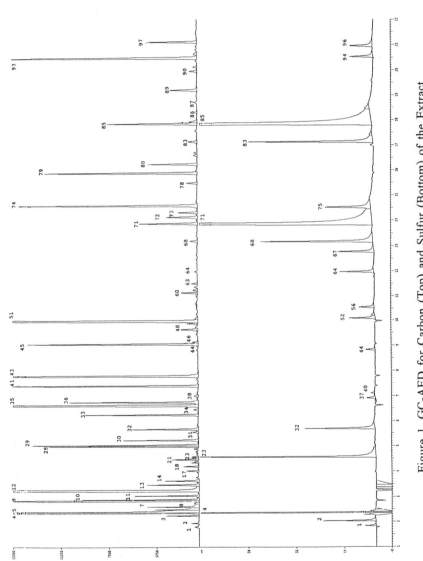

Figure 1. GC-AED for Carbon (Top) and Sulfur (Bottom) of the Extract of Makdimon.

Table I. Volatile Constituents of Makdimon

Peak No.	Kovats Lit. a	Kovats	Compound	Conc. b (ug/kg)	ID
1	nd		unknown c	nd	
1a	nd		sulfur	nd	AED
2	nd		unknown c	nd	
2a	nd		sulfur	nd	AED
3	nd		unknown c	nd	
4	nd		dimethyl sulfide	nd	MS
5	nd		pentane	nd	
6	nd		unknown c	nd	
7	nd		unknown c	nd	
8	nd		unknown c	nd	
9	nd		cyclopentane c	nd	MS
10	nd		methylpentane c	nd	MS
11	nd	575	3-methylpentane c	nd	MS
12	nd	595	ethyl acetate	2906	MS,RT
13	615	611	methyl propanoate	155	MS,RT
14	627		methylcyclopentane c	tr	MS
15	629		unknown	tr	
16	647		unknown	7	
17	649	645	methylethyl acetate	35	MS,RT
18	659		cyclohexane c	49	MS
19	666		trans-2,4-dimethyltetrahydrofuran*	20	MS
20	669		unknown	tr	
21	671	673	methyl 2-methylpropanoate	90	MS,RT
22	676		3-methylhexane c	19	MS
23	678		S-methyl thioethanoate	38	MS
24	682		unknown	tr	
25	687		unknown	tr	
26	689		unknown	12	
27	692		unknown	13	
28	696	691	ethyl propanoate	409	MS,RT
29	699	694	propyl acetate	653	MS,RT
30	706	705	methyl butanoate	312	MS,RT
31	719		methylcyclohexane c	29	MS
32	722	719	2-methylbutanol (+dimethyldisulfide)	234	MS,RT
33	745	746	ethyl 2-methylpropanoate	409	MS,RT
34	750		methylbenzene	25	MS
35	759	758	2-methylpropy acetate	2307	MS,RT
36	765	765	methyl 2-methylbutanoate	618	MS,RT
37	738		unknown (contains sulfur)	22	MS,AED
38	776		2-methylprop-2-enyl acetate *	19	MS
39	778		unknown (contains sulfur)	14	MS,AED
40	780		unknown	12	MS
41	784	784	ethyl butanoate	1109	MS,RT
42	792	785	propyl propanoate*	tr	MS
43	796	793	butyl acetate	1586	MS,RT
44	807		S-methyl thiobutanoate	tr	MS
45	838	837	ethyl 2-methylbutanoate	989	MS,RT
46	843	844	ethyl 3-methylbutanoate	tr	MS,RT
47	851		unknown	tr	
48	854	858	hexanol (+ 2-methylpropyl propanoate)	105	MS,RT
49	862		ethylbenzene c	12	MS
50	866	867	3-methylbutyl acetate	53	MS,RT
51	868	869	2-methylbutyl acetate	2865	MS,RT

Continued on next page

Table I. Continued

52	874	3-(methylthio)propanal	21	MS
53	878	unknown (contains sulfur)	tr	AED
54	880	unknown	tr	
55/56	882	881 methyl (methylthio)acetate+propyl butanoate	23	MS,RT
57	885	884 ethyl pentanoate	16	MS,RT
58	887	unknown	tr	
59	891	butyl propanoate	tr	MS,RT
60	895	895 pentyl acetate	102	MS,RT
61	901	901 2-methylpropyl 2-methylpropanoate + unknown	35	MS
62	904	902 3-methyl-2-butenyl acetate*	18	MS
63	907	906 methyl hexanoate	48	MS,RT
64	928	S-methyl thiopentanoate	18	MS
65	934	933 propyl 2-methylbutanoate*	tr	MS,RT
66	939	unknown	tr	MS,RT
67	958	952 dimethyl trisulfide	tr	MS,RT
68	958	ethyl (methylthio)acetate	80	MS,RT
69	975	unknown	tr	
70	977	unknown	tr	
71	979	2-(methylthio)ethyl acetate	615	MS,RT
72	983	983 ethyl hexanoate	328	MS,RT
73	988	987 cis-3-hexenyl acetate	155	MS,RT
74/75	996	1004 hexyl acetate+methyl 3-(methylthio)propanoate	1388	MS,RT
76	1001	unknown	25	
77	1003	unknown	tr	
78	1026	1030 limonene	102	MS,RT
79	1032	2,3-butanediol diacetate	1595	MS,RT
80	1043	2,3-butanediol diacetate	558	MS,RT
81	1054	1061 1-octanol + unknown	32	MS,RT
82	1058	unknown	48	MS
83	1074	1075 ethyl 3-(methylthio)propanoate	124	MS,RT
84	1082	unknown	28	
85	1095	1091 3-(methylthio)propyl acetate	1325	MS,RT
86	1103	unknown	36	
87	1118	unknown	40	MS
88	1127	unknown	13	MS
89	1141	1144 benzyl acetate	423	MS,RT
90	1160	unknown	108	MS
91	1165	unknown	58	MS
92	1179	unknown	tr	
93	1180	1177 butyl hexanoate	3143	MS,RT
94	1182	unknown (contains sulfur)	44	AED
95	1185	1180 ethyl octanoate	57	MS
96	1190	1186 dimethyltetrasulfide	29	RT
97	1194	1193 octyl acetate	684	MS,RT

a: from references
b: as determined from TIC data
nd: not determined by GCMS
c: solvent and contaminants therein
tr: trace <10ug/kg
*:tentative identification

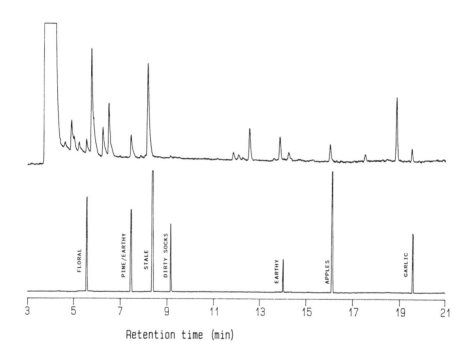

Figure 2. FID (Top) and Aromagram (Bottom) of the Most Dilute Makdimon Extract by AEDA.

Table II. Key Aroma Compounds of Makdimon by AEDA

Compound	SDE/20	SDE/10	SDE/3	SDE	Descriptor
unknown				++	pumpkin
dimethylsulfide*		+	++	+++	rubber/sulfur
ethyl acetate*			+	++	butterscotch
S-methyl thioethanoate			++	+++	garlic toast
ethyl propanoate			+	+++	vinyl
propyl acetate				+	floral
methyl butanoate				+++	sick
ethyl 2-methylpropanoate	+	+	++	+++	floral
methyl 2-methylbutanoate			++	+++	medicinal/floral
unknown				++	stale
ethyl butanoate				++	medicinal/sick
butyl acetate				++	toast
S-methyl thiobutanoate	+	+	++	+++	pine/earthy
ethyl 2-methylbutanoate		+	++	+++	strawberry
hexanol + unknown		++	++	+++	stale
unknown		+		+	pine/skunk
2-methylbutyl acetate		++	++	+	ether
3-(methylthio)propanal	++	+++	+++	+++	stale
pentyl acetate*			++	++	toast
unknown	++	+++	+++	+++	dirty socks
S-methyl thiopentanoate			+	++	garlic toast
unknown				+++	fruity
2-(methylthio)ethyl acetate				++	skunk
ethyl hexanoate			+	++	floral
hexyl acetate				+	pine
methyl 3-(methylthio)propanoate				+	stale
limonene		+++	++	+++	stale/dirty socks
2,3-butanediol diacetate	+	++	++	+++	earthy
S-methyl thiopentanoate			+	++	garlic toast
octanol + unknown				+	pine
dimethyltrisulfide			++	++	garlic
ethyl 3-(methylthio)propanoate			+	++	fruity/stale
unknown				+	raw potato
3-(methylthio)propyl acetate	++	++	+++	+++	apples
unknown			+	++	stale
benzyl acetate			++	+	pine
dimethyltetrasulfide*	+	+			garlic
unknown			++	+	garlic
octyl acetate*				+	pine

* tentative identification
+, ++, +++ describe intensity of response

tentatively, dimethyltetrasulfide. Only two of the odorants, ethyl 2-methylpropanoate and 3-(methylthio)propyl acetate, contribute to the classical fruity notes while the other five odorants provide the musk overtones to this melon's aroma. There is an earthy note associated with one of the pair of peaks from the diastereomers of 2,3-butanediol diacetate, however, additional work on pure enantiomers using chiral phase GC analysis must be completed to confirm this contribution. The identity of the dirty socks aroma is unknown.

At increased concentrations of the extract the olfactory impact of the esters notably, ethyl 2-methylbutanoate, methyl 2-methylbutanoate and 2-methylbutyl acetate become significant and presumably enhance the fruity characters that are not so dominant in the most dilute extract. The aromagram of the SDE extract itself contains forty discernable aromas and most intriguingly none of the descriptors are melon-like.

An alternative approach to assessing the sensory significance of aroma compounds is through the use of odor units (21). Compounds with odor units greater than unity (Table III) correlated well with those components identified as significant by AEDA. Unfortunately odor values could only be calculated for those compounds for which odor thresholds are available and that of the key compounds S-methyl thiobutanoate and 2,3-butanediol diacetate do not appear to have been determined. The dominance of the typical esters by this method is at odds with the GCO aromagram data that suggests 3-(methylthiopropyl) acetate should have a higher ranking, and the esters, ethyl and methyl 2-methylbutanoate should be detectable in the most dilute extract. These anomalies most probably result from the known inconsistencies of odor threshold data (22) that may be further compounded by the current lack of knowledge regarding the chiral composition of some of these key components to melon aroma.

The lack of C_6 and C_9 aldehydes and alcohols in our aroma extracts prompted an investigation of the influence of blending samples prior to the extraction of volatiles. The chromatograms of SDE extracts of the same Makdimon melons with and without blending are shown in Figure 3. The TICs clearly indicate the dramatic loss of volatiles resulting from blending the flesh for one minute prior to extraction. The identity of all lipoxygenase products has yet to be confirmed, however, the appearance of new compounds, notably (Z)-2-hexenol, was accompanied by a substantial loss of esters that can only be attributed to hydrolases activated by blending the sample. The increased number of peaks in the latter part of the chromatogram are responsible for the sensory responses in the aromagram for this extract (Figure 4). The aromagram also confirms the significant loss of esters from the front portion of the profile.

Conclusion

Sulfur volatiles play a significant role in the perceived aroma of *Cucumis melo* cv. Makdimon. In particular 3-(methylthio)propanal and S-methyl thiobutanoate and dimethyltetrasulfide (tentative) convey a musky overtone to the classical fruity aromas associated with 3-(methylthio)propyl acetate, ethyl 2-methylpropanoate, ethyl 2-methylbutanoate and methyl 2-methylbutanoate. Further work is required to confirm the contribution of the enantiomers of 2,3-butanediol diacetate and the unknown compound that conveys the sweaty, dirty socks component to the aroma of this melon.

Aroma volatiles of this Makdimon melon are dramatically altered by homogenizing the flesh prior to SDE. The increase in some lipoxygenase derived volatiles, namely (Z)-2-hexenol along with hitherto unidentified compounds is accompanied by a substantial loss of esters. More importantly, the sensory properties of the blended melon flesh extract indicate that these changes have a significant impact on the aroma perception. Much of the earlier work that has recorded lipoxygenase products as key contributors to melon aroma have blended samples prior to extraction.

Table III. Key Aroma Compounds of Makdimon by Odor Units (Uo)

Compound	Amount (ppb)	Threshold (ppb) a	Odor Unit
ethyl 2-methylpropanoate	409	0.1	4093.18
ethyl 2-methylbutanoate	989	0.3	3297.47
methyl 2-methylbutanoate	619	0.25	2474.00
ethyl butanoate	1109	1	1108.77
hexyl acetate *	1388	2	694.24
ethyl acetate	2906	5	581.24
ethyl hexanoate	328	1	327.78
2-methylbutyl acetate	2865	11	260.42
benzyl acetate	423	2	211.72
dimethyl tetrasulfide	29	0.2	143.25
3-(methylthio)propanal	21	0.2	105.73
ethyl 3-methylbutanoate	6	0.1	63.48
octyl acetate	684	12	56.99
ethyl propanoate	409	10	40.93
2-methylpropyl acetate	2307	65	35.49
3-methylbutyl acetate	53	2	26.56
butyl acetate	1586	66	24.03
ethyl 3-(methylthio)propanoate	124	7	17.67
methyl butanoate	312	59	5.29
ethyl pentanoate	16	5	3.25
ethyl (methylthio)acetate	80	25	3.21
3-(methylthio)propyl acetate	1325	600	2.21
2-methylbutanol*	234	300	0.78
ethyl octanoate	57	92	0.62
methyl hexanoate	48	84	0.57
dimethyl trisulfide	5	10	0.50
limonene	102	210	0.48
1-octanol	32	110	0.29
propyl butanoate*	23	124	0.18
hexanol*	105	2500	0.04
butyl propanoate	7	180	0.04
phenylethyl acetate	43	1800	0.02

a: from references 1,11,12,14
*:coelute with other compounds on BP1

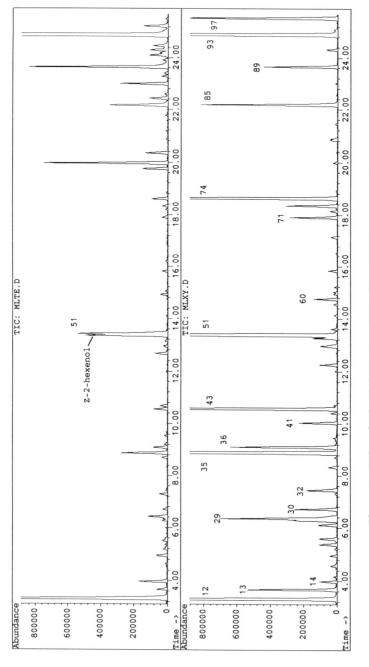

Figure 3. TICs of Blended (Top) and Nonblended (Bottom) Makdimon Flesh from SDE.

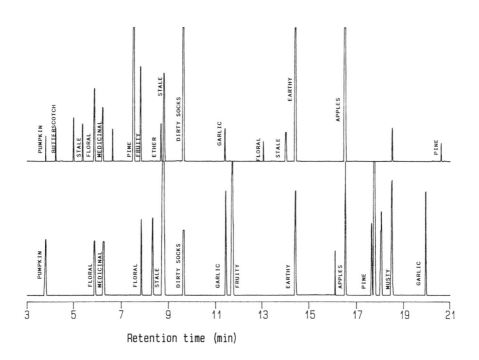

Figure 4. Aromagrams of Nonblended (Top) and Blended (Bottom) Makdimon Flesh from SDE.

Acknowledgments

We would like to acknowledge the financial support of the American Chemical Society and the University of Western Sydney, Hawkesbury that enabled us (SGW and DNL) to attend the 209th Annual ACS Conference, Chicago, August 23-27th, 1993. We would also like to thank the Hawkesbury Foundation and the UWS, Hawkesbury Research and Development Committee for financial support of the research work. Without the expertise of our hydroponics grower Mr Brian Corliss, Conisseur's Choice, the horticultural expertise of Drs Barry McGlasson and Tony Haigh in the School of Horticulture, UWS, Hawkesbury and the cooperation of the University of Georgia, this work could not have been completed. The GC-AED work was done at the Research Institute for Chromatography, Kortrijk, Belgium and their cooperation is greatly acknowledged.

Literature Cited

1. *Volatile Compounds in Foods and Beverages*, Maarse, H.,Ed.; Marcel Dekker Inc., New York, 1991; pp 295.
2. Schreier,P.; *Chromatographic Studies of the Biogenesis of Plant Volatiles*; Huthig, Heidelberg, 1984; pp 64-76.
3. Wyllie, S. G. and Leach, D. N. *J. Agric. Food Chem.*, **1990**, *38*, 2042-2044.
4. Wyllie, S. G. and Leach, D. N. *J. Agric. Food Chem.*, **1992**, *40*, 253-256.
5. Homatidou, V.; Karvouni, S. and Dourtoglou, V. In *Flavors and Off-Flavors;* Charlambous, G., Ed.; Proceedings of the 6th International Flavor Conference; Elsevier, Amsterdam, **1989**; pp 1011-1023.
6. Homatidou, V.; Karvouni, S.; Dourtoglou, V. and Poulos, C.N. *J. Agric. Food Chem.* **1992**, 40, 1385-1388.
7. Buttery, R. G.; Seifert, R. M.; Ling, L. C.; Soderstom, E. L.; Ogawa, J. M. and Turnbaugh, J. G. *J. Agric. Food Chem.* **1982**, *30*, 1208-1211.
8. Horvat, R. J. and Senter, S. D. *J. Food Sci.* **1987**, *52(4)*, 1097-1098.
9. Dirinck, P.; De Pooter, H.; Willaert, G.; Schamp, N. In *Analysis of Volatiles*; Schreier, P.; Ed.; W.de Gruyter, Berlin, 1984; pp 387-390.
10. LeQuere, J-L. and Latrasse, A. *J. Agric. Food Chem.* **1990**, *38*, 3-10.
11. Takeoka, G.; Buttery, R. G.; Flath, R. A.; Teranishi, R.; Wheeler, E. L.; Wieczorek, R. L. and Guntert, M. In *Flavor Chemistry, Trends and Developments*; Teranishi, R., Buttery, R.G., Shahidi, F., Eds.; ACS Symposium Series 388; American Chemical Society: Washington, DC, 1989; pp 223-237.
12. Takeoka, G.;Buttery, R. G.;Teranishi, R.; Flath, R. A.; Guntert, M. *J. Agric. Food Chem..*, **1991**, *39*, 1848-1851.
13. Umano, K.; Hagi, Y.; Nakahara, K.; Shoji, A. and Shibamoto, T. *J. Agric. Food Chem.*, **1992**, *40*, 599-603.
14. Takeoka, G. R.; Buttery, R. G. and Flath, R. A. *J. Agric Food Chem.*, **1992**, *40*, 1925-1929.
15. Engel, K. and Tressl, R. . *J. Agric. Food Chem.*, **1991**, *39*, 2249-2252.
16. Schieberle, P.; Ofner, S. and Grosch, W. *J. Food Sci.*,**1990**, *55*, 193-195.
17. Lester, G. *J. Amer. Soc. Hort. Sci.* **1990**, *115(4)*, 612-615.
18. Miranda-Lopez, R.; Libbey, L. M.; Watson, B. T. and McDaniel, M. R. *J. Food Sci.* **1992**, *57(4)*, 985-1019.
19. Yabumoto, K.; Jennings, W. G.; Yamaguchi, M. *J. Food Sci.* **1977**, *42*, 32-37.
20. Jennings, W. G. and Shibamoto, T. Qualitative Analysis of Flavor and Fragrance Volatiles by Glass Capillary Gas Chromatography, Academic Press, New York, 1980, p29-85.
21. Guadagni, D. G.; Buttery, R. G. and Harris, J. *J. Sci. Food Agric.* **1966**, *17*, 142-144.
22. Drawert, F. and Christoph, N. In *Analysis of Volatiles*; Schreier, P., Ed.; Proceedings International Workshop; de Gruyter, Berlin, **1984**; pp 269-291.

ECEIVED March 23, 1994

Chapter 5

Sulfur-Containing Flavor Compounds in Beef: Are They Really Present or Are They Artifacts?

Arthur M. Spanier, Casey C. Grimm, and James A. Miller

Southern Regional Research Center, Agricultural Research Service, U.S. Department of Agriculture, 1100 Robert E. Lee Boulevard, New Orleans, LA 70122

>Several analytical procedures are available for extracting and identifying the flavor components of food. Unfortunately, these same analytical procedures may, in themselves, alter the composition of the food being analyzed. This paper demonstrates that the chromatographic profiles of volatiles from cooked and from cooked-stored ground beef are directly affected by and related to the temperature used to purge the volatiles from the sample, the temperature of analysis, and the end-point cooking temperature of the sample. Analytical efficacy and recovery is minimal when the samples are purged at 50°C due to inefficient extraction of the flavor volatiles. Misleading chromatographic profiles occur when samples are purged at temperatures over 90°C due to conversion of one volatile form to another. Optimal volatile extraction/purging and limited conversion of volatiles to other forms is observed at temperatures of 70-75°C. The data suggest that a thorough examination of the effect of temperature on volatile production must be performed prior to analytical appraisal of the food's flavor-volatiles if the true flavor-profile or flavor-picture of the food is to be known.

The flavor quality of a meat is dependent upon several key antemortem and postmortem factors (1). These factors include the animal's age, breed, sex and nutritional status as well as all of the postmortem handling and cooking protocols. While both the antemortem and postmortem factors are involved in the development of the food's final flavor and texture, the most important factors are those which develop during the postmortem aging period and during postmortem handling, cooking and storage (1-4).

The study of meat flavor is complicated further, and in major part, by its structure which is a highly complex fibrous network of interrelated organelles and organelle-systems (1). Perhaps the most important organelle-system impacting the flavor of a muscle food is the sarcotubular/lysosomal (SR/L) system (*4-7*). This organelle system is multifunctional and is involved in both excitation/contraction coupling and in normal-and-pathological protein turnover. Its involvement in excitation contraction coupling makes it a repository for large quantities of free calcium ions. These calcium ions are released from the SR/L during the postmortem aging period leading to the tenderization of the beef by virtue of activation of the calcium-dependent proteinases (*8*). The SR/L is also the repository of lysosomal hydrolases such as the proteinases, glycosidases, and lipases (*5-7*). These enzymes, specifically the proteinases, are responsible for the generation of other flavor products and precursors in meat (*1, 4*).

The origin of meat flavor has been shown to arise from the combination of two primary sources. The first is the tissue fat, both extracellular and intracellular, which produces carbonyl and other lipid and lipid-oxidation products. The fat component of meat flavor is viewed as being responsible for the species specific flavor in meat (*9*). The second major component of meat flavor is the lean portion. The proteins, peptides, and amino acids of the lean, add not only to the muscle food's general meaty flavor, but also undergo Maillard reactions with sugars to produce Amadori and Heyns compounds having meat flavor characteristics.

In 1986, a review of the analysis of meat volatiles by Shahidi and colleagues (*10*) listed 995 compounds that have been found in meat. Mechanistic studies that have combined various amino acids and sugars have predicted the presence of even more compounds that have yet to be observed (*11*). The formation of flavor-producing compounds results from the complex interaction of numerous precursors and treatments. Some factors involved in beef flavor production are diet, postmortem aging, storage time and temperature, and cooking method. Additionally, it has recently been shown by Block et al. (*12, 13*) and others (*14*) that the method used to analyze flavor compounds may, in the process, create new flavor compounds.

As meat ages and as it is cooked or stored, it shows a significant alteration in the level of numerous endogenous chemical components such as sugars, organic acids, peptides and free amino acids, and products of adenine nucleotide metabolism (adenosine triphosphate, ATP). Independent of the history of the meat, as meat is stored and heated, both desirable and undesirable flavors develop from proteolysis, thermal degradation, and interaction of sugars, amino acids, and nucleotides (*15*). These chemical modifications to the muscle-food serve as a pool of reactive flavor compounds and flavor intermediates which later interact to form additional flavor notes during cooking such as the Maillard reaction products formed during the heating of sugars and amino acids (*16, 17*). It becomes immediately apparent that the development of flavor in a muscle food is an extremely complicated process which occurs continuously from moment of slaughter, continues through cooking and storage and ends when the food is eaten and the flavor perceived.

Sulfur-containing flavor compounds in meat can be categorized into aliphatic, thiamines, thiazoles, and all others. Model studies have shown that cysteine reacts to produce thiazoles that have some meaty odors (15). However, the thiazoles are generally seen only in grilled meat, but this may be dependent upon the method of analysis (2). Sources of sulfur-containing flavor compounds include, but are not limited to thiamine degradation. MacLeod's review (11) listed seven aliphatic sulfur compounds, 65 heterocyclic sulfur compounds, and six non-sulfur heterocyclic compounds as containing meat flavors. Only 25 of these compounds have been identified in meat.

There have been suggestions that the oxidation of sulfhydryl groups by aldolase could be the initial event in the degradation of proteins to amino acids (18). It is the reaction of sugars with these amino acids that has become known as Maillard reaction products (MRP) (19). These MRP are considered to have a vital role in the production of flavor in meat (15). A significant proportion of the organic compounds, which are known to produce meaty-like flavors, contain sulfur. The specific aldehydes produced during the Maillard reaction are controlled mainly by the specific amino acid used in the reaction (19), whereas the amount of a specific aldehyde produced is determined mostly by the type of sugar used in the reaction (19). With this information in hand, it is reasonable to expect that through control of reactants and reaction conditions, the MRP can be controlled; furthermore, through the use of heuristic models, the MRP can be predicted. The literature contains a notable amount of information relating to synthetic Maillard Reaction Products. There is an ever growing list of MRP found in synthetic mixtures and in meat, but we were unable to find published quantitative correlations between the levels of MRPs and the sensory response of these compounds in meat. Although much is known about the formation of MRP *in vitro*, little is know of their formation and influence on the overall sensory perception and flavor of meat *in vivo*. Thus, the mechanism of MPR formation, the specific content of MRP in meat, and the influence of MRP on sensory perception need to be major research areas in food science. Recognition of these facts, as seen by the development of this symposium, is the subject of this text.

Analytical Methodology.

Sample Preparation. *Semimembranosus muscle* (top round) from angus-cross steers was trimmed of all excess fat and connective tissue and ground twice (3). Samples (100g) were packed into 4 oz glass jars and covered with aluminum foil. Samples were baked in a Jenn-Air convection oven at 177°C (350°F) until a 70°C end-point in the center of the sample was measured using a type-J probe with a digital thermometer (Cole-Palmer). Cooking time was approximately 20 min. Samples were removed, allowed to reach ambient temperature and the liquid portion (gravy) decanted and discarded. Samples were stored at 4°C for two (2d) and four (4d) days. At the appropriate time, they were removed and processed as described below. The cooked/stored ground-beef was finely minced to enhance the sample's homogeneity. Portions of 10g of the minced cooked/stored ground beef were purged with nitrogen, sealed with a nitrogen purge in vacuum (Reiser Manufacturing) impermeable bags (Koch Supplies, Inc.) and then stored frozen at -20°C until analysis.

Direct Thermal Desorption. A short-path thermal desorbtion (SPTD) device (Scientific Instrument Services, New Jersey; Figure 1) was used for volatile concentration and loading. A 100 mg sample of freshly cooked or cooked-and-stored beef was packed into a 10cm x 4mm I.D. tube prepacked with a glass wool plug. The glass wool was placed in the tube to hold the sample in place. One μl of a standard solution of benzothiophene (40ng/μl) was injected into the upper plug of glass wool. An injection needle was screwed into place at the lower end of the tube and the assembly then attached to the SPTD. A prepurge time of 0.1 sec preceeded a pressure equilibration time of 1 min. During this time carrier gas was rerouted from the injector through the SPTD. After 1 min purge with helium the sample tube was maintained in the heated block of the injector for 5 min. During this period, volatiles are carried from the sample to a splitless injector that is heated to the same temperature as the desorber. The split vent on the injector was capped off to prevent loss of volatiles.

The SPTD was reprogrammed to send a start signal to the GC at the beginning of the desorption. A 60 m, 0.75 mm I.D., 5% phenyl, 95% dimethyl siloxane capillary column was cooled to 1°C to cryofocus the desorbed meat volatiles. The initial temperature was held at 1°C during the desorption 5 min period and then ramped to 150°C at 5°C per min. The temperature ramp was then increased to 250°C at 10°C per min. The effluent from the capillary was split to allow simultaneous detection of carbonyls by a flame ionization detector (FID) and the sulfur-containing components via a flame photometric detector (FPD).

Results and Discussion

Several methods have been employed (*20*) in the extraction of flavor volatiles from their chemical and physical bonds in the complex muscle matrix. These methods include steam distillation, solvent extraction, headspace analysis and supercritical fluid extraction. Each method had specific advantages and shortcomings when used in the isolation of numerous compound types such as carbonyls, sulfur-containing compounds such as thiols and thiazoles, pyrazines, furans, and pyrroles (*21, 22*). The mechanisms responsible for the formation of these flavor volatiles have been proposed and developed based on such experimentation (*23, 24*).

Almost all methods for extraction/isolation of flavor volatiles involve some form of heating to volatilize the compounds from the sample into a headspace collection region or trapping matrix such as Tenax or for elution and movement of the compounds from the adsorbent column (capillary, megabore, or packed). The temperatures used for sparging, purging and thermal desorbing may result in either thermal degradation of some of the compounds present or in generation of new compounds from existing and cooking-derived precursors. Since a detailed gas chromatographic analysis may result in 200 to 1000 peaks (*25, 26*), the investigator must use caution in judging the validity of the results and interpreting the data. Deciphering the true volatile profile of the food being analyzed is a complex problem further confounded by choice of extraction method. The experiments described here were designed to analyze the effect of temperature on the development and/or degradation of the volatile flavor precursors and products of cooked and cooked/stored ground beef.

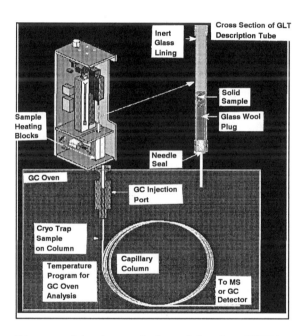

Figure 1. Diagram of the short path thermal desorber (SPTD) used for isolation and separation of sulfur-containing compounds in beef.

Unlike most environmental samples, in which the analyte does not change and severe matrix disruption is permitted, the volatiles released above the food are in a dynamic state. No single analytical technique has been devised that is useful for determining all the volatile components in beef (*23, 27-30*). Scores of collection schemes, in combination with a single separation technique (GC) having a number of detectors available, have been employed.

Earlier investigations using an external closed loop inlet device (ECID) in conjunction with a packed column and a sulfur specific flame photometric detector, indicated that the concentration of H_2S varied inversely with temperature (*14, 24*). This resulted in a temperature-dependent variability in the concentrations of the dimethyl sulfides (*14*). In order to determine whether the dimethyl sulfides were being created during desorption, meat samples were cooked to three end-point temperatures and then desorbed at these same three temperatures. Additionally, the meat was stored for periods of 0, 2, and 4 days before being sealed and analyzed.

The short path thermal desorber (SPTD; Figure 1) was used to isolate the volatiles from cooked and cooked/stored ground beef. The SPTD was chosen for use as it enabled us to maintain accurate control of the sparge, purge, and desorption temperatures, and to have minimal space between the sample and the capillary column. Since metal ions catalyze the process of lipid oxidation (*31*), use of the SPTD also minimized the amount of metal the sample and volatiles would contact.

Examination of the effect of storage on the generation or loss of sulfur-containing volatiles indicates that some compounds are lost while others are increased. Ground beef cooked to a final end-point temperature of 70°C was the sample used on a 60 meter, DB-5 column; this column was chosen because of the enhanced capacity of wide- or megabore columns to handle the larger volatile load and the large quantity of water associated with muscle. Compounds were identified by running known standards and recording the retention times (RT). The water content of the meat samples did not permit us to identify the compounds directly on the mass spectrometer. Comparison of the samples purged at 50°C at 0 days and 4 days storage (Figure 2, top left and top right, respectively) indicated that both methyl trisulfide (RT=39.0 min) and hydrogen sulfide (RT=10.0 min; predominant sulfur-containing compound) were found at lower concentrations after storage. Changes with storage are also seen when the sample is purged at 100°C. This temperature proved to be quite unreliable as the elevated temperature produced more volatiles than were present initially (see *22, 32*).

An examination of the effect of purge temperature on the production and analysis of sulfur-containing compounds can be seen by comparing the upper to the lower graph in Figure 2. The top-left graph in Figure 2, shows the volatile profile of cooked ground beef purged at 50°C. This is compared to the graph on the bottom-left of Figure 2 which shows the same sample purged at 100°C. As predicted and as reported previously (*4, 22, 32*) there is a marked increase in total sulfur volatiles as the purge temperature is increased from 50°C to 100°C. The same temperature effect is observed in stored meat (see Figure 2 top-right vs bottom-right). A more complete examination of the effect of purge temperature on

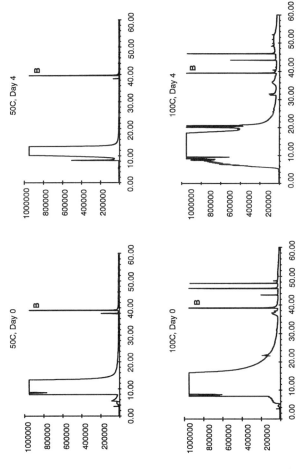

Figure 2. Flame photometric detection of sulfur-containing volatiles from 0.1g of cooked ground-beef. (***top left***), beef cooked to an end-point temperature (EPT) of 70°C. Sample was purged at 50°C in the SPTD the same day it was prepared. (***top, right***), same as the top, left but the sample was stored for 4 days. (***bottom, left***), beef cooked to an EPT of 70°C. Sample was purged at 100°C in the SPTD the same day as it was prepared. (***bottom, right***), same as the bottom, left but the sample was stored for 4 days. B = benzothiophene external standard.

the analysis of sulfur-containing compounds is seen in Figure 3. It is evident that as the purge temperature is increased from 50°C to 75°C to 100°C, the number of detectable sulfur-containing compounds also increases.

Since a temperature-dependent difference in the number of detectable sulfur-containing compounds was observed and since the temperature at the sample holding block and the GC injector port (Figure 1) had been maintained constant in the previous experiments, it was decided to see whether these two heating compartments (sample holding block and injector port) had an independent effect on the determination of sulfur-containing volatiles. As seen in Figure 4, there was little or no difference in the volatiles and external standard measured as the temperature of the injector port is changed. Lower temperatures resulted in some of the volatiles condensing on the septum, resulting in carryover and skewing reproducibility (not shown). These data were similar to those reported previously (32) for carbonyl compounds of meat.

Discussion of the data among the authors led to some concern regarding the conclusions drawn from a sample of such small size (0.1 g cooked ground beef). Conclusions drawn from data obtained from sample of such small size could lead, potentially, to a high degree of sample heterogeneity. Thus, instrumentation was modified to permit the use of a larger sample size (1.0g). An external closed inlet device (ECID) was used with a packed column to isolate and identify the sulfur-containing compounds of ground meat (3, 14, 32). The packed column had the advantage of permitting larger samples to be focused onto the column and would handle large quantities of water. A disadvantage of the ECID is that there is more area for contact of metal surfaces with the sample and the volatiles. Samples included fresh cooked ground beef and cooked ground beef stored for 1, 2, and 4 days. The slopes for the rate of change in the sulfur-containing compounds were determined such that positive slopes represented compounds being created during storage while negative slopes represented compounds being lost during storage.

Unlike the previous ground beef samples that were cooked in a jar, these samples were cooked on an open-top Farberware grill for 7 minutes/side to reach a final average end-point temperature of 66-68°C. Purge and concentration of the volatiles took place on a 10 ft x 1/8" Tenax GC - polymetaphenyl ether column for 30 minutes at room temperature. Elution from the column was with nitrogen flow at 20 mL/m. The desorption ramp was 25°C at 3 °C/min. Sample identity was confirmed via Mass Spectroscopy. Six major sulfur-containing compounds were identified by this method and include hydrogen sulfide (H_2S), carbonyl sulfide (CS), methane thiol (MT), dimethyl sulfide (DMS), dimethyl disulfide (DMDS), and dimethyl trisulfide (DMTS; Figure 5).

At 50°C only dimethyl disulfide and dimethyl trisulfide were seen with some methane thiol and hydrogen sulfide. There appears to be a very minor storage-dependent decrease in dimethyl disulfide and dimethyl trisulfide and an increase in the hydrogen sulfide and methane thiol (Figure 5).

At 75°C all six major sulfur-containing volatiles are observed. These include hydrogen sulfide, carbonyl sulfide, and methane thiol as low molecular weight components and dimethyl sulfide, dimethyl disulfide, and dimethyltrisulfide as the higher molecular mass sulfur-containing compounds. During refrigerated

5. SPANIER ET AL. *Sulfur-Containing Flavor Compounds in Beef* 57

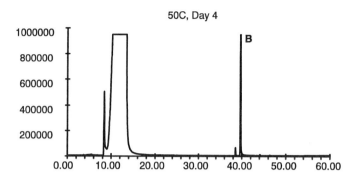

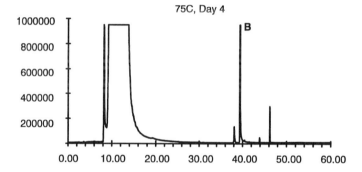

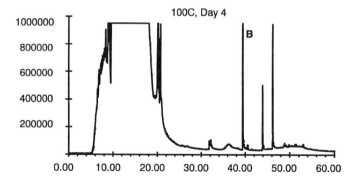

Figure 3. Flame photometric detection of sulfur-containing volatiles from 0.1g of ground-beef cooked to an end-point temperature of 70°C, then purged at 50°C, 75°C, and 100°C after 4 days of refrigerated storage. B = benzothiophene external standard.

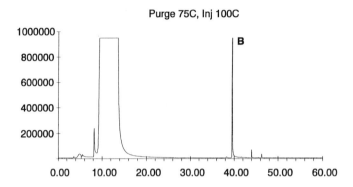

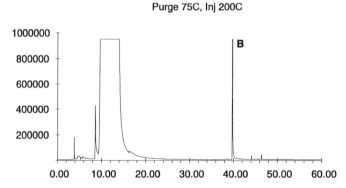

Figure 4. Flame photometric detection of sulfur-containing volatiles from 0.1g of beef cooked to an end-point temperature of 70°C and purged at 75°C with the injection port at 100°C (***top***) or 200°C (***bottom***). B = benzothiophene external standard.

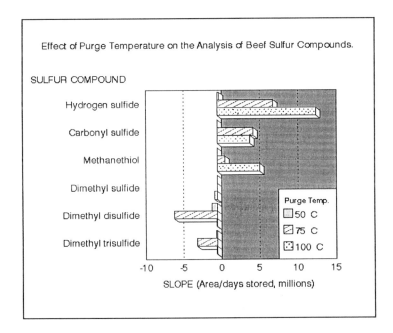

Figure 5. Effect of purge temperature on the analysis of beef sulfur-containing compounds. Detection of the sulfur-containing volatiles was by flame photometric detection. Samples were analyzed using the external closed inlet device (ECID) described previously (*3*). Rates of change in compond composition during storage over a 4 day period was determined from the data in ref. *14*.

storage there are clear increases in the low molecular mass components with a concomitant decrease in the higher molecular mass components.

When the purge temperature is brought to 100°C it appears that the elevated purge temperature has caused the conversion of the higher molecular mass sulfur-containing compounds into the lower mass compounds (*14*). Indeed the carbonyl sulfide peak is almost totally masked by the tremendous increase in the H_2S peak (see *14, 32*). The hydrogen sulfide peak masks most of the other peaks making it difficult to determine the slope of the remaining components hidden under it.

In summary, these data indicate that increased purging temperature leads to an increase in identifiable sulfur-containing compounds. These data also indicate that 75°C, the final temperature of a medium to well-done ground beef patty (*33*), appears to be the optimal temperature for identifying meat sulfur-containing compounds (*14, 32*). These experiments indicate that sulfur-compounds not present in the original food sample can be created and identified in the sample due to the temperature used in the analytical procedure. Finally, some sulfur-compounds are lost and others are generated during the refrigerated storage of ground beef.

CONCLUSION

Analytical techniques can actually create compounds in the sample which may not be there originally. Therefore, we leave the reader with the following thoughts and questions to ask when selecting an analytical method.

- Is sample size a limitation, i.e., do you have mg, g or kg of sample to be analyzed?
- Do you want qualitative or quantitative data and results? Do you want to examine relative changes or values? Do you want to know absolute values? Do absolute values actually exist?
- What type of resolution of compounds do you desire? Will you use a capillary column, a widebore column or a packed column.
- What are the compounds you are trying to isolate? Are they carbonyl and/or sulfur-containing?
- What type of sample collection method are you planning on using, i.e. distillation, direct purge-and-cryofixation?
- What method of column desorption will you use, i.e. thermal desorption or rate of removal.

It is only by knowing the true picture of the volatiles that compose a food that we can take steps to learn how these compounds interact within the food to form our final flavor perception.

LITERATURE CITED

1. Spanier, A.M.; J.A. Miller The role of proteins and peptides in meat flavor. In *Food Flavor and Safety: Molecular Analysis and Design.* Spanier, A.M., Okai, H., Tamura, M., Eds.; ACS Symposium Series No. 528; American Chemical Society: Washington, D.C.; **1993** pp **78-97.**
2. Spanier, A.M. Current Approaches to the Study of Meat Flavor Quality.

In. *Food Science and Human Nutrition.* G. Charalambous, Ed..; Elsevier Science Publishers, B.V.; **1992**. pp **695-709**.
3. Spanier, A.M.; St. Angelo, A.J.; and Shaffer, G.P *J. Agric. Food Chem.* **1992**, *40*,1656-1662.
4. Spanier, A.M.; McMillin, K.W.; Miller, J.A. *J. Food Sci.* **1990**, *55(2)*,318-326.
5. Bird, J.W.C.; In, *Lysosomes in Biology and Pathology.* Dingle, J.T.; Dean, R.T.; Ed.; American Elsevier Publishing Co.: New York, NY, **1975**, Vol. 4; pp **75-109**.
6. Bird, J.W.C.; Schwartz, W.N.; Spanier, A.M. *Acta Biol. Med. Germ.* **1977** *36*,1587-1604.
7. Bird, J.W.C.; Spanier, A.M.; Schwartz, W.N. In, *Protein Turnover and Lysosomal Function.* Segal, H.; Doyle, D.; Eds.; Academic Press: New York, NY; **1978** pp **589-604**.
8. St. Angelo, A.J.; Koohmaraie, M.; Crippen, K.L.; Crouse, J. *J. Food Sci.* **1991** *56(2)*,359-362.
9. Wasserman, A.E. *J. Food Sci.* **1979** *44*,5-11.
10. Shahidi, F.; L.J. Rubin; L.A. D'Souza. *CRC Crit. Rev. Food Sci. Nutr.* **1986** *24*,141-243.
11. MacLeod, G.; In *The scientific and Technological basis of Meat Flavors, Development in Food Flavors.* Birch, G.G.; Lindley, M.G.; Eds.; **1986** pp 191-223.
12. Block, E., Naganathan, S., Putman, D., and Zhao, S-H. *J. Agric. Food Chem.* **1992** *40*, 2418-2430.
13. Block, E.; Putman, D; Zhao, S-H *J. Agric. Food Chem.* **1992** *40*, 2431-2438.
14. Spanier, A.M.; Boyleston, T.D. *Food Chemistry.* **1994** IN PRESS
15. Dwivedi, B.K.; CRC Critical Reviews in Food Technology. **1975** *5*,487-535.
16. Bailey, M.E. *Food Technol.* **1988** *52*,123-126.
17. Bailey, M.E., Shin-Lee, S.Y., Dupuy, H.P., St. Angelo, A.J. and Vercellotti, J.R. In *Warmed-over flavor of meat.* St. Angelo, A.J.; Bailey, M.E.; Eds.; Academic Press; Orlando, Florida **1987** pp **237-266**.
18. McKay, M.J.; Bond, J.S. In, *Intracellular protein catabolism. Progress in Clinical and Biological Research.* Vol. 180. Khairallah, E.A.; Bond, J.S.; Bird, J.W.C.; Eds. Alan R. Liss, Inc.: New York, NY Vol. 180. **1985** pp. **351-361**.
19. Danehy, J.P; In, *Advances in Food Research.* Chichester, C.O.; Mrak, E.M.; Schweigert, B.S., Eds. Academic Press: New York, NY **1986**.
20. Risch, S.J.; Reineccius, G.A. Isolation of thermally generated aromas. In, *Thermal Generation of Aromas.* ACS Symposium Series 409. Parliment, T.H., McGorrin, R.J.; Ho, C-T; Eds.; ACS Books, Inc.: Washington, DC, **1989** p. 42-50.
21. Teranishi, R.; Flath, R.A.; Sugisawa, H. In, *Flavor research, recent advances.* Marcel Dekker, Inc.: New York, NY **1981**
22. Drumm, T. D.; Spanier, A. M. *J. Agric. Food Chem.* **1991** *39(2)*, 336-343.

23. Parliment, T.H.; McGorrin, R.J.,; Ho, C-T (Eds.) In *Thermal Generation of Aromas.* ACS Symposium Series 409; ACS Books, Inc.: Washington, DC **1989**.
24. Manley, C.H. Progress in the science of thermal generation of aromas. In, *Thermal Generation of Aromas.* ACS Symposium Series 409. Parliment, T.H., McGorrin, R.J. and Ho, C-T (Eds.); ACS Books, Inc.: Washington, DC, **1989** p. **12-22**.
25. Gasser, U.; Grosch, W. *Z. Lebensm. Unters. Forsch.* **1988** *86*, 489-494.
26. MacLeod, G.; Ames, J.M. *Flavour Fragrance J.* **1986** *1*, 91-104.
27. Chang, S.S.; Peterson, R.J. *J. Food Sci.* **1977** *42*,298-305.
28. Reineccius, G. A.; S. Anandarman Analysis of volatile flavors. In, *Food Constituents and Food Residues: Their Chromatographic Determination.* Lawrence, J.F., Ed.; Marcel Dekker, Inc.: New York, NY **1984** pp **195-293**.
29. Vercellotti, J.R., Kuan, J.W.; Spanier, A.M.; St. Angelo, A.J. Thermal generation of sulfur-containing flavor compounds in beef. In *Thermal Generation of Aromas.* ACS Symposium Series 409 Parliment, T.H., McGorrin, R.J. and Ho., C-T (Eds.); ACS Books, Inc.: Washington, DC, **1989** pp **452-459**.
30. Zlatkis, A.; Lichtenstein, H.A.; Tishbee, A. *Chromatographia* **1973** 6(2): 67-76.
31. Spanier, A.M.; Miller, J.A.; Bland, J.M. Lipid Oxidation. Effect on Meat Proteins. In *Lipid Oxidation in Food.* St. Angelo, A.J., Ed. ACS Symposium Series 500 ACS Books: Washington DC; **1992** pp. **104-119**.
32. Spanier, A.M.; St. Angelo, A.J.; Grimm, C.C.; Miller, J.A. The relationship of temperature to the production of lipid volatiles from beef. In *Lipids in Food Flavors.* Ho, C.T. and Hartman, T.A., Eds. ACS Symposium Series ACS Books: Washington DC; **1994** Chapter 10. In press.
33. Cross, H.R.; Bernholdt, H.F.; Dikeman, M.E.; Greene, B.E.; Moody, W.G.; Staggs, R.; West, R.L. In *Guidelines for Cookery and Sensory Evaluation of Meat.* American Meat Science Association. Chicago, Illinois **1978** pp **4-6**.

RECEIVED March 23, 1994

Chapter 6

Facts and Artifacts in *Allium* Chemistry

Eric Block[1] and Elizabeth M. Calvey[2]

[1]Department of Chemistry, State University of New York at Albany, Albany, NY 12222
[2]Center for Food Safety and Applied Nutrition, U.S. Food and Drug Administration, Washington, DC 20204

> Problems associated with distinguishing genuine genus *Allium* flavorants from artifacts formed by analytical methods and sampling techniques are considered. For flavorant isolation and extraction, methods such as supercritical carbon dioxide extraction and room temperature steam distillation are recommended. For determination of thermally unstable thiosulfinate esters and their breakdown products, methods such as gas chromatography (GC), high-performance liquid chromatography (HPLC), and supercritical fluid chromatography (SFC), as well as GC-mass spectrometry (MS), LC-MS and SFC-MS, each have strengths and weaknesses. These techniques have been used to study the stereospecificity associated with biogenesis of *Allium* sulfinyl flavorants; the effects of sulfur fertility on the pungency of hydroponically cultivated onions; and the decomposition of garlic and onion distilled oil components, bis(2-propenyl) disulfide and bis(1-propenyl) disulfide, respectively.

The "natural flavors" of garlic (*A. sativum*), onion (*A. cepa*), and other *Allium* spp., like those from many other common vegetables and fruits, are not present as such in the intact plants but are formed by enzymatic processes when the plants are chewed or cut. Under these circumstances the cells rupture, and the released flavor precursors and enzymes commingle (*1*). Additional flavors, also considered to be natural, are formed during food preparation by heating the thermal breakdown of the initial enzymatically produced flavorants in either an aqueous or nonaqueous (e.g., cooking oil) medium; heating in an aqueous medium can lead to products of hydrolysis. If the breakdown products are unstable, further compounds can be formed, which could also contribute to the aroma and taste of the food, particularly if the unstable substances have low taste thresholds. The U.S. Code of Federal Regulations recognizes all of these possibilities in defining natural flavors as "the essential oil, oleoresin, essence or extractive, protein hydrolysate, distillate or any product of roasting, heating or enzymolysis, which contains the flavoring constituents derived from a spice, fruit juice, vegetable or vegetative juice, edible yeast, herb, bark, bud, root, leaf or similar plant material, meat, seafood, poultry, eggs, dairy products, or fermentation products thereof, whose significant function in food is flavoring rather than nutrition (*2*)." The complex sequence of events that can occur when garlic, onion and other *Allium* spp.

are processed provides a special challenge to those concerned with identifying the natural flavors. Qualitative and quantitative analyses of *Allium* organosulfur flavorants take on added significance because of the current interest in the reported health benefits of consumption of fresh *Allium* spp. (*3, 4*) and the claims of numerous health food products to duplicate these benefits in pill form.

The problems addressed in this chapter stem from the instability (thermodynamic) and reactivity (kinetic) of many *Allium*-derived organosulfur compounds, which make it difficult to distinguish the genuine flavorants, as defined above, from artifacts formed by analytical methods and sampling techniques (*5*). Nobel Laureate Artturi I. Virtanen notes, "literature contains much data concerning substances in plants which in reality are not present at all (*1*)." Jennings and Takeoka emphasize that in the analysis of terpenoid flavorants, "great care must be taken to avoid the formation of artifacts during sample storage, sample preparation, and subsequent chromatographic analysis (*6*)." In our opinion, there can be no substitute for synthesizing suspected flavorants and studying their behavior under well-defined analysis and sampling conditions. Furthermore, rigorous application of modern mechanistic theory to considerations of flavorant formation and decomposition can aid evaluation of previous research in this area and can assist in the identification of new flavorants. In this chapter these various points will be illustrated with examples from the authors' work and from the literature.

Background

The first reports concerning isolation of organic compounds from garlic and onion by Wertheim in 1844 and Semmler in 1892 in Germany identified diallyl sulfide (corrected in 1892 to the disulfide) and a propenyl propyl disulfide as the principal components of the distilled oils ("ätherische Öl") of garlic and onions, respectively (*7*). In the 1940s these and other volatile sulfur compounds were found to be secondary compounds formed by enzymatic action on precursors in the intact bulb. Thus in 1945, Cavallito et al. in Rennselaer, New York, reported that when garlic cloves were frozen in dry ice, pulverized, and extracted with acetone, "the acetone extracts upon evaporation yielded only minute quantities of residue and no sulfides, indicating the absence of free sulfides in the plant.... The [white garlic] powder had practically no odor, but upon addition of small quantities of water, the typical odor was detected and the antibacterial principle [**1**, Scheme 1] could be extracted and isolated. This demonstrates that neither I [**1**] nor the allyl sulfides found in 'Essential Oil of Garlic' are present as such in whole garlic.... When the powder was heated to reflux for thirty minutes with a small volume of 95% ethanol, no activity could be demonstrated by addition of water to the insoluble residue.... When, however, a small quantity (1 mg per cc) of fresh garlic powder was added to the alcohol insoluble fraction in water (20 mg per cc), the activity of the treated sample was shown to be equal to that of the original untreated powder.... The 95% ethanol treatment has inactivated the enzyme required for cleavage of the precursor and addition of a small quantity of fresh enzyme brought about the usual cleavage (*8*)."

Pioneering studies in the 1940s by Stoll and Seebeck (*7*) in Basel demonstrated that the stable precursor of Cavallito et al.'s antibacterial principle of garlic (**1**, Scheme 1) is (+)-*S*-2-propenyl L-cysteine *S*-oxide (**2**, alliin). In the intact cell, alliin and related *S*-alk(en)yl-L-cysteine *S*-oxides (aroma and flavor precursors) are located in the cytoplasm and the C-S lyase enzyme allinase in the vacuole. Disruption of the cell releases allinase, which causes subsequent α,β-elimination of the sulfoxides, ultimately affording volatile and odorous low-molecular-weight organosulfur compounds such as S-2-propenyl 2-propenethiosulfinate (**1**, allicin) by a process thought to involve an intermediary sulfenic acid, 2-propenesulfenic acid (**3**). Allicin and related symmetrically substituted thiosulfinates are easily synthesized via peracid oxidation of the corresponding disulfide, e.g. diallyl disulfide (**4**). Four sulfoxides occur in *Allium* spp.: *S*-2-propenyl, *S*-(*E*)-1-propenyl, *S*-methyl and *S*-propyl L-cysteine *S*-

oxides (**2, 5-7** respectively, Scheme 2). Onions contain compounds **5-7** while garlic contains compounds **2, 5** and **6**; the ratio of the amounts of compounds **2:5:6** in garlic is 86:3:10 (*4*). Scheme 2 shows the distribution of the four sulfoxides in common *Allium* spp.

The distinctive flavors of the various *Allium* spp. reflect the levels of compounds **2, 5-7**, particularly **2**, which is the precursor to the characteristic allylic thiosulfinates of garlic, and **5** (isoalliin), which is the precursor to the onion lachrymatory factor (LF) propanethial *S*-oxide (**8a**, Scheme 7 below) bis-(thial) **8b**, α,β-unsaturated thiosulfinates, and onion flavorant 2,3-dimethyl-5,6-dithiabicyclo[2.1.1]hexane 5-oxide (zwiebelane) **9**, which is thought to be formed from 1-propenyl 1-propenethiosulfinate (*7*). Also present in *Allium* spp. are 24 γ-glutamyl peptides of sulfur amino acids. These peptides function as storage compounds of nitrogen and sulfur (which may be precursors to plant chemical defense agents) and may have a role in transport of amino acids across cell membranes (*7*). Recently Sendl et al. reported that γ-glutamyl-*S*-alkylcysteines inhibit the blood pressure-regulating hormone (*9*). It should also be noted that garlic's fresh-weight total sulfur content of 3.5 mg/g (*4*) is higher than that of any other common vegetable or fruit.

Allicin (**1**), which possesses the characteristic aroma and taste of fresh garlic, is quite reactive and unstable. It hydrolyses on heating in water to give diallyl disulfide (**4**), diallyl trisulfide, and the corresponding polysulfides. (Allicin is moderately stable in water at room temperature because of the protective action of hydrogen bonding.) As for possible health benefits associated with consumption of garlic, diallyl trisulfide is said to possess antithrombotic activity and, at high doses, diallyl sulfide and disulfide are reported to be antitumorigenic (*4*). Furthermore, allicin shows significant antibacterial activity, and along with diallyl disulfide, inhibits enzymes and cofactors involved in cholesterol and fatty acid biosynthesis (acetyl-CoA synthetases and 3-hydroxy-3-methylglutaryl CoA, respectively)(*7*). The *in vivo* antibiotic activity of allicin and related thiosulfinates toward microorganisms capable of reducing nitrate to nitrite could explain the activity of *Allium* spp. toward stomach cancer, suggested by epidemiological evidence (*3, 4*).

Attempts by Brodnitz et al. in 1971 to determine allicin by gas chromatography-mass spectrometry (GC-MS) provided evidence for an unusual mode of decomposition (*10*). They indicated that GC caused diallyl thiosulfinate to dehydrate according to the reaction $C_6H_{10}S_2O \rightarrow C_6H_8S_2 + H_2O$, affording a 2.4:1 mixture of two compounds, claimed (incorrectly as we shall see below) to be 3-vinyl-4*H*-1,2-dithiin (**10**, equation 1) and 3-vinyl-6*H*-1,2-dithiin (**10a**, equation 1) (*10*). Analytical methods for allicin quantitation that rely on GC detection of compounds **10** and **10a** are still used (*11*), even though serious problems exist with this method and more accurate high-performance liquid chromatographic (HPLC) methods have been developed (see below).

Recent Work on *Allium* spp. Flavorants

In our work, chromatography of solutions prepared by soaking chopped garlic in methanol afforded allicin (**1**); diallyl di-, tri-, and tetrasulfide; allyl methyl trisulfide; 3-vinyl-4*H*-1,2-dithiin (**10**) and 2-vinyl-4*H*-1,3-dithiin (**11**); along with two isomeric polar compounds termed (*E*)- and (*Z*)-ajoene (based on the Spanish word for garlic, ajo, pronounced "aho"), ultimately characterized as (*E,Z*)-4,5,9-trithiadodeca-1,6,11-triene 9-oxide (**12**) by spectroscopic and synthetic methods (Scheme 3)(*12*). Compounds **10-12** showed antithrombotic activity in vitro; ajoene also displayed antifungal properties (*3, 4, 7, 12*). Compound **12** had previously been misidentified as allyl-1,5-hexadienyltrisulfide (*13*). When synthetic allicin was refluxed with acetone:water (3:2) for 4 hours, compound (*E,Z*)-**12** was isolated in 34% yield (*12*). Other products included compounds **10** and **11**, in a 1:4.4 ratio, and diallyl di- and trisulfide. The structures of compounds **10** and **11** were established by comparison with the

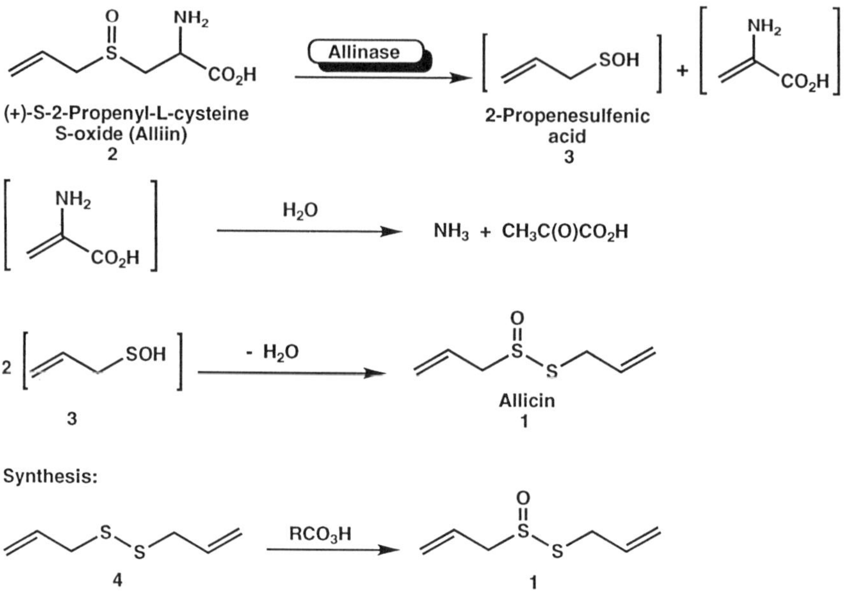
Allinase
(+)-S-2-Propenyl-L-cysteine S-oxide (Alliin)
2
2-Propenesulfenic acid
3
H2O
NH3 + CH3C(O)CO2H
- H2O
Allicin
1
Synthesis:
RCO3H
4
1

Scheme 1

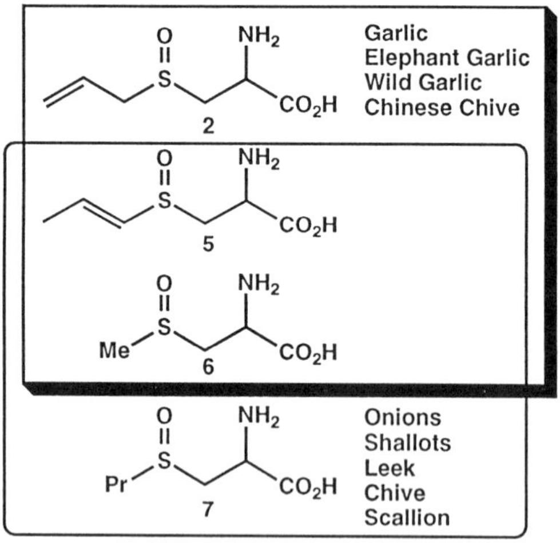
Garlic
Elephant Garlic
Wild Garlic
Chinese Chive
2
5
6
7
Onions
Shallots
Leek
Chive
Scallion

Scheme 2

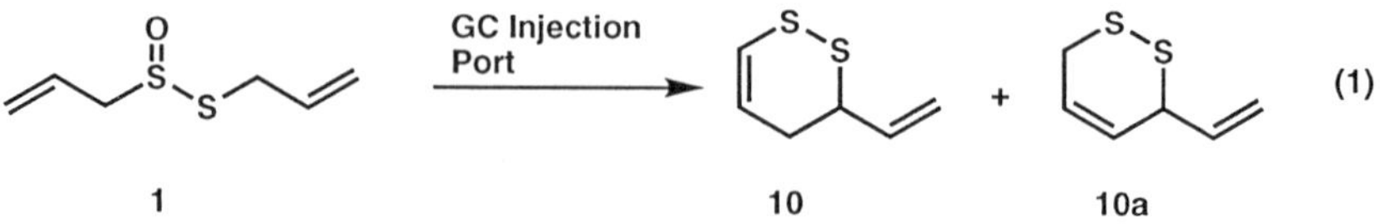
GC Injection Port
1
10
10a

Equation 1.

known compounds, *also formed in a 1:4.4 ratio* from dimerization at -180 °C of thioacrolein, $CH_2=CHCH=S$ (13), from flash vacuum pyrolysis of diallyl sulfide (*12*).

Our isolation of compounds 10 and 11 from decomposition of allicin and the comparison of their abundance ratio with that observed from dimerization of thioacrolein (13, Scheme 3) suggests that the first step in the decomposition of allicin involves Cope-elimination to 2-propenesulfenic acid (3) and thioacrolein (13). This postulate, which follows our proposal for decomposition of methyl methanethiosulfinate (Scheme 4), is also supported by the observation of the characteristic sapphire blue color of thioacrolein (13) when allicin is distilled into a liquid nitrogen cooled trap and our isolation of 2- and 3-carboethoxy-3,4-dihydro-2*H*-thiopyran and (*Z*)-2-carbethoxy-4-pentenethial *S*-oxide when solutions of allicin in ethyl acrylate and ethyl propiolate, respectively, are kept overnight at room temperature (Scheme 5). Presumably, (*Z*)-2-carbethoxy-4-pentenethial *S*-oxide results from sulfoxide thio-Claisen rearrangement of the initial adduct of ethyl propiolate and thioacrolein (13). The earlier work of Brodnitz et al. (*10*) (Scheme 2) on the decomposition of allicin in a gas chromatograph must be corrected by replacing his incorrect structure 10a with that of 11. Because other thiosulfinates of structure $RS(O)SCH_2CH=CH_2$, such as $MeS(O)SCH_2CH=CH_2$ and $MeCH=CHS(O)SCH_2CH=CH_2$, have been found in garlic extracts (see below), and these thiosulfinates can also give thioacrolein by the mechanism of Scheme 3, then formation of the two dithiins 10 and 11 cannot be used to assay for allicin alone.

Support for our mechanism of thermal decomposition of allicin (1) was obtained by supercritical fluid chromatography-MS (SFC-MS) of a methylene chloride extract of fresh garlic (*14*). Initial open-tubular SFC data showed two chromatographic peaks. Using an interface to the mass spectrometer equipped with a variable temperature restrictor tip through which the supercritical carbon dioxide expands into the MS source, it was observed that at a restrictor tip temperature of 115 °C the most abundant ion of the major chromatographic peak was *m/z* 163, which is the MH$^+$ for allicin. The minor chromatographic peak was identified by its mass spectrum as a mixture of unresolved isomers of allyl methyl thiosulfinate ($MeS(O)SCH_2CH=CH_2$ and $MeS-S(O)CH_2CH=CH_2$). When the restrictor tip temperature was increased to 165 °C, the *m/z* 163 ion of allicin disappeared and was replaced by the base ion *m/z* 73 and another major ion at *m/z* 91, in addition to ions at *m/z* 145, 129, 115 and 99, which are indicative of formation of thioacrolein dimers. The *m/z* 73 and 91 ions correspond to protonated thioacrolein (13-H$^+$) and protonated 2-propenesulfenic acid (3-H$^+$), respectively, both originating from the thermal decomposition of allicin shown in Scheme 3. With a different SFC-MS interface, the abundance of the *m/z* 145 ion relative to the *m/z* 73 base peak was greatly reduced, suggesting that in this case sufficient energy was provided in the interface to completely decompose allicin under unimolecular conditions before substantial amounts of the thioacrolein dimers could form.

We sought to answer the question "What compounds are primarily responsible for the characteristic flavor of freshly cut members of the genus *Allium*?" Because of their excellent resolution and mass identification capabilities, GC and GC-MS have figured prominently in the effort to characterize *Allium* volatiles. This characteristic has occurred despite the previous warning that many of the compounds from *Allium* spp. by GC may be "artifacts of analysis" (*15*) and despite more recent work suggesting that the resolution of HPLC may be better. For example, the leek (*A. porrum* L.) thiosulfinate, propyl propanethiosulfinate (PrS(O)SPr), is the most attractive compound for the leek moth *Acrolepiopsis assectella*. Auger et al., however, concluded that "the large number of other sulfur-containing substances identified in these odors are thus only analytical artifacts created by sample manipulation, passage to the liquid state, heating, etc., except perhaps for a certain proportion of disulfides, which are always present. The disulfides could arise directly from ruptured cells where a small proportion of thiosulfinate could be degraded to disulfide before entering the vapor phase (*16*)." They further stated that "on long gas chromatographic

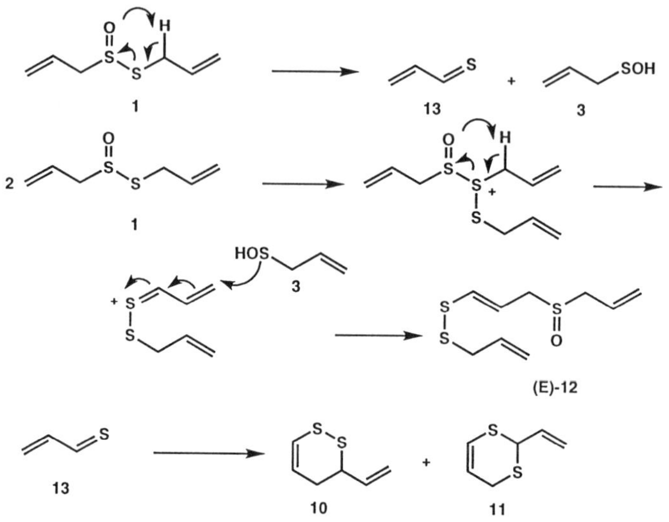

Scheme 3

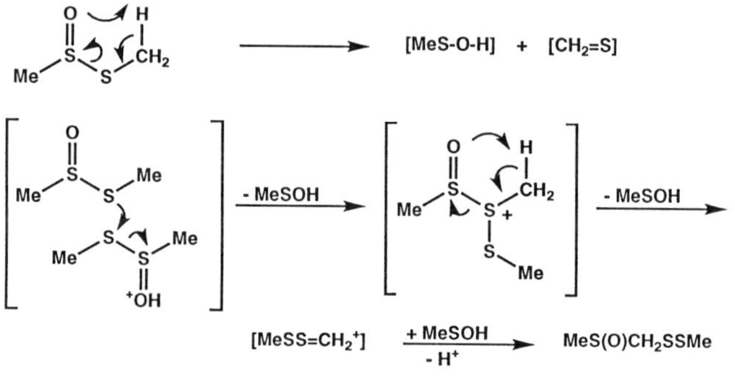

Scheme 4

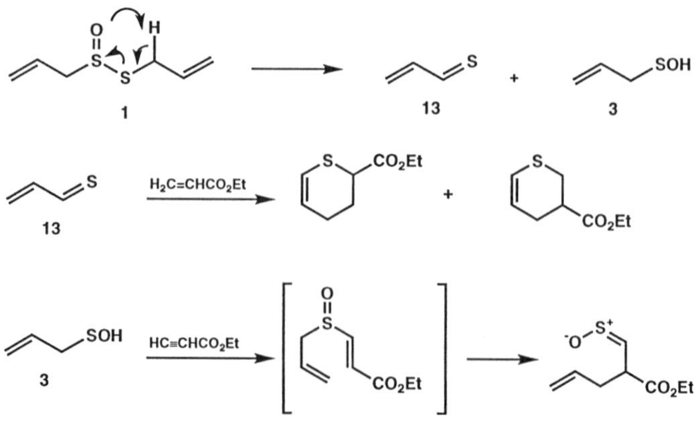

Scheme 5

columns, propyl propanethiosulfinate is broken down, possibly by mechanisms other than simple disproportionation, into dipropyl disulfide since propanethiol and dipropyl trisulfide appear, apart from dipropyl disulfide and propyl propanethiosulfonate. In rapid chromatography (very short column), the thiosulfinates do not break down. This technique allows one to conclude that leek odor just emitted contains primarily propyl propanethiosulfinate and perhaps a small amount of dipropyl disulfide. The majority of sulfur volatiles identified by GC-MS in *Allium* spp. are thus artifacts produced during the isolation of the sample and during chromatography (*17*)." Other classes of natural products found in fruits and vegetables, such as terpenes and retinoids, are well known to be particularly labile under conventional GC conditions. They are sensitive to hot injector surfaces and partly inactivated column packings and therefore undergo acid-catalyzed rearrangment and thermal degradation (*18*).

In 1974, Block and O'Connor established that aliphatic thiosulfinates with up to eight carbon atoms survive GC on 0.32 cm x 1.8 m packed columns (*19*). Because synthetic specimens of most possible thiosulfinates resulting from allinase cleavage of precursors 2, 5-7 were available (Scheme 6)(*7*), extracts and room-temperature steam distillates (e.g., steam distillation under vacuum) of various *Allium* spp. were examined using LC-MS, GC-MS and UV and ^{1}H NMR spectroscopy to confirm product identities. On the basis of this work we concluded that GC, as typically performed with elevated injector and column temperatures, presents an erroneous picture of the composition of both headspace volatiles and room-temperature extracts from *Allium* spp. and that HPLC provides a reliable qualitative and quantitative measure of what is actually present (*20*). Thus, analysis of *Allium* extracts or vacuum distillates by both normal (Si) and reversed phase (C_{18}) HPLC, using diode-array UV detection and LC-MS, accompanied by ^{1}H NMR analysis of these same materials, indicated that the predominant constituents were thiosulfinates. We found no evidence from HPLC results for the presence in these materials of significant quantities of the polysulfides and thiophenes claimed by previous GC-MS studies!

By using Si-HPLC and benzyl alcohol as internal standard, almost all of the thiosulfinates from *Allium* spp. can be separated and quantified. The thiosulfinates and related flavorants found in garlic and onion extracts are shown in Scheme 7. Contrary to earlier reports (*21, 22*) but consistent with the observations of Lawson et al. (*23*), the 1-propenyl and not the *n*-propyl group is present in garlic as well as in wild garlic (*A. ursinum*), elephant garlic (*A. ampeloprasum* L. var. *ampeloprasum* auct.), and Chinese chive (*A. tuberosum* L.). In the case of onion, shallot (*A. ascalonicum* auct.), scallion (*A. fistulosum* L.), chive (*A. schoenoprasum* L.), Chinese chive, and leek, the flavorant profiles could be established using cryogenic GC-MS methods (fused silica capillary column, on-column injection with injection port and column initially at 0 °C followed by slow simultaneous warming of injector and oven to 200 °C, Figure 1) (*24*).

We have used chiral γ-cyclodextrin GC columns under isothermal GC-MS conditions (column temperature 90-120 °C) and found that enantiomers of *trans*-zwiebelane (9a, Figure 2) and the thiosulfinates MeS(O)SMe, MeS(O)SPr-*n*, and MeSS(O)Pr-*n* (Figure 3) can be resolved and that individual enantiomers are stable under the analytical conditions (*25*). However, analysis of an onion extract on the chiral column showed that all of these compounds were present as racemic mixtures, suggesting that asymmetric induction is not involved in their formation from achiral sulfenic acids.

Even under the mildest GC-MS conditions, the allylic thiosulfinates were completely decomposed, indicating that GC-MS is unsuitable for separation and identification of garlic flavorants. Among the types of thiosulfinates studied, MeCH=CHS-S(O)R undergoes rapid E-Z interconversion (*7*). Clearly, great caution should be used in determining "natural" ratios of easily isomerized compounds of this type.

The constituents of SC-CO_2 extracts of fresh garlic were separated by reversed-phase HPLC, and the effluent was directly analyzed by HPLC-MS using thermospray and tandem MS (MS-MS) methods (*26*). Products included allicin (MH+, *m/z* 163),

HC≡CCH₂Br + PrSNa → 94% → HC≡CCH₂SPr → NaOMe 85% →

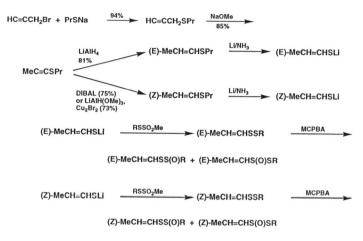

(E)-MeCH=CHSLi → RSSO₂Me → (E)-MeCH=CHSSR → MCPBA →

(E)-MeCH=CHSS(O)R + (E)-MeCH=CHS(O)SR

(Z)-MeCH=CHSLi → RSSO₂Me → (Z)-MeCH=CHSSR → MCPBA →

(Z)-MeCH=CHSS(O)R + (Z)-MeCH=CHS(O)SR

Scheme 6

C2, C4 and C6 Thiosulfinates in Garlic Extracts

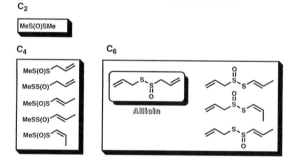

C2, C4 and C6 Sulfur Compounds in Onion Extracts

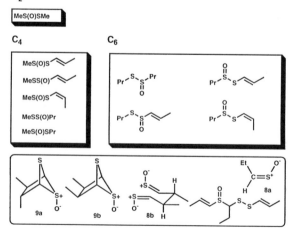

Scheme 7

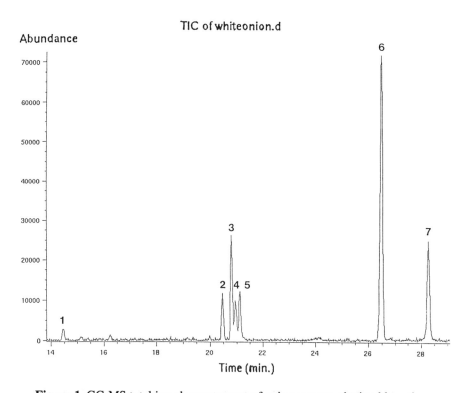

Figure 1. GC-MS total ion chromatogram of onion extract, obtained by using a 30 m x 0.53 mm methyl silicone gum capillary column, injector and column programmed from 0 to 200 °C. Peak identification: 1-MeS(O)SMe; 2-MeS(O)SPr-n; 3-(E,Z)-MeCH=CHSS(O)Me; 4-MeSS(O)Pr-n; 5-(E)-MeCH=CHS(O)SMe; 6-(E,Z)-MeCH=CHSS(O)Pr-n, *trans*-zwiebelane (**9a**); 7-*cis*-zwiebelane (**9b**).

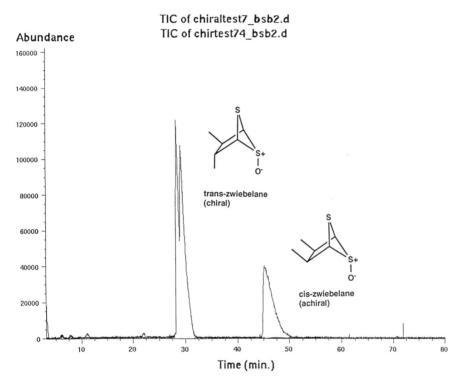

Figure 2. GC-MS total ion chromatogram showing chiral *trans*-zwiebelane (**9a**) and achiral *cis*-zwiebelane (**9b**) obtained by using a 30 m x 0.32 mm γ-cyclodextrin capillary column at 120 °C.

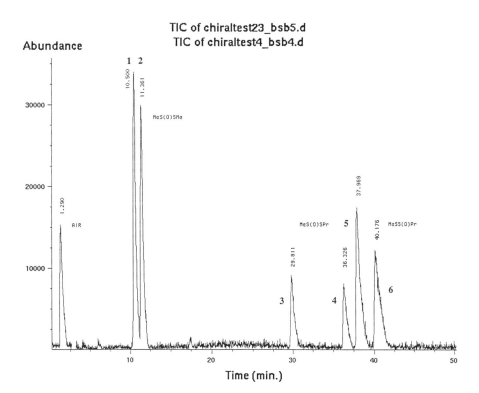

Figure 3. GC-MS total ion chromatograph obtained by using a 30 m x 0.32 mm γ-cyclodextrin capillary column at 80 °C. Peak identification: 1,2-enantiomers of MeS(O)SMe; 3,4-enantiomers of MeS(O)SPr-n; 5,6-enantiomers of MeSS(O)Pr-n.

isomeric 1-propenyl 2-propenyl thiosulfinates (MH+, m/z 163), and isomeric allyl methyl and allyl 1-propenyl thiosulfinates (both MH+, m/z 137). The HPLC chromatograms of the 35-36 °C SC-CO_2 extract and a CH_2Cl_2 extract showed excellent agreement. The use of a solid-phase SC-CO_2 trap (e.g., condensation on glass beads) led to extraction of 25% *more* allicin than could be extracted with CH_2Cl_2. An increase in the amount of thermal decomposition products with respect to allicin was evident when garlic was extracted with SC-CO_2 at temperatures greater than 36 °C.

Sinha et al. reported that GC-MS analysis of the SC-CO_2 extract of onion showed the presence of "28 sulfur-containing compounds including diallyl thiosulfinate (or its isomer, di-1-propenyl thiosulfinate), propyl methanethiosulfonate, dithiin derivatives, diallyl disulfide," and diallyl trisulfide along with 13 other compounds also found in steam-distilled onion oil (27). None of these compounds were previously seen by us using HPLC, cryogenic GC-MS, or NMR analysis of CH_2Cl_2 extracts of several varieties of onion (19), nor is there any evidence in the literature for the presence of allylic compounds in onion. No standards were used by Sinha et al. to verify assignments. Because the GC-MS conditions used by Sinha et al. included GC injector temperatures of 280 °C, which is high enough to cause immediate rearrangement of "di-1-propenyl thiosulfinate [1-propenyl 1-propenethiosulfinate]" or decomposition of the very labile allicin, it is unlikely that their claims are correct (see below).

We have independently examined the SC-CO_2 extraction of onion, which judged by GC-MS, was ca. 69% as efficient as extraction with diethyl ether. The SC-CO_2 mixture was identical to that of the ether extract, containing thiosulfinates, zwiebelanes (9), and propanethial S-oxide (8a)(26). Because our cryogenic GC-MS conditions do not permit analysis of allicin (1), an SC-CO_2 extract of onion was also subjected to Si-HPLC. The extract was then spiked with a standard solution of allicin containing the internal standard benzyl alcohol. The chromatogram of the spiked onion extract clearly shows that allicin was not present in the original extract. LC-MS shows that although a chromatographic component eluting near allicin had an MH+ ion of m/z 163, its spectrum and retention time did not match those of the allicin standard. The difference in the LC-MS-MS spectra further confirmed that this compound was not allicin (28). These results are quite different from those reported by Sinha et al. (27).

We have used the cryogenic GC-MS procedure to examine the effects of sulfur fertility on pungency of hydroponically cultivated Southport white globe onions (29). For particular levels of sulfur (S) fertility the corresponding values of combined thiosulfinates and zwiebelanes (9, T + Z) and the levels of enzymatically released pyruvate (P)(given as meq/L S), nMol T+Z))/g F.W., and nMol P/g F.W. P are the following: 0.10 S, 43.2 T+Z, 1400 P; 0.48 S, 56.3 T+Z, 2000 P; 0.85 S, 96.6 T+Z, 4300 P; 1.60 S, 171 T+Z, 5000 P; and 3.10 S, 297 T+Z, 9700 P. A second series of analyses for onions cultivated a year later under similar conditions gave similar results, i.e., an almost linear dependency of levels of organosulfur flavorants on the level of sulfur fertility. On a molar basis the amount of pyruvate formed averages 35 times that of combined thiosulfinates and zwiebelanes. This result is similar to that reported by Thomas et al. (30), who used N-ethylmaleimide to measure thiosulfinate (but not zwiebelane) levels in onion as well as in garlic homogenates. They suggest that the low thiosulfinate:pyruvate ratio in onion homogenates reflects conversion of much of the intermediate sulfenic acid to the LF, propanethial S-oxide (8a).

Although thiosulfinates, zwiebelanes (9), and thial S-oxides, such as compound 8 are most likely the *primary* flavorants, disulfides and other compounds formed by hydrolysis or other reactions of the primary compounds during cooking or other processing may still be important *Allium* spp. aroma and flavor components. Particularly interesting chemistry is associated with the major disulfides from garlic and onion, bis(2-propenyl) disulfide (diallyl disulfide) (4) and bis(1-propenyl) disulfide (14), respectively. Upon heating, the former compound undergoes a complex sequence of reactions leading both to diallyl polysulfides as well as to a series of acyclic and heterocyclic compounds (Scheme 8) resulting from generation and reaction of thioacrolein (13) and radical species $CH_2=CHCH_2S_n\bullet$ with precursor diallyl polysulfides (*31*). Most of these compounds have been identified in distilled oil of garlic and may be assumed to arise from thermal breakdown of diallyl disulfide (4). A comparison of the results of GC and HPLC analyses of distilled oil of garlic leads to an unusual observation: Simply because a product has been distilled does not guarantee that all distilled components will survive GC analysis. Thus, Lawson et al. has reported that a number of the higher molecular weight polysulfide components, RS_nR' (n = 5, 6), of garlic oil that are detected by C_{18}-HPLC decompose during GC (*32*).

The subtleties associated with *Allium* spp. flavorants are also illustrated by the thermal chemistry of bis(1-propenyl) disulfide (14). Isomers of compound 14 and 3,4-dimethylthiophene (15) have both been frequently identified in distillates and extracts of *Allium* spp. by GC-MS and therefore might be considered "natural flavors." Because compound 14 was reported to decompose in the inlet system of a mass spectrometer to produce compound 15 (*33*), the thermal chemistry of 14 was examined in dilute solution. We discovered that compound 14 undergoes facile conversion at 80 °C to *cis*- and *trans*-2-mercapto-3,4-dimethyl-2,3-dihydrothiophene (16, Scheme 9) (*34*) which in turn decomposes at slightly higher temperatures to 15. We therefore suggested that in *Allium* distilled oils, compound 15 originates, at least in part, from 16 by loss of hydrogen sulfide (*34*). Shortly after publication of our work, Kuo et al. reported the presence of compounds 15 and 16, as well as methyl, *n*-propyl, and 1-propenyl 3,4-dimethyl-2-thienyl disulfides (17a-c, respectively), in distilled oils and extracts of Welsh onion (*A. fistulosum* L. var. maichuon) and scallion (*35-37*) while Sinha et al. (*27*) reported that compounds 15 and 17a are significant components of SC-CO_2 extracts of onion. Investigators (*33,35*) have suggested that compounds 15 and 17 are formed by a process involving radical addition of MeCH=CHS• to MeCH=CHSSR followed by ring-closure and aromatization (Scheme 10). Dihydrothiophenes, such as compound 18, are probable intermediates in the aromatization steps and might result from reaction of 16 with RSX (R = Me, n-Pr, MeCH=CH-, X = SR' or S(O)R').

We found that compound 16 could be aromatized at room temperature by direct treatment of an unpurified solution in benzene with a slight excess of 2,3-dichloro-5,6-dicyano-1,4-benzoquinone (DDQ) giving 3,4-dimethyl-2-thiophenethiol (19a) in fair yield. Thioalkylation of the corresponding lithium salt (19b) with methanesulfonothioic acid *S*-methyl ester (MeSSO$_2$Me, 20a; Scheme 11) affords compound 17a (*38*). A more satisfactory procedure involves conversion of compound 16 to bis(1,5-dihydro-3,4-dimethyl-2-thienyl) disulfide (21) by using methanesulfonyl chloride, DDQ aromatization of 21 to bis(3,4-dimethyl-2-thienyl) disulfide 22, reductive cleavage of 22 with lithium triethylborohydride, and thioalkylation of the resultant lithium salt (19b) with RSSO$_2$Me (R = Me, *n*-Pr, or MeCH=CH; 20a-c, respectively) giving 17a-c (*38*). These compounds possess mass spectra identical with those reported for 17a-c found by GC-MS in distilled oils from Welsh onion and scallion (*35-37*), confirming the identity of the naturally-derived compounds. On the other hand, comparing cryogenic GC-MS (*23*) data for SC-CO_2 extracts of onion (*26*) with data obtained under identical conditions using synthetic 17a, we were unable to reproduce the claims of Sinha et al. (*27*) for the presence of either 17a or major (i.e., 55%) amounts of an "isomer" of 17a in onion SC-CO_2 extracts. We suspect that in this latter work, compound 17a and its isomer are artifacts whose formation is associated with the 280 °C GC injector temperature used in the analysis and that neither these compounds nor their immediate precursors (14 or 16) are actually present in onion SC-CO_2 extracts (*37*). Because compound 17a had a nutty flavor with a 5-ppm detection threshold, 17b had a green vegetative flavor at the 10-ppm detection threshold and a cabbage flavor at higher levels, and 17c had a green, slightly sulfury and nutty flavor at a 20-ppm detection threshold, we suggest that compounds 17a-c do not have characteristic fresh onion flavors but may contribute to the flavor of roasted onion or other *Allium* spp.

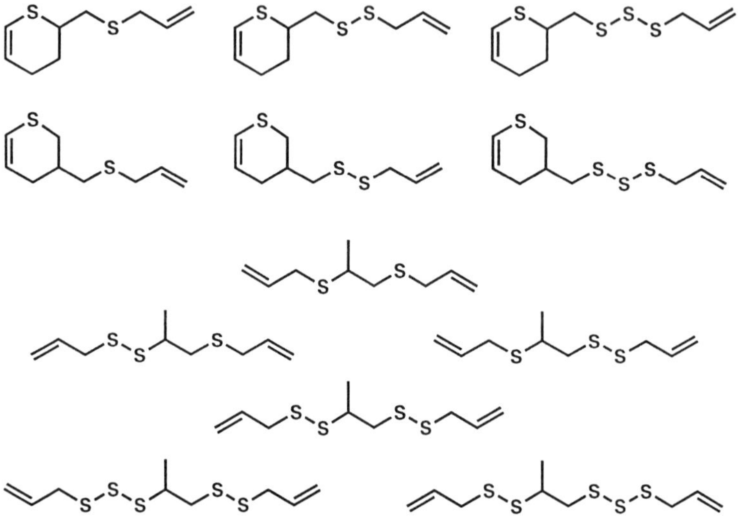

Scheme 8

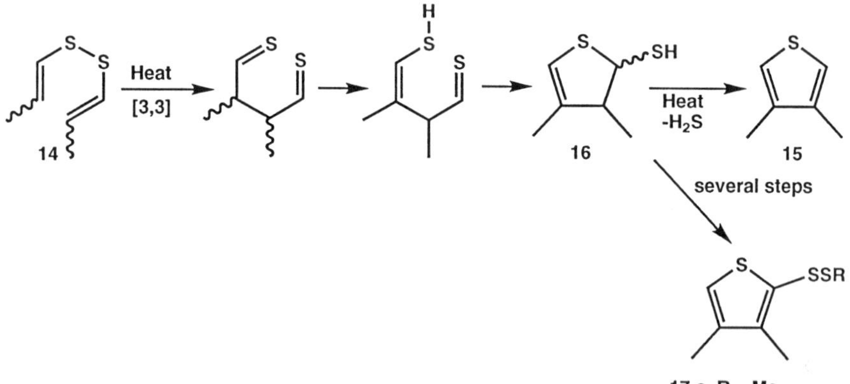

Scheme 9

17 a, R = Me,
b, R = n-Pr
c, R = CH=CHMe

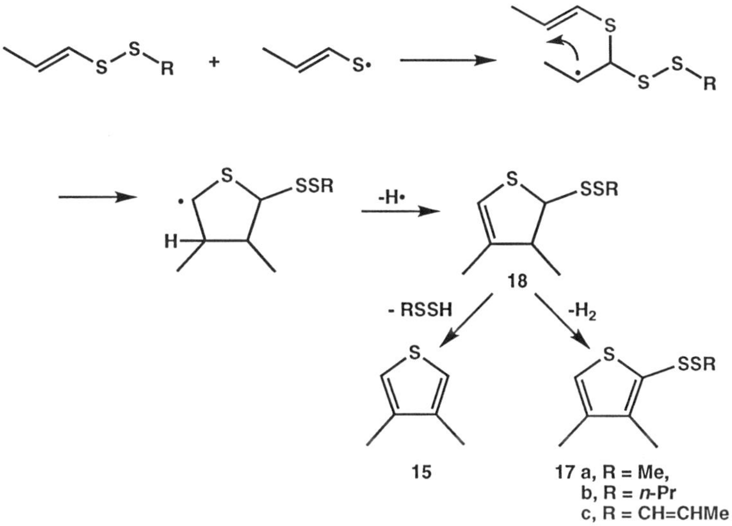

6. BLOCK AND CALVEY *Facts and Artifacts in* Allium *Chemistry*   77

Scheme 10

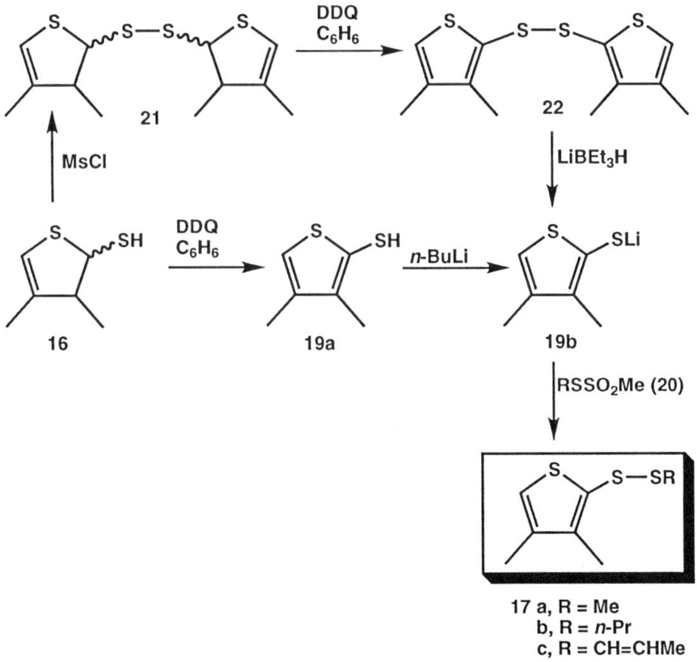

Scheme 11

17 a, R = Me
b, R = *n*-Pr
c, R = CH=CHMe

One final example of artifact formation in *Allium* chemistry can be cited. Along with a variety of biologically active α-sulfinyl disulfides (RS(O)CHEtSSR') isolated from onion (*39, 40*) and termed "cepaenes" (*39*), a methanolic extract of onion contained 3-ethyl-6-methoxy-2,4,5-trithiaoctane 2-*S*-oxide (MeS(O)CHEtSSCH(OMe)Et), which was assumed to arise from addition of methanol to the onion LF (**8**)(*40*).

Acknowledgments

We thank our skilled colleagues, whose names are listed in the references, for contributing to the work reported herein. We gratefully acknowledge support from the National Science Foundation, the NRI Competitive Grants Program/U.S. Department of Agriculture (Award No. 92-37500-8068), and McCormick & Company Inc. The State University of New York at Albany NMR and MS facilities are funded in part by instrument grants from the National Science Foundation.

Literature Cited

1. Virtanen, A.I. *Phytochem.* **1965**, *4*, 207.
2. Code of Federal Regulations *21*, 101.22.a.3.
3. Block, E. In *Food Phytochemicals for Cancer Prevention*; C.T. Ho, Ed.; ACS Symposium Series 546; American Chemical Society: Washington, DC, 1994; Chapter 5, pp 84-96.
4. Lawson, L.D. In *Human Medicinal Agents from Plants*; A.D. Kinghorn and M.F. Balandrin, Eds.; ACS Symposium Series 534; American Chemical Society: Washington, DC, 1993, p. 306.
5. Block, E. *J. Agric. Food Chem.* **1993**, *41*, 692.
6. Jennings, W.; Takeoka, G. In *Methods in Enzymology*, Vol. 111; J.H. Law and H.C. Rilling, Eds.; Academic Press: Orlando, FL, 1985; p. 149.
7. Block, E. *Angew. Chem., Int. Ed. Engl.* **1992**, *31*, 1135.
8. Cavallito, C.J.; Bailey, J.H.; Buck, J.S. *J. Am. Chem. Soc.* **1945**, *67*, 1032.
9. Sendl, A.; Elbl, G.; Steinke, B.; Redl, K.; Breu, W.; Wagner, H. *Planta Med.* **1992**, *58*, 1.
10. Brodnitz, M.H.; Pascale, J.V.; Van Derslice, L. *J. Agric. Food Chem.* **1971**, *19*, 273.
11. Saito, K.; Horie, M.; Hoshino, Y.; Nose, N.; Mochizuki, E.; Nakazawa, H.; Fujita, M. *J. Assoc. Off. Anal. Chem.* **1989**, *72*, 917.
12. Block, E.; Ahmad, S.; Catalfamo, J.; Jain, M.K.; Apitz-Castro, R. *J. Am. Chem. Soc.*, **1986**, *108*, 7045.
13. Apitz-Castro, R.; Cabrera, S.; Cruz, M.R.; Ledezma, E.; Jain, M.K. *Thrombosis Res.* **1983**, *32*, 155.
14. Calvey, E.M.; Roach, J.A.G.; Block, E. *J. Chromatogr. Sci.*, in press.
15. Saghir, A.R.; Mann, L.K.; Bernhard, R.A.; Jacobsen, J.V. *Proc. Am. Soc. Hort. Sci.* **1964**, *84*, 386.
16. Auger, J.; Lalau-Keraly, F.X.; Belinsky, C. *Chemosphere* **1990**, *21*, 837.
17. Auger, J.; Lecomte, C.; Thibout, E. *J. Chem. Ecol.* **1989**, 15, 1847.
18. Furr, H.C.; Clifford, A.J.; Jones A.D., In *Methods in Enzymology*, Vol. 213; L. Packer, Ed.; Academic Press: Orlando, FL, **1992**; p. 281.
19. Block, E.; O'Connor, J. *J. Am. Chem. Soc.* **1974**, *96*, 3921.
20. Block, E.; Naganathan, S.; Putman, D.; Zhao, S.H. *J. Agric. Food Chem.* **1992**, *40*, 2418.
21. Fujiwara, M.; Yoshimura, M.; Tsuno, S. *J. Biochem.* **1955**, *42*, 591.
22. Whitaker, J.R. *Adv. Food Res.* **1976**, *22*, 73.
23. Lawson, L.D.; Wood, S. G.; Hughes, B.G. *Planta Med.* **1991**, *57*, 263.
24. Block, E.; Putman, D.; Zhao, S.H. *J. Agric. Food Chem.* **1992**, *40*, 2431.
25. Block, E.; Putman, D.; Littlejohn, M. Unpublished results.

26. Calvey, E.M.; Betz, J.M.; Matusik, J.E.; White, K.D.; Block, E.; Littlejohn, M.; Naganathan, S.; Putman, D., *J. Agric. Food Chem.*, manuscript submitted.
27. Sinha, N.K.; Guyer, D.E.; Gage, D.A.; Lira, C.T. *J. Agric. Food Chem.* **1992**, *40*, 842.
28. Matusik, J.E.; Calvey, E.M.; Block, E., paper presented at 41st ASMS Conference on Mass Spectrometry, May 30-June 4, 1993, San Francisco, CA.
29. Randall, W.; Block, E.; Putman, D.; Litlejohn, M., *J. Agric. Food Chem.*, manuscript submitted.
30. Thomas, D.J.; Parkin, K.L.; Simon, P.W. *J. Sci. Food Agric.* **1992**, *60*, 499.
31. Block, E.; Iyer, R.; Saha, C.; Grisoni, S.; Belman, S.; Lossing, F. *J. Am. Chem. Soc.* **1988**, *110*, 7813.
32. Lawson, L.D.; Wang, Z.-Y.J.; Hughes, B.G. *Planta Med.* **1990**, *56*, 363.
33. Boelens, H.; Brandsma, L. *Rec. Trav. Chim. Pays-Bas* **1972**, *91*, 141.
34. Block, E.; Zhao, S.-H. *Tetrahedron Lett.* **1990**, *31*, 4999.
35. Kuo, M.-C.; Chien, M.; Ho, C.-T. *J. Agric. Food Chem.* **1990**, *38*, 1378.
36. Kuo, M.-C. Ph.D. Thesis, Rutgers University, 1991.
37. Kuo, M.-C.; Ho, C.-T. *J. Agric. Food Chem.* **1992**, *40*, 111, 1906.
38. Block, E.; Thiruvazhi, M. *J. Agric. Food Chem.*, **1993**, *41*, 2235.
39. Bayer, T.; Breu, W.; Seligmann, O.; Wray, V.; Wagner, H. *Phytochemistry* **1989**, *28*, 2373.
40. Morimitsu, Y.; Kawakishi, S. *Phytochemistry* **1990**, *29*, 3435.

RECEIVED March 23, 1994

Chapter 7

Effect of L-Cysteine and N-Acetyl-L-cysteine on Off-Flavor Formation in Stored Citrus Products

M. Naim[1], U. Zehavi[1], I. Zuker[1], R. L. Rouseff[2], and S. Nagy[3]

[1]Department of Biochemistry, Food Science and Nutrition, Faculty of Agriculture, Hebrew University of Jerusalem, Rehovot 76–100, Israel
[2]University of Florida, Citrus Research and Education Center, Lake Alfred, FL 33850
[3]Florida Department of Citrus, Citrus Research and Education Center, Lake Alfred, FL 33850

> Fortification by low concentrations (up to 2.5 mM) of L-cysteine or N-acetyl-L-cysteine of orange juice stored for 14 days at 45°C, reduced browning and formation of 2,5-dimethyl-4-hydroxy-3-(2H)-furanone (DMHF) and p-vinyl guaiacol (PVG), two major detrimental volatiles in citrus products. These low levels of thiol compounds were also effective in reducing PVG formation in orange juice stored at 15, 25 and 35°C for 45 days. Results of sensory-aroma similarity analyses indicated results that were compatible with the chemical analyses data suggesting that low doses of L-cysteine may improve the aroma quality of stored orange juice. Thus, it appears that low doses of L-cysteine and N-acetyl-L-cysteine may reduce DMHF and PVG formation in citrus products without significant production of sulfur-containing off-flavors.

Non-enzymatic reactions that occur during processing and storage of food products are detrimental to color (browning), aroma and taste (*1-4*). Sodium sulfite has been used to minimize Maillard type browning in a variety of foods including citrus. The use of sodium sulfite, however, has been restricted or eliminated in most foods due to sensitivity of humans to sulfite (*5,6*). Suggestions for the use of L-cysteine to inhibit non-enzymic browning (e.g., *7,8*) have not attracted further research since cysteine, in itself, is a source for flavors in some foods (*9,10*) and can be objectional in fruit juice (*11*). Nevertheless, recent studies (*11*) suggest that thiols should be considered as additives for browning prevention. Some of these thiols are natural components of human diets and play significant physiological roles in vivo as nucleophiles and scavengers of free radicals (*12*).

Limited information is available on the nature and mechanisms of off-flavors formation during citrus juice processing and storage (*13*). Some of the objectional volatiles can significantly affect acceptance since they reach their taste threshold levels under typical processing and storage conditions while other volatiles may be irrelevant as off-flavors (4). In stored orange and grapefruit juices, α-terpineol, 2,5 - dimethyl-4-hydroxy-3-(2H)-furanone (DMHF) and p-vinyl guaiacol (PVG) are believed to be the main detrimental off-flavor compounds with taste threshold levels of 2.5, 0.1 and 0.075 ppm, respectively (1). Evidently, the common use of the term "taste" rather than smell (or aroma) threshold is scientifically inappropriate. In most cases, during sensory evaluation of food, panel members are not requested to distinguish between taste and smell sensations and they actually judge aroma properties rather than taste. Taste threshold is usually much higher than the aroma threshold for volatile compounds. Therefore, the term flavor threshold, although measured during the usual taste testing, will be used throughout this paper. PVG contributes "old fruit" or "rotten" flavor to the juice, DMHF produces a pineapple-like odor typical of aged juice, and α-terpineol imparts a stale, musty or piney odor. When the above three compounds were added collectively to freshly processed juice they gave an aged, off-flavor aroma, similar to that observed in stored juice (1).

DMHF, one of the putative degradation products of sugars, is believed to be obtained by either acid catalysis or through an Amadori rearrangement during Maillard reactions (Figure 1, 14,15). It produces a characteristic malodorous, pineapple-like flavor (1) and can mask the fresh orange juice aroma at levels above 0.05 ppm (16). DMHF can be further degraded to additional flavors (17). PVG has been proposed to be formed in citrus products from free ferulic acid due to non-enzymic decarboxylation following the release from its bound forms (Figure 2; 18,19). PVG formation increased under practical storage conditions of orange juice (20,21) and is accelerated when the juice is fortified by free ferulic acid resulting in inferior aroma quality (20). Although ferulic acid occurs mainly in bound forms, the amount of free ferulic acid present in citrus fruit before processing exceeds the amount needed to form an above-flavor threshold level of PVG during processing and storage (22).

Effect of L-Cysteine and N-Acetyl-L-Cysteine on Browning, and DMHF Formation

The formation of DMHF, 5-hydroxymethyl furfural (HMF) and browning in orange juice during accelerated non-enzymic aging was monitored. Pasteurized, commercial single-strength orange juice (SSOJ) was preserved with 500 ppm sodium benzoate. Twenty mL SSOJ samples with and without added L-cysteine and N-acetyl-L-cysteine were transferred to 20 mL glass vials, sealed and then stored at 45°C for 14 days. Experiments employed levels of 0.5, 1.0 and 2.5 mM of either L-cysteine or N-acetyl-L- cysteine (23). Following storage, DMHF and HMF were extracted and analyzed by HPLC (23). Browning was determined in 10 mL stored SSOJ samples after centrifugation (4°C, 5 min, 500 X g) and ethanol extraction of the supernatant.

In line with recent data on browning inhibition by thiol compounds (11), the present results indicated that the use of a variety of concentrations of L-cysteine inhibited browning (Figure 3) and reduced ascorbic acid degradation in a dose-

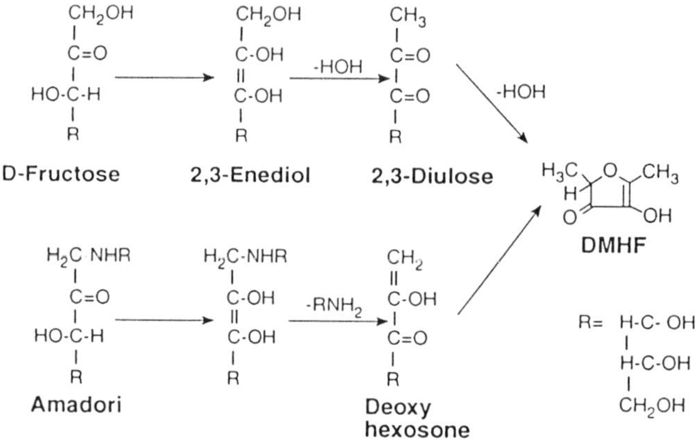

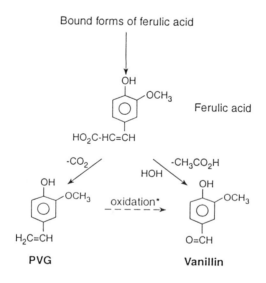

Figure 1: Possible pathways for DMHF formation following sugar degradation.

Figure 2: PVG and vanillin formation from ferulic acid following its release from bound forms. An asterisk indicates that the oxidation pathway was not identified in our experiments.

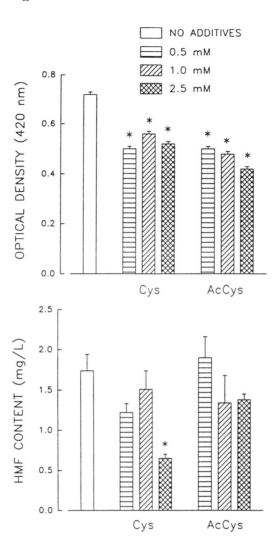

Figure 3: Effects of L-cysteine (Cys) and N-acetyl-L-cysteine (AcCys) on browning and HMF formation in SSOJ samples stored for 14 days at 45°C. Values are the mean and SEM of 3 samples. Asterisks indicate significant ($p < 0.05$) lower value than that observed in samples stored with no thiol fortification. Adapted from ref. 23.

dependent manner (23). Our study was directed to the effects of low levels of thiols as high levels may, in themselves, lead to the formation of off-flavor in fruit juices (11). N-Acetyl-L-cysteine was used since it is believed to be an effective inhibitor of browning with less off-flavor formation (11). These experiments demonstrated that low levels of both thiols (below 2.5 mM) can significantly reduce browning and DMHF formation in stored SSOJ (Figures 3 and 4). A level of 2.5 mM L-cysteine was also effective in reducing HMF content (Figure 3). Furthermore, low levels of L-cysteine reduced ascorbic acid degradation by 40% (23). In contrast, the same levels of N-acetyl-L-cysteine were ineffective in preventing ascorbic acid degradation. This may be related to the difference between these two compounds in their ability to form complexes with heavy metal ions such as copper. Complexes between copper and amino acids or peptides are well recognized (24,25) and chelates (e.g. EDTA) have been reported to reduce browning and ascorbic acid degradation in grapefruit juice (26). Cysteine, may be more effective than N-acetyl-L-cysteine in forming metal complexes due the presence of a free amino group. Furthermore, L-cysteine and apparently also N-acetyl-L-cysteine, being in great excess compared to the content of Cu^{2+}, is expected to reduce Cu^{2+} to Cu^{+} which in turn form a particularly stable complex with L-cysteine (27).

Effect of L-Cysteine and N-Acetyl-L-Cysteine on Formation of PVG

Since ferulic acid (free or bound), the precursor of PVG, contains an activated double bond conjugated to a carboxylic acid and to a phenyl ring, we hypothesized that it may also be sensitive to the nucleophilic attack by thiols, thereby inhibiting PVG formation. Pasteurized SSOJ samples (20 mL) were fortified with L-cysteine or N-acetyl-L-cysteine and stored at 45°C for 14 days (28). In addition, SSOJ samples were also stored for six weeks at 4, 15, 25 and 35°C. Extraction of PVG from the incubated SSOJ samples was performed, and samples were analyzed for PVG by HPLC.

The presence of L-cysteine and N-acetyl-L-cysteine reduced the accumulation of PVG during storage of SSOJ samples (Figure 4, Table I). Similar effects were observed in model solutions of orange juice containing ferulic acid (28), and when a higher range of L-cysteine concentrations was used in model solutions, the effect of L-cysteine was clearly dose-dependent. The results in Figure 4 show that fortification of SSOJ by 0.5 mM L-cysteine reduced the PVG accumulation following storage by 50%, resulting in a value close to its flavor threshold level. N-Acetyl-L-cysteine (2.5 mM) was also effective in reducing PVG level in stored SSOJ. The accumulation of PVG in stored SSOJ samples was accelerated as temperature increased (Table I). Six-week storage of unadulterated SSOJ at 25 and 35°C already induced levels of PVG above its flavor threshold. PVG levels under storage at 15°C for six weeks, however, appear to be below the flavor threshold level, especially when fortified with L-cysteine. There was a clear inhibition of PVG accumulation by L-cysteine, although not always did it reach statistical significance.

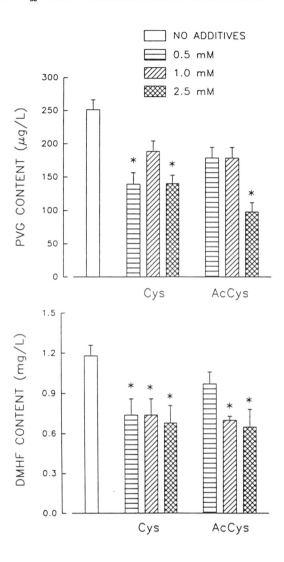

Figure 4: Effects of L-cysteine (Cys) and N-acetyl-L-cysteine (AcCys) on DMHF and PVG formation in SSOJ samples stored for 14 days at 45°C. Values are the mean and SEM of 3 samples. Asterisks indicate significant ($p < 0.05$) lower value than that observed in samples stored with no thiol fortification. Adapted from refs. 23 and 28.

Table I. PVG Content (μg/L) in Orange Juice Fortified by L-Cysteine and Stored for Six Weeks

L-Cysteine Level	Storage Temperature			
	4°C	15°C	25°C	35°C
Control (none)	17.1±2.9	54±1	103±5	292±8
0.5 mM	11.7±1.2	30±4*	85±3	245±15
1.0 mM	12.4±0.6	41±1	72±5*	231±19
2.5 mM	16.6±0.9	32±3*	73±1*	210±17*

Values are the mean ±SEM of 4 samples. Asterisks indicate significant difference from control juice samples.

Aroma Similarity Evaluation of Stored Orange Juice Fortified by L-Cysteine and N-Acetyl-L-Cysteine

The significance of thiols to aroma characteristics was studied with SSOJ samples stored at 45°C for 14 days. Naive taste-panel members evaluated the aroma similarity for eight orange juice treatments. During smell sessions, on a verbal signal from the experimenter, each panelist opened the vials of 2 samples at 15 sec intervals, smelled them, and was requested to rate the similarity level of aroma of the 2 samples on a 0 to 20 scale (0 for no similarity, 20 for identical). Each panelist did so for all 64 treatment combinations presented in a coded randomized order. A data matrix representing the similarity results for each of the sensory tests was obtained, where each cell in the matrix represents the mean similarity for all panelists for the corresponding comparison. This proximity matrix was then analyzed by a clustering program (ADDTREE, 29), which yields a tree structure of branches and subdivisions with aromas located at the ends of the branches (Figure 5). Sets of aromas which are connected to the same node with relatively short (horizontal) distances from each other, represent high similarity.

The proportion of variance explained (0.677, r = 0.84) suggested that distances in the clustering structure are highly correlated with the original aroma similarity data (Figure 5). Two major clusters were obtained. The most intriguing finding was that the addition of 0.5 mM L-cysteine (60 ppm) to stored SSOJ resulted in a branch (No. 2) connected to the same node as that of control juice kept at 4°C (No. 8) indicating that both exhibited similar aroma. This suggests that relatively minor amounts of L-cysteine added to juice stored at an elevated temperature were able to either retain the aroma of the original juice and/or to reduce off-flavor formation. Other treatments (higher doses of L-cysteine or the use of N-acetyl-L-cysteine) resulted in either aromas similar to that of the stored-untreated juice (No. 1) or produced completely different aromas. The aroma of N-acetyl-L-cysteine-fortified SSOJ samples (branches No. 5, 6 and 7) grouped together, separately from the L-

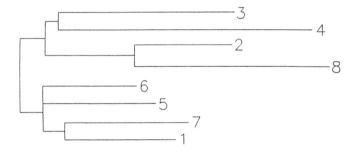

Figure 5: Cluster analysis by ADDTREE of aroma similarity values. Legends: 1 - SSOJ stored at 45°C for 14 days; 2 - with added 0.5 mM L-cysteine; 3 - with 1 mM L-cysteine; 4 - with 2.5 mM L-cysteine; 5 - with 0.5 mM N-acetyl-L-cysteine; 6 - with 1 mM N-acetyl-L-cysteine; 7 - with 2.5 mM N-acetyl-L-cysteine; 8 - control SSOJ, stored at 4°C. Each value in the original matrix represents the mean of 8 panelists. Reproduced from ref. 23. Copyright 1993 ACS.

cysteine-fortified samples (No. 2, 3 and 4) which were connected to the aroma of control SSOJ samples (No. 8). This indicates that the aroma of N-acetyl-L-cysteine-fortified SSOJ was inferior compared with that of L-cysteine-fortified SSOJ.

Concluding Remarks

The demand for fresh tasting (premium) citrus juices has increased significantly in recent years. Since chemical rather than enzymic processes are usually responsible for reduced quality of stored citrus products, storage at low temperature and the exclusion of oxygen and light are critical factors. Various juice extraction and concentration techniques along with diverse container and storage condition combinations have been evaluated in order to maximize fresh flavor retention and minimize off-flavors. The most effective packaging materials and low temperature storage are economically prohibitive except for premium juices. Therefore, most current commercial juices represent compromises either in packaging or in storage.

The present studies suggest that browning and ascorbic acid degradation as well as the formation of two major off-flavor compounds in citrus products, PVG and DMHF, can be significantly reduced by low doses (0.5 mM) of L-cysteine, an amino acid found in almost all proteins. The aroma similarity experiments were compatible with the chemical analyses suggesting that such low doses of L-cysteine may significantly improve the aroma quality of stored orange juice. It is thus proposed that although PVG and DMHF are formed by different pathways, objectional aroma may be reduced by a common approach; using naturally occurring thiols, such as L-cysteine. Therefore, the consideration of the use of low levels of L-cysteine (and perhaps similar thiols) in order to retain the fresh aroma of orange juice should be further tested under practical processing and storage conditions.

Acknowledgments

This research was supported by Grant No. I-1967-91 from BARD, The United States-Israel Binational Agricultural Research & Development Fund. The excellent technical assitance by Mrs. M. Levinson and the statistical advise by Dr. H. Voet are highly appreciated. We also thank Mr. S. Bernhardt for assisting us with computerized drawings.

Literature Cited

1. Tatum, J. H.; Nagy, S.; Berry, R. E. *J. Food Sci.* **1975**, *40*, 707-709.
2. Shaw, P. E.; Tatum, J. H.; Berry, R. E. (1977). *Dev. Food Carb.* **1977**, *1*, 91-111.
3. Meydav, S.; Berk, Z. *J. Agric. Food Chem.* **1978**, *26*, 282-285.
4. Handwerk, R. L.; Coleman, R. L. *J. Agric. Food Chem.* **1988**, *36*, 231-236.
5. Brown, J. L. *J. Am. Med. Assoc.* **1985**, *254*, 825
6. Taylor, S. L.; Bush, R. K. *Food Technol.* **1986,** *40*, 47-51.
7. Arnold, R. G. (1969). *J. Dairy Sci.* **1969**, *52*, 1857-1859.

8. Montgomery, M. W. *J. Food Sci.* **1983**, *48,* 951-952.
9. Hurrell, R. F. In *Food Flavors Part A, Introduction*; Morton, I D.; Macload, A J. Eds; Elsevier Publishing Co.: Amsterdam, Oxford and New York, 1982; pp 399-437.
10. Shu, C- K.; Hagedorn, M. L.; Mookherjee, B. D.; Ho, C- T. *J. Agric. Food Chem.* **1985**a, *33,* 442-446.
11. Molnar Perl, I; Friedman, M. *J. Agric. Food Chem.* **1990**, *38,* 1648-1651.
12. Friedman, M. In: *Nutritional and Toxicological Consequences of Food Processing.* Friedman, M., ed., Plenum Press: New York, 1991; pp. 171-215.
13. Naim, M.; Zehavi, U.; Nagy, S.; Rouseff, R.L. In *Phenolic Compounds in Food and their effects on Health I, Analysis Occurrence and Chemistry;* Ho, C- Tand Lee, C. Y. Eds.; American Chemical Society Symposium Series No. 506; ACS: Washington, D.C., 1992, Chap. 14; pp. 180-191.
14. Shaw, P. E. Tatum, J. H.; Berry, R. E. *Carb. Res.* **1967,** *5,* 266-273.
15. Baltes, W. *Food Chem.* **1982**, *9,* 59-95.
16. Nagy, S.; Rouseff, R. L.; Lee, H. S. In *Thermal Generation of Aromas,* Parliment, T. H.; McGorrin, R. J.; Ho, C- T., Eds. American Chemical Society Symposium Series No. 409, ACS: Washington, D.C., 1989, Chap. 31; pp 331-345.
17. Shu, C- K.; Mookherjee, B. D.; Ho, C-. T. *J. Agric. Food Chem.* **1985**b, *33,* 446-448.
18. Peleg, H.; Striem, B. J.; Naim, M.; Zehavi, U. In: *Proc. Int. Citrus Congress;* Goren, R., Mendel, K., Eds.; Balaban: Rehovot and Margraft, Weikershim, 1988; Vol 4, pp. 1743-1748.
19. Peleg, H.; Naim, M.; Zehavi, U.; Rouseff, R. L.; Nagy, S. *J. Agric. Food Chem.* **1992**, *40,* 764-767.
20. Naim, M.; Striem, B. J.; Kanner, J.; Peleg H. *J. Food Sci.* **1988**, *53,* 500-503, 512.
21. Lee, H. S.; Nagy, S. *J. Food Sci.* **1990**, *55,* 162-163 & 166.
22. Peleg, H.; Naim, M.; Rouseff, R. L.; Zehavi, U. *J. Sci. Food Agric.* **1991**, *57,* 417-426.
23. Naim, M.; Wainish, S.; Zehavi, U.; Peleg, H.; Rouseff, R. L.; Nagy, S. *J. Agric. Food Chem.* **1993**, *41,* 1355-1358.
24. Liang, Y- C.; Olin, A. *Acta Chemica Scand.* **1984**, *A38,* 247-252.
25. Thomas, G.; Zacharias, P. S. *Indian J. Chem.* **1984,** *23,* 929-932.
26. Kanner, J.; Shapira, N. In *Thermal Generation of Aromas,* Parliment, T. H.; McGorrin, R. J.; Ho, C- T., Eds. American Chemical Society Symposium Series No. 409, ACS: Washington, D.C., 1989, Chap. 5; pp. 55-64.
27. Levitzki, A. *Copper complexes of peptides and polypeptides as a model for copper containing proteins.* 1964. Ph.D. Thesis, The Weizmann Institute of Science, pp. 4-5.
28. Naim, M.; Zuker, I.; Zehavi, U.; Rouseff, R. L. *J. Agric. Food Chem.* **1993**, *41,* 1359-1361.
29. Sattath, S.; Tversky, A. *Psychometrika* **1977**, *42,* 319-345.

RECEIVED May 3, 1994

Chapter 8

Modulation of Volatile Sulfur Compounds in Cruciferous Vegetables

Hsi-Wen Chin and Robert C. Lindsay

Department of Food Science, University of Wisconsin, Madison, WI 53706

Methanethiol is produced in cruciferous vegetable tissues following injury via the action of cysteine sulfoxide lyase, and various other volatile sulfur compounds are subsequently formed via chemical reactions which also contribute sulfurous off-flavors. Using headspace gas chromatography and sulfur-selective detection, the effects of several conditions, including oxygen and ascorbic acid, upon the production of dimethyl disulfide and dimethyl trisulfide were determined. The suppression of sulfurous off-flavors in cruciferous vegetables was investigated and found to involve a thiol oxidase-like activity and a chemical reactivity involving quinone-like substances. Various strategies were evaluated for the control of off-flavors caused by methanethiol-derived compounds in cruciferous vegetables.

Cruciferous vegetables (cabbage, broccoli, cauliflower, and Brussels sprouts) are characterized by sulfurous aromas and flavors following tissue injury or cooking. Isothiocyanates derived from glucosinolate precursors via the action of thioglucosidase (Figure 1) generally contribute to the desirable, pungent flavors (1,2), while methanethiol-related volatile sulfur compounds produced from $\underline{S}$-methyl-L-cysteine sulfoxide following cooking or via the action of cysteine sulfoxide lyase (C-S lyase; Figure 2) cause objectionable odors in cruciferous vegetable foods (3,4,5).

Cruciferous vegetables have recently been identified as possessing anticarcinogenic properties (6). For example, indole-3-carbinol derived from indolylmethyl glucosinolate which is widely present in cruciferous vegetables has been shown to induce estradiol 2-hydroxylation in mice, and thereby reduce the risk of estradiol-linked mammary cancer (7,8). Moreover, 4-methylsulfinyl isothiocyanate (Sulforaphane) isolated from broccoli has been shown to induce several anticarcinogenic protective enzymes (9). However, development of excessive methanethiol-related off-flavors in cruciferous foods (e.g., sauerkraut) detracts from consumer acceptance of these cancer-preventive foods.

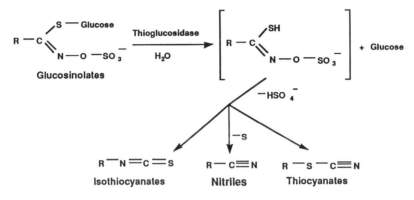

Figure 1. Formation of isothiocyanates from glucosinolate precursors via the action of thioglucosidase in cruciferous vegetables.

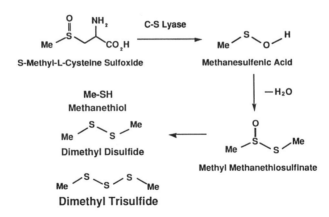

Figure 2. Formation of methanethiol-related volatile sulfur compounds from S-methyl-L-cysteine sulfoxide via the action of cysteine sulfoxide lyase (C-S lyase) in cruciferous vegetables.

Minimally processed vegetables which were first developed in Europe have become increasingly popular in North American markets (10). Minimally processed foods have the attributes of convenience and fresh-like quality, but in most cases they are more perishable than unprocessed raw materials (11). A number of methods have been developed to control physiological and microbiological changes in minimally processed fruits and vegetables (12). Furthermore, the use of controlled and modified atmosphere storage technologies has been shown to extend the shelf life of fresh broccoli florets (10,13). The benefit of controlled and modified atmosphere storage of broccoli florets included delayed yellowing, reduced development of mold, and prolonged retention of chlorophyll and ascorbic acid (10,13).

The atmosphere within a package can be modified by retaining the respiratory gases of living plant tissues or by injection of gases, and generally carbon dioxide concentrations elevated to 5-10% and oxygen lowered to 2-5% extend the shelf life of fruits and vegetables (12). However, broccoli florets stored under reduced oxygen levels have been found to develop offensive odors (4,14,15). Similarly, shredded cabbage packaged in bags having low oxygen transmission rates has resulted in significantly greater fermentation and other off-odors than similar cabbage packaged in bags with higher oxygen transmission rates after 19 days storage at 11°C (16).

Methanethiol has been identified as a major contributor to the objectionable odors in broccoli florets held under anaerobic conditions for 7 days at 15°C (4). Additionally, dimethyl disulfide and dimethyl trisulfide were found in significant amounts in broccoli samples stored under controlled atmospheres (17). Because the oxidized forms (dimethyl disulfide and dimethyl trisulfide) of methanethiol also possess unpleasant aromas and flavors (18), it appears that methanethiol-related volatile sulfur compounds are major limiting factors in preventing the use of controlled and modified atmosphere storage technologies. Thus, suppression of formation of these malodorous compounds should extend the shelf life of cruciferous vegetables maintained by controlled and modified atmosphere storage technologies. Furthermore, suppression of unpleasant methanethiol-related volatile sulfur compounds in cruciferous vegetables foods should also enhance consumer acceptance and consumption of these cancer-preventive foods.

Targets for Modulating the Formation of Volatile Sulfur Compounds

Methyl Methanethiosulfinate. Methanethiol-related compounds are secondary products of the primary C-S lyase action on S-methyl-L-cysteine sulfoxide (19). The primary product, presumably methanesulfenic acid, is very unstable and facilely converted to methyl methanethiosulfinate which has been found in macerated Brussels sprouts (20). Methyl methanethiosulfinate has also been reported to be present in a model system composed of S-methyl-L-cysteine sulfoxide and partially purified cabbage C-S lyase (20).

Methanethiol-related volatile sulfur compounds are derived from methyl methanethiosulfinate via nonenzymic chemical reactions. Methanethiol and dimethyl trisulfide are readily formed in a model system composed of methyl methanethiosulfinate and hydrogen sulfide (Figure 3; 21,22) which is also a volatile component in disrupted tissues of cruciferous vegetables (4,5,17). Additionally, dimethyl disulfide has also been shown to be formed facilely from methyl methanethiosulfinate through chemical disproportionation (Figure 4; 23,24). Thus, methyl methanethiosulfinate provides a potential target when designing strategies for controlling the formation of methanethiol-related volatile sulfur compounds in cruciferous vegetables.

Since methyl methanethiosulfinate is a product of C-S lyase-catalyzed degradation of S-methyl-L-cysteine sulfoxide, it follows that selection of plant cultivars that are low in the amino acid precursor or C-S lyase activity should afford a means to provide cruciferous vegetable foods with less malodorous volatile sulfur compounds and more consumer acceptance. Variations in concentrations of S-methyl-L-cysteine sulfoxide among different cruciferous vegetables have been observed (20). However, suitable studies on the content of S-methyl-L-cysteine sulfoxide in different cultivars of the same cruciferous vegetable have not been carried out. On the other hand, volatile sulfur compounds formed in disrupted tissues of different cabbage cultivars have been analyzed (5), and wide variations in ability to produce volatile sulfur compounds were observed for the different cultivars. Figure 5 shows a comparison of formation of volatile sulfur compounds in two cabbage cultivars analyzed by headspace gas chromatography with a flame photometric detector (GC-FPD) according to the method described by Chin and Lindsay (5).

Methanethiol. Methanethiol was a major volatile sulfur compound formed in broccoli florets stored in sealed jars for 7 days at 15°C which developed severe off-flavors, and it was considered the most important contributor to the objectionable odors that render the modified atmosphere stored broccoli florets unmarketable (14). Similarly, we observed that broccoli florets stored in heat-sealed barrier pouches (15 cm x 25 cm o.d.; 70 g/bag; Curlon X/K-28; Grade 850; Curwood, New London, WI) with oxygen transmission rates less than 1 cc $O_2/645$ $cm^2/24$ hr developed offensive sulfurous odors within a few days storage at 4°C.

We also analyzed the volatile sulfur compounds within the pouches containing chopped broccoli using GC-FPD according to the method described by Chin and Lindsay (5). Headspace samples (0.5 mL) were withdrawn with gas-tight syringes through silicone septa which had been previously fixed onto the exterior of pouches by silicone sealant. Methanethiol was confirmed as the most abundant volatile sulfur compound produced by the injured broccoli tissues in the pouches (Figure 6). Hydrogen sulfide, dimethyl sulfide (CH_3-S-CH_3) and dimethyl disulfide were also detected in the pouches which confirmed the findings of Forney et al. (4). Although dimethyl trisulfide has been commonly encountered in cruciferous vegetables (5,6,17), it was not detected in the broccoli held in barrier pouches held at 4°C for 8 days (Figure 6). However, significant amounts of dimethyl trisulfide was formed in broccoli florets which were stored in the barrier pouches held at an elevated temperature (21°C) for 2 days (data not shown).

Methanethiol has a very low flavor threshold value (0.02 ppb in water; 15), and possesses a strong, offensive, fecal-like odor (18). It also is readily converted to very unpleasant oxidized sulfurous off-flavor compounds with low detection thresholds, such as dimethyl disulfide and dimethyl trisulfide (18,25). Thus, suppression or modulation of methanethiol formation in cruciferous vegetables would lead to a reduction of off-flavors associated with methanethiol-related compounds, and enhance the utility of controlled and modified atmosphere storage technologies for these vegetables.

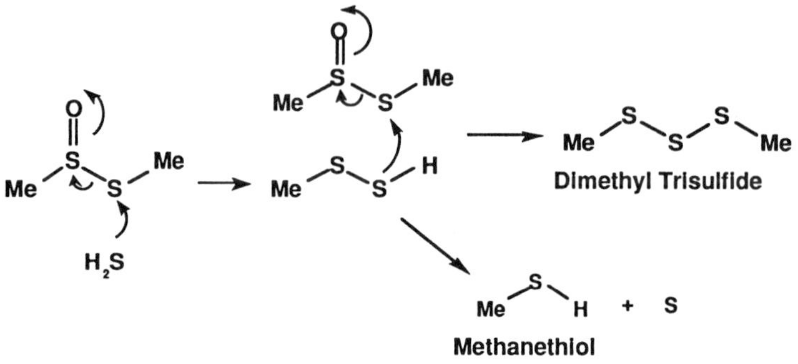

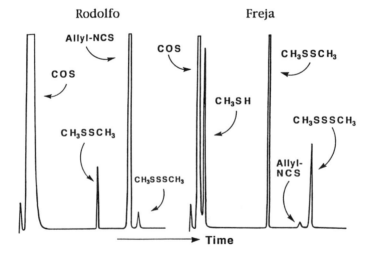

Figure 3. Formation of methanethiol and dimethyl trisulfide in cruciferous vegetables involving methyl methanethiosulfinate and hydrogen sulfide.

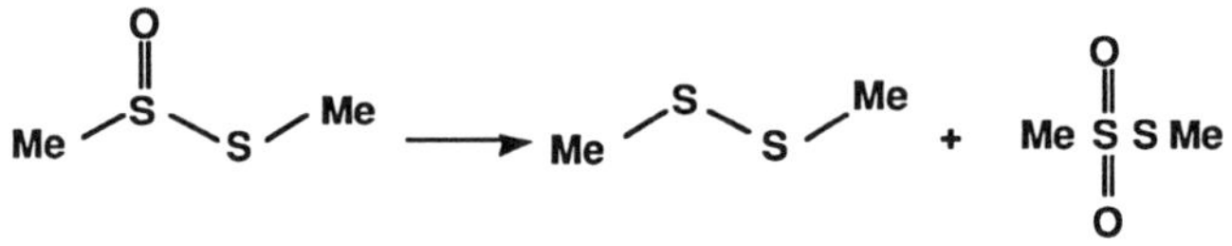

Figure 4. Formation of dimethyl diisulfide in cruciferous vegetables involving disproportionation of methyl methanethiosulfinate (Adapted from references 23 and 24).

Figure 5. Comparison of formation of volatile sulfur compounds in two different cabbage cultivars.

Modulation of Sulfurous Off-Flavors in Cruciferous Vegetable Foods

Suppression of Undesirable Sauerkraut Flavors by Caraway Seed. More than 200,000 metric tons of fresh cabbage are processed into sauerkraut annually in the United States (26). However, the aroma of sauerkraut detracts from consumer acceptance (27). Sulfur-like aromas have been long recognized as contributors to the undesirable flavors in sauerkraut (28), and dimethyl trisulfide has been identified as one of the dominant odor components in sauerkraut (30).

Several spices have been sensorily evaluated for their modification effects on sauerkraut flavors, and the addition of umbelliferous seeds, including caraway, celery, and dill weed, increased the flavor acceptance of sauerkraut (Table I; 26). Further, sauerkraut seasoned with caraway seed and known as Bavarian-style, has been marketed for some time now. Bavarian-style sauerkraut products are generally accepted as exhibiting lower intensity, less sulfurous, and more desirable flavors than comparable unspiced sauerkraut. The less intense flavors found in Bavarian-style sauerkraut have been attributed to the effects of added spice flavoring and to the ability of caraway seeds to absorb unpleasant sauerkraut aromas (31).

Table I. Odor assessment of spiced-sauerkrauts[1]

Sample	Acceptance Scores[2]
Caraway Seed	+4
Dill Weed	+3.5
Celery Seed	+3
Plain	0
Anise	-2

(Adapted from reference 27).

[2] +: makes the flavor more acceptable;
0 : gives a flavor as acceptable as plain sauerkraut;
- : makes the flavor less acceptable

GC-FPD analysis of freshly-opened commercial caraway-spiced and comparable unspiced sauerkraut samples (70 g drained sauerkraut) were conducted. These samples were contained in 120-mL serum-type vials (Supelco Co., Bellefonte, PA) closed with Mininert valves (Supelco). Samples were warmed to 30°C in a water bath for 10 min before withdrawing headspace gases (4 mL) for GC-FPD analysis. Data in Table II show that caraway-spiced sauerkraut had lower quantities of dimethyl disulfide than the unspiced sauerkraut, whereas dimethyl sulfide was present in comparable amounts in both samples. Methanethiol and dimethyl trisulfide were not detected in the caraway-spiced sauerkraut, but they were present in significant amounts in the unspiced samples. These results showed that caraway seed modified the flavor of unspiced sauerkraut by suppression of the formation of methanethiol-related volatile sulfur compounds rather than simply masking the undesirable flavors in sauerkraut with spice flavoring.

Table II. Concentrations of volatile sulfur compounds in commercial unspiced and caraway-spiced sauerkrauts

Sauerkraut	Concentration (ppb)			
	CH_3SH	CH_3SCH_3	CH_3SSCH_3	CH_3SSSCH_3
Unspiced	15	56	63	150
Caraway-Spiced	0	59	25	0

Nature of the Suppression Effect of Caraway Seed. Since analytical data demonstrated that the addition of caraway seeds depleted volatile sulfur compounds in sauerkraut, this effect was studied using aqueous extracts of caraway seeds and authentic volatile sulfur compounds in model systems. Cell-free crude caraway seed extracts were prepared by blending the unheated spice seeds with chilled (ice water) potassium phosphate solutions (50 mM, pH 7) containing 0.1 M potassium chloride (1:5, w/w) in a cold room (4°C). Homogenates were then clarified by filtration and centrifugation as described by Chin (21). Data in Figure 7 shows that unheated and heat-treated crude caraway seed extracts (1 mL in a 3 mL buffered solution, pH 8) removed methanethiol (10 µg) from the headsapce in closed serum-type vials (120 mL) at 37 °C which indicated that both heat-stable and heat-labile entities in caraway seed were involved in the depletion of methanethiol.

However, unheated and heat-treated crude caraway seed extracts did not affect the concentrations of dimethyl disulfide and dimethyl trisulfide in model systems over an extended period (5 hr). Thus, it appeared that the overall suppression of methanethiol-related compounds in sauerkraut by caraway seed (Table I) was dependent on methanethiol which has been shown to be a precursor for both dimethyl disulfide and dimethyl trisulfide under oxidative conditions (18,25). Recently, Chin and Lindsay (25) have shown that methanethiol was readily converted to dimethyl disulfide and dimethyl trisulfide in an aerobic cabbage model system (pH 6.3) consisting of 1 ppm hydrogen sulfide, 450 ppm ascorbic acid and 4 ppm Fe(III).

Nature of Caraway Seed Components Responsible for Methanethiol Depletion. The nature of the methanethiol-depleting heat-stable and heat-labile entities in caraway seed were further studied. Flavonoid compounds, which are naturally-occurring plant phenolics, appeared to comprise the heat-stable methanethiol-depleting entity based on evidence obtained from absorption spectra and color changes at different pH values (21). The caraway seed flavonoid components removed methanethiol from the headspace of closed model systems, and the formation of oxidized forms of methanethiol (dimethyl disulfide and dimethyl trisulfide) did not account for its disappearance (Table III).

Tertiary butylhydroquinone (TBHQ), a commonly-employed synthetic antioxidant in foods, was also included in the studies as a model phenolic compound, and it had the same effect on depletion of methanethiol as caraway seed flavonoids (Table III). These results indicated that TBHQ could be used also for suppressing unpleasant flavors in foods that are caused by methanethiol and its oxidative derivatives. Because the methanethiol-depleting effect of caraway seed flavonoids and TBHQ increased with increasing oxygen tensions and pH values in the model systems

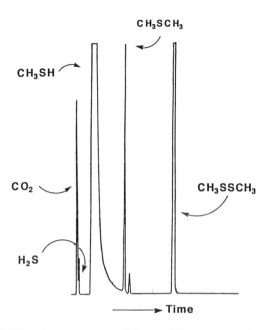

Figure 6. GC-FPD chromatogram of broccoli florets stored under modified atmospheres at 4°C for 8 days.

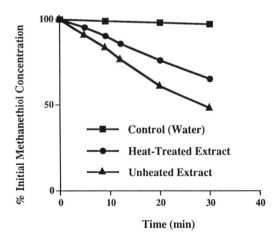

Figure 7. Depletion of methanethiol by cell-free crude caraway seed extracts in closed model systems at 37°C.

(21), and oxidation of methanethiol did not explain its disappearance (Table III), it was concluded that the depletion effect involved oxidation of flavonoids to quinoid compounds which then formed adduct compounds with methanethiol (Figure 8).

Table III. Effects of isolated caraway seed flavonoid components and TBHQ on methanethiol concentrations in closed model systems after 5 hr at 30°C

Samples	Loss of CH_3SH[1]	Amounts (μg) Production of CH_3SSCH_3	Production of CH_3SSSCH_3
Caraway Flavonoids[2]	3.23 ± 0.04	0.48 ± 0.01	0
Caraway Flavonoids (N_2-saturated)	1.23 ± 0.04	0.16 ± 0.01	0
TBHQ[3]	5.07 ± 0.01	0.75 ± 0.02	0
TBHQ (N_2-saturated)	1.84 ± 0.10	0.27 ± 0.01	0

[1] Ten μg initial amount; equivalent to 2 ppm of solution in sample vials.
[2] Five mL aliquots from a Sephadex G-10 gel filtration fraction that had a high methanethiol-depleting activity.
[3] Equivalent to 100 ppm of solution (5 mL volume).

The methanethiol-depleting heat-labile entity in caraway seed was partially characterized by studying the effects of various assay conditions in closed model systems on the rate of methanethiol depletion by cell-free crude caraway seed extracts (21). The effects of pH, temperature, concentrations of the heat-labile entity, and initial concentrations of methanethiol are shown in Figure 9, and these data indicated that an enzyme comprised the heat-labile entity. This methanethiol-depleting enzyme did not require oxygen in removal of methanethiol because varying oxygen tensions in the enzyme assay systems did not alter the rate of methanethiol depletion. Also, dimethyl disulfide and dimethyl trisulfide were not end products of the caraway seed enzyme-catalyzed depletion of methanethiol.

Further studies showed that the caraway seed enzyme lost the methanethiol-depleting activity following acetone precipitation, but reconstitution of the acetone precipitate with the soluble caraway seed cellular substances reinstated the methanethiol-depleting activity. Thus, it appeared that a cofactor or a second substrate was required by the enzyme to function in removal of methanethiol.

Modulation of Objectionable Flavors in Broccoli. Since caraway seed has been utilized successfully in commercial Bavarian-style sauerkraut for suppressing undesirable sulfurous flavors, other applications to cruciferous vegetables seemed

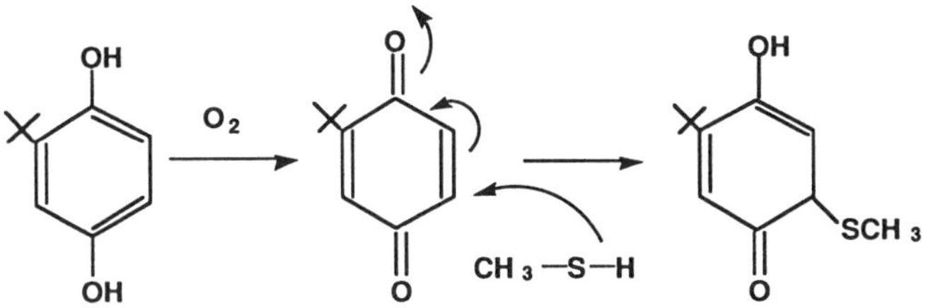

Figure 8. Proposed mechanism for the formation of an adduct compound from the reaction of methanethiol with TBHQ.

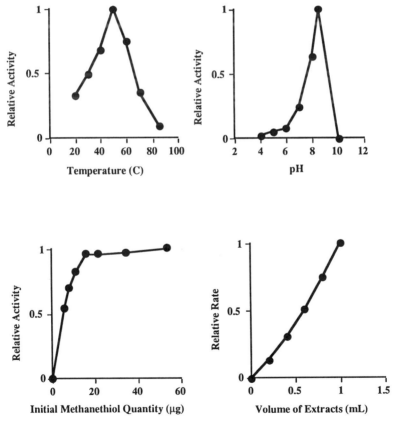

Figure 9. Partial characterization of the methanethiol-depleting enzyme in caraway seed.

feasible. Therefore, studies were carried to evaluate the efficacy of caraway seed for modulating undesirable broccoli flavors.

Modified Atmosphere Stored Broccoli Florets. Since TBHQ and caraway seed flavonoid components exhibited the same effect on depletion of methanethiol in model systems, they were examined for efficacy in suppressing methanethiol-related off-flavors in broccoli florets stored under modified atmospheres. Freshly prepared broccoli florets (ca 3 cm long) were treated by dipping them in cell-free crude caraway seed extracts or aqueous TBHQ solutions (100 ppm), and allowing excess liquids to drain off. Control samples were dipped in distilled water. Broccoli samples were placed in the heat-sealed Curlon 850 pouches described earlier (70 g/pouch), and then held under refrigeration at 4°C until used for GC-FPD and descriptive sensory analysis 8 days later.

Sensory evaluations were carried out in the University of Wisconsin Sensory Analysis Laboratory under conditions prescribed for sensory analysis (32). Panelists were instructed to cut-open sealed pouches, smell samples, and then record observations on ballots which contained horizontal line scales (1=weak, 7=strong) for each attribute (33). Data in Table IV shows that both TBHQ- and crude caraway seed extracts-treated broccoli florets received significantly lower sensory scores for the unpleasant sulfurous aroma. However, there was no significant difference in the sulfurous aroma between TBHQ- and caraway-treated samples. Additionally, both caraway seed and TBHQ treatments resulted in higher scores for green vegetable aromas and overall pleasantness.

Table IV. Descriptive sensory analysis of broccoli dipped in crude caraway seed extracts and held at 2°C for 2 days

	Attribute		
Sample	Broccoli Flavor Intensity[1]	Flavor Quality[2]	Overall Preference[3]
	(------------------------ Mean Scores --------------------------)		
Untreated	4.38[a]	3.57[a]	3.16[a]
Caraway-Treated	3.74[b]	4.56[b]	3.70[a]

[1] Scale: 1 = very weak; 7 = very strong.
[2] Scale: 1 = staling, broccoli flavor; 7 = green vegetable flavor.
[3] Scale: 1 = dislike extremely; 7 = like extremely.
[a,b] Mean scores in the same column with the same superscript are not significantly different at 5% level.

Data from GC-FPD analysis of the broccoli florets are shown in Table V, and these data corresponded directionally with the sensory data (Table IV). Although TBHQ and crude caraway seed extracts showed suppression effects on the formation of methanethiol-related compounds, the performance of TBHQ and caraway seed extracts in depletion of methanethiol within the packages was not as high as expected from studies employing model systems. This was probably caused by the anaerobic conditions developed by respiration of broccoli tissues within the pouches which retarded the rates of oxidation of flavonoids and TBHQ to quinoid compounds which are the binding components for methanethiol. Thus, it is likely that chemical quinoid binding systems for methanethiol will function more efficiently in situations that provide more aerobic or oxidative conditions.

Table V. Descriptive sensory analysis scores for broccoli florets treated with water (control), cell-free crude caraway seed extracts, or TBHQ solutions (100 ppm), and stored under modified atmospheres at 4°C for 8 days

Attribute	Sample		
	Control	Caraway Extract	TBHQ
	(------------------ Mean Scores ------------------)		
Sulfurous Aroma[1]	4.30[a]	2.75[b]	2.50[b]
Green Vegetable Aroma[1]	2.19[a]	2.63[a]	3.48[a]
Overall Pleasantness[2]	2.18[a]	2.75[a]	3.42[a]

[1] Scale: 1 = very weak; 7 = very pronounced.
[2] Scale: 1 = very unpleasant; 7 = very unpleasant.
[a,b] Mean scores in the same row with the same superscript are not significantly different at the 5% level.

Some panelists noted that when pouches containing control or treated samples were opened, offensive odors were perceived which dissipated rapidly. Presumably, this effect would be caused by methanethiol which has a b.p. 6°C. However, panelists noted that they could still differentiate the odor characteristics between the samples receiving different treatments. Thus, it appears that the concentrations of higher boiling volatile sulfur compounds (e.g., dimethyl disulfide) were also influential in the flavor quality and consumer acceptance of broccoli florets stored under modified atmospheres.

Analysis of samples showed that concentrations of methanethiol and dimethyl disulfide were lower in caraway- and TBHQ-treated broccoli florets than in control samples (Table V). Dimethyl trisulfide was not detected in the refrigerated samples which indicated that its contribution to the objectionable odors of modified atmosphere stored broccoli florets was minimal. Dimethyl sulfide which has a precursor (S-methyl methionine sulfonium salt; 34) different from that for methanethiol-related compounds was present in all samples.

The influence of dimethyl disulfide on flavor quality of modified atmosphere stored broccoli florets was further demonstrated by data shown in Table VI which revealed that aroma assessments of undesirable sulfurous aromas in broccoli florets more closely corresponded with concentrations of dimethyl disulfide than with methanethiol. In this experiment broccoli florets were treated by dipping in various solutions containing either ascorbic acid (500 ppm), sodium hydroxide (0.01 M), or phosphoric acid (0.1 M). Broccoli florets dipped in distilled water were used as the control samples. Samples held in sealed Curlon 850 pouches were analyzed after storage for 4 days at 10°C. Ascorbic acid was included in this study because it acts as a reducing agent for methanethiol under anaerobic conditions in model systems (25).

Alkaline and acidic solutions were employed in the experiment because Marks et al. (20) reported that the formation of S-methyl-L-cysteine sulfoxide breakdown products was dependent on pH of the vegetable tissues. Table VII shows that the alkaline treatment had the most pronounced effect among the treatments on suppressing the formation of dimethyl disulfide which appears to contribute strongly to the unpleasant volatile sulfur compound aroma in stored broccoli. This effect probably results from the destruction of methyl methanethiosulfinate by the hydroxide ion (35). Thus, strategies for suppressing undesirable volatile sulfur compounds in cruciferous vegetables and other foods need to incorporate means to minimize contributions by oxidized forms of methanethiol as well as methanethiol itself.

Table VI. Concentrations of volatile sulfur compounds in broccoli florets treated with water (control), cell-free crude caraway seed extracts, or TBHQ solutions (100 ppm), and stored under modified atmospheres at 4°C for 8 days

Sample	Concentration (ppm)		
	CH_3SH	$CH_3\text{-}S\text{-}CH_3$	$CH_3\text{-}S\text{-}S\text{-}CH_3$
Control	25.5 ± 0.5	0.32 ± 0.05	2.5 ± 0.4
Caraway Extract	23.8 ± 0.8	0.23 ± 0.04	2.0 ± 0.1
TBHQ	22.8 ± 0.3	0.24 ± 0.05	1.5 ± 0.3

Table VII. Odor assessment and concentrations of volatile sulfur compounds in broccoli florets treated with water (control), or solutions of ascorbic acid (500ppm), phosphoric acid (0.1 M), or sodium hydroxide (0.01 M), and stored under modified atmospheres at 10°C for 4 days

Sample	Concentration (ppm)			Sulfurous Aroma Intensity[1]
	CH_3SH	$CH_3\text{-}S\text{-}CH_3$	$CH_3\text{-}S\text{-}S\text{-}CH_3$	
Control (Water)	5.6 ± 0.2	0.10 ± 0.02	2.01 ± 0.28	5.9
Acid Treatment	5.4 ± 0.1	0.08 ± 0.01	1.49 ± 0.35	2.6
Ascorbic Acid	5.7 ± 0.3	0.12 ± 0.01	1.47 ± 0.15	2.1
Alkali Treatment	5.5 ± 0.2	0.10 ± 0.03	1.34 ± 0.21	1.3

[1] Scale: 1 = very weak; 7 = very pronounced.

Conclusions

Unpleasant sulfurous odors associated with methanethiol-related volatile sulfur compounds limit consumer acceptance and the use of modified atmosphere technologies to extend the shelf life of cruciferous vegetables. Studies have revealed that both enzymic and nonenzymic systems which are present in caraway seed extracts are capable of depleting methanethiol in model systems and experimental vegetable packs. Additionally, the widely-used synthetic phenolic antioxidant, TBHQ, exhibits active methanethiol-depleting activity through adduct formation. Thus, the results of these studies have provided encouragement toward a goal of expanding new product opportunities in the cruciferous vegetable industry, but additional research on volatile sulfur compound-binding in practical applications with cruciferous vegetables is needed. Further, strategies examined in this study might be applied to the control of noxious sulfurous odors in the environment that have been generated from other industrial sources, such as pulp and paper mills.

Literature Cited

1. Schwimmer, S. *J. Food Sci.* **1963**, 28, 460-466.
2. Fenwick, G. R.; Heaney, R. K.; Mullin, W. J. *CRC Crit. Rev. Food Sci. Nutr.* **1983**, 18, 123-201.
3. Maruyama, F. T. *J. Food Sci.* **1970**, 35, 540-543.
4. Forney, C. F.; Matteis, J. P.; Austin, R. K. *J. Agric. Food Chem.* **1991**, 39, 2257-2259.
5. Chin, H.-W.; Lindsay, R. C. *J. Food Sci.* **1993**, 58: 835-839.
6. Bailey, G. S.;Williams, D. E. *Food Technol.* **1993**, 47(2), 105-118.
7. Bradlow, H. L.; Michnovicz, J. J.; Telang, N. T.; Osborne, M. P. *Carcinogenesis* **1991**, 12, 1571-1574.
8. Michnovicz, J. J.; Bradlow, H. L. *Nutr. Cancer* **1991**, 16, 59-66.
9. Zhang, Y.; Talalay, P.; Cho, C.-G.; Posner, G. H. *Proc. Natl. Acad. Sci. USA.* **1992**, 89, 2399-2403.
10. Bastrash, S.; Makhlouf, J.; Castaigne, F.; Willemot, C. *J. Food Sci.* **1993**, 58, 338-341.
11. Huxsoll, C. C; Bolin, H. R. *Food Technol.* **1989**, 43(2), 124-128.
12. King, A. D.; Bolin, H. R. *Food Technol.* **1989**, 43(2), 132-135.
13. Barth, M. M.; Kerbel, E. L.; Perry, A. K.; Schmidt, S. J. *J. Food Sci.* **1993**, 58, 140-143.
14. Lipton, W. J.; Harris, C. M. *J. Am. Soc. Hortic. Sci.* **1974**, 99, 200-205.
15. Ballantyne, A.; Stark, R.; Selman, J. D. *Int. J. Food Sci. Technol.* **1988**, 23, 353-360.
16. Botero-Omary, M., Testin, R. F., Barefoot, S. F., Halpin, E. S., and Rushing, J. W. Effect of packaging on the sensory characteristics and consumer acceptibity of shredded cabbage. Presented at **1991** IFT Annual Meeting and Food Expo, Dallas, TX.
17. Hansen, M.; Buttery, R. G.; Stern, D. J.; Cantwell, M. I.; Ling, L. G. *J. Agric. Food Chem.* **1992**, 40, 850-852.
18. Lindsay, R. C., Josephson, D. B., and Olafsdottir, G. In *Seafood Quality Determination*, D. E. Krammer and J. Liston (Ed.), **1986**, pp. 221-234. Elsevier Science Publishers, Amsterdam.
19. Whitaker, J. R. *Adv. Food Res.*, **1976**, 22, 73-133.
20. Marks, H. S.; Hilson, J. A., *J. Agric. Food Chem.* **1992**, 40, 2098-2101.
21. Chin, H.-W. **1993**, Ph.D. Thesis, University of Wisconsin-Madison.
22. Chin, H.-W.; Lindsay, R. C. *J. Agric. Food Chem.*, **1994**, accepted for publication.
23. Ostermayer, F.; Tarbell, D. S. *J. Am. Chem. Soc.* **1960**, 82, 3752-3755.

24. Block, E.; O'Connor, J. *J. Am. Chem. Soc.* **1974**, 96, 3929-3944.
25. Chin, H.-W.; Lindsay, R. C. *Food Chem.*, **1994**, 49, 387-392.
26. Stamer, J. R. In *Biotechnology. Vol. 5*, H.-J. Rehm and G. Reed (Ed.), **1983**, pp. 365-378. Verlag Chemie GmbH, Weinheim.
27. Meyer, W. Using spices and their flavors to provide attractive new flavors for kraut products. Presented at Winter Conference, National Kraut Packers Association. Dec. 13, **1984**, Chicago, IL.
28. Pederson, C. S. and Albury, M. N. The sauerkraut fermentation. Bull. 824, *N.Y. State Agric. Exp. Stn.*, **1969**, Geneva, NY.
29. Lee, C. Y.; Acree, T. E.; Butts, R. M.; Stamer, J. R. *Proc. Int. Congr. Food Sci. Technol., 4th,* **1974**,1, 175-178.
30. Dahlson, A. **1986**, M.S. Thesis, University of Wisconsins-Madison.
31. Price Waterhouse Associates. A marketing study of sauerkraut and related cabbage products in the Atlantic Region. *Price Waterhouse Associates,* **1985**, Nova Scotia.
32. IFT Sensory Evaluation Division. *Food Technol.* **1981**, 35(11), 50.
33. Stone, H.; Sidel, J. Sensory Evaluation Practices. *Academic Press* **1985**, New York.
34. McRorie, R. A.; Sutherland, G. L.; Lewis, M. S.; Barton, A. D.; Glazener, M. R.; Shive, W. *J. Am. Chem. Soc.* **1954**, 76, 115.
35. Kice, J. L.; Rogers, T. E.; Warheit, A. C. *J. Am. Chem. Soc.* **1974**, 96, 8020-8026.

RECEIVED March 23, 1994

Formation

Chapter 9

Thioglucosides of *Brassica* Oilseeds and Their Process-Induced Chemical Transformations

Fereidoon Shahidi

Department of Chemistry, Memorial University of Newfoundland, St. John's, Newfoundland A1B 3X7, Canada

Glucosinolates are sulfur-containing compounds found in canola and other species of *Brassica* oilseeds and are considered as antinutritional factors. Glucosinolates and their degradation products are associated with various flavor, off-flavor and toxic effects in products derived from such seeds. Consequently, their removal from oilseeds is necessary for upgrading of the quality of their protein meals. Extraction of crushed canola seeds with methanol/ammonia/water removed 50-100% of individual glucosinolates present in the samples. However, removal of glucosinolates from seeds also coincided with the formation of degradation products such as epithionitriles, nitriles, isothiocyanates, sulfinylnitriles, sulfinylisothiocyanates, glucose and thioglucose. Although most glucosinolate degradation products were extracted into the polar methanol/ammonia/water phase, some residues were detected in the oil and protein meal fractions. Ultrafiltration may offer a means to further detoxify the resultant meals.

Plants of the genus *Brassica*, including canola, contain glucosinolates which are considered as the source of goitrogens. Goiter enlargement and improper functioning of the thyroid gland due to iodine deficiency is consistently observed when *Brassica* seeds are fed to experimental animals (1-3). Sulfur-containing compounds in canola oil, derived from degradation of glucosinolates, have also been implicated as hydrogenation catalyst poisons (4). The basic chemical structure of glucosinolates which was established by Ettlinger and Lundeen in 1956 (5) is shown in Figure 1.

More than one hundred glucosinolates are known to occur in nature (6), and these differ from one another due to differences in the side chain R-group originating from their respective amino acid precursors; indole glucosinolates being apparent exceptions (7-9). Table I provides a list of common glucosinolates of *Brassica* oilseeds. The intact glucosinolates are apparently free from toxicity. However, on hydrolysis by an endogenous enzyme myrosinase (thioglucose glucohydrolase (E.C.3.2.3.1) (10), which co-occurs in the seed and unheated meal, their thioglucosidic bond cleaves, giving rise to a variety of potentially toxic products. These include isothiocyanates, thiocyanate, nitriles (Figure 2) and oxazolidinethiones formed by cyclization of hydroxyisothiocyanates (Figure 3).

Degradation products of glucosinolates are associated with various flavor, off-flavor, antinutritive and toxic effects (7). The pungent flavor of mustard and its biting taste as well as the characteristic flavors of radish, broccoli, cabbage and cauliflower originate from degradation of their glucosinolates. Some breakdown products of glucosinolates such as isothiocyanates are fairly reactive and produce substituted thioamides and oxazolidine-2-thiones upon reaction with amines or cyclization. The thiocyanate ion and some phenolics may be produced from sinalbin. In acidic solutions, however, nitriles are dominant degradation products including cyanoepithioalkanes from progoitrin.

The commercial processing of rapeseed and its canola varieties includes a heat treatment step which facilitates oil release and deactivates the enzyme myrosinase. However, the intact gluosinolates left in the meal may still produce toxic aglucons in the lower gastrointestinal tract. Oginsky *et al.* (11) have shown that some bacteria, especially *para colobatrum*, common to the digestive system of man, have myrosinase activity. Therefore, it is prudent to remove glucosinolates and/or their degradation products from canola meal in order to upgrade the quality of the final protein meal.

The present paper reports on the glucosinolate content of several *Brassica* oilseeds and examines the effects of various detoxification methods, including alkanol/ammonia/water-hexane as an extraction medium for canola. The fate of glucosinolates in this process and possible methods for their elimination from the reaction medium is also presented.

Toxicity, Antinutritional and Possible Beneficial Effects of Glucosinolates

The traditional varieties of rapeseed contain 1-7% glucosinolates after oil extraction. Consequently, only a small portion of rapeseed meal is used at reduced levels in animal feed formulations and a major portion of it is used as a fertilizer. Some seeds and meals are also used as fish feed in small aquaculture operations in China. However, growth response of livestock and other animals upon ingestion of such meals is generally poor and even at low levels, problems with breeding of animals have been noted (12, 13). Goitrin, a cyclization product of progoitrin (Figure 3), has been shown to interfere with the function of the thyroid gland by reducing the uptake of iodine, thus adversely affecting the growth of experimental animals. Incorporation of larger amounts of iodine in the diet did not circumvent this problem. Therefore, no rapeseed or canola preparation is currently used in food for human consumption.

Table I. Common glucosinolates of *Brassica* oilseeds

No.	Trivial Name	Side Chain, R	
		Name	Formula
1	Sinigrin	Allyl	$CH_2=CH-CH_2-$
2	Gluconapin	3-Butenyl	$CH_2=CH-(CH_2)_2-$
3	Progoitrin	2-Hydroxy-3-butenyl	$CH_2=CH-CHOH-CH_2-$
4	Glucobrassicanapin	4-Pentenyl	$CH_2=CH-(CH_2)_3-$
5	Gluconapoleiferin	2-Hydroxy-4-pentenyl	$CH_2=CH-CH_2-CHOH-$
6	Gluconasturtin	Phenylethyl	$C_6H_5-CH_2-CH_2-$
7	Glucobrassicin	3-Indolylmethyl	3-indolylmethyl structure
8	4-Hydroxyglucobrassicin	4-Hydroxy-3-indolylmethyl	4-hydroxy-3-indolylmethyl structure
9	Neo(Glucobrassicin)	1-Methoxy-3-indolylmethyl	1-methoxy-3-indolylmethyl structure
10	Glucoerucin	4-Methylthiobutyl	$CH_3-S-(CH_2)_4-$
11	Gluco(Sinalbin)	4-Hydroxybenzyl	$p-HO-C_6H_4-CH_2-$

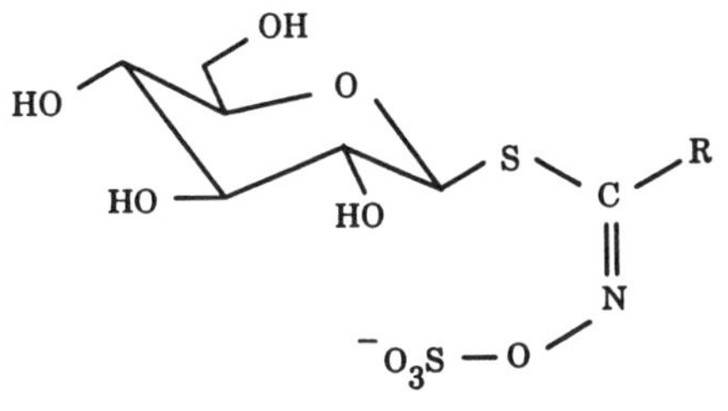
OH
HO
O
HO
HO
S
C
R
N
$^{-}O_3S$ — O

Figure 1. General chemical structure of glucosinolates.

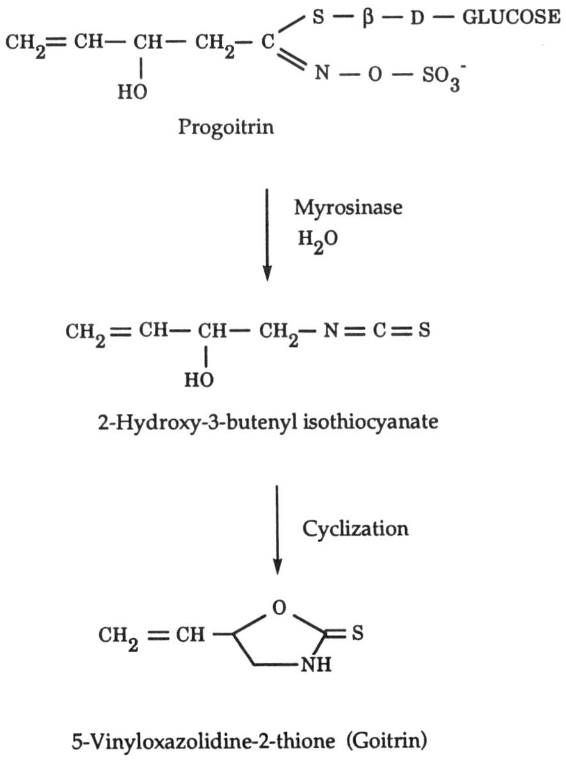
S — β — D — GLUCOSE
CH_2= CH— CH— CH_2— C
N — O — SO_3^-
HO
Progoitrin
Myrosinase
H_2O
CH_2= CH— CH— CH_2— N=C=S
HO
2-Hydroxy-3-butenyl isothiocyanate
Cyclization
CH_2 = CH
O
S
NH
5-Vinyloxazolidine-2-thione (Goitrin)

Figure 2. Enzymatic hydrolysis products of glucosinolates.

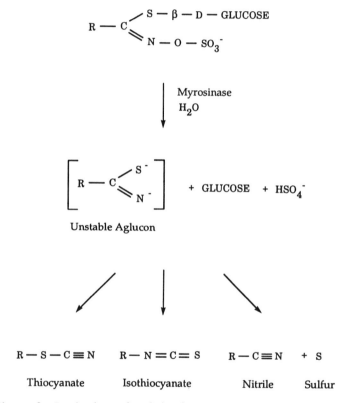

Figure 3. Production of goitrin from progoitrin upon enzymatic hydrolyis.

The toxic effect of nitriles from decomposition of glucosinolates has been thoroughly studied (*14*). For example, 3-hydroxy-4-pentane nitrile and 3-hydroxy-4,5-epithiopentane nitrile (threo) have LD_{50} values of 170 and 240 mg/kg, respectively. Acute toxicities of these nitriles exceed those of oxazolidine-2-thiones (*14, 15*). At a 0.1% level of addition, mixed nitriles brought about poor growth as well as liver and kidney lesions in rats after a 106 days of feeding (*14*). However, despite severe effects of 4, 5-epithiopentane nitrile on animals, no toxic effects were registered after the administration of its polymerized form (*16*). The toxic and goitrogenic effects of glucosinolate breakdown products were generally more pronounced in non-ruminants.

In contrast to aliphatic glucosinolates, indolyl glucosinolates such as glucobrassician and its myrosinase-catalyzed transformation products have been shown to inhibit chemical and other types of carcinogensis induced by polycyclic aromatic hydrocarbons and other initiators (*17*). Indolyl glucosinolates are found abundantly in cruciferae vegetables and are present in relatively large amounts in the canola varieties of rapeseed.

Processing of *Brassica* Oilseeds

The conventional extraction process for rapeseed and canola is an adaption of soybean technology adjusted for the small seed size, high oil content, and presence of glucosinolates (Figure 4). The seeds are crushed to fracture the seed coat and to rupture oil cells. This increases the surface-to-volume ratio. The crushed seeds are then cooked to 90-110°C for 15 to 20 min to facilitate good oil release and to inactivate the enzyme myrosinase (*18*). The crushed and cooked seeds are then prepressed, which reduces their oil content from 42 to about 20% and compresses the material into large cake fragments. These fragments are then flaked and hexane-extracted, using percolating bed extractors, to remove most of the remaining oil. The formation of toxic degradation products of glucosinolates is prevented, but the thermal processing adversely affects the color and the functional properties of the resultant meal (*19*).

If the hydrolysis of glucosinolates occurs during crushing before oil extraction, the sulfur-containing products such as isothiocyanates and oxazolidinethiones enter the oil. Industrially processed crude rapeseed and canola oils contain 10-57 ppm of sulfur which can not be totally removed by conventional refining and bleaching operations (*20*). Consequently, as high as 3-5 ppm of sulfur may remain in refined oils. In addition to interference of sulfur-containing materials during the hydrogenation process, sulfurous compounds also cause certain unpleasant odors in heated canola oil. Therefore, adequate processing and proper heat treatment is required to prevent sulfur contamination of the oil.

Glucosinolate Content of Canola Varieties

The name "canola" refers to low-glucosinolate, low-erucic acid rapeseed cultivars which were initially developed in Canada. They are also known as "double-

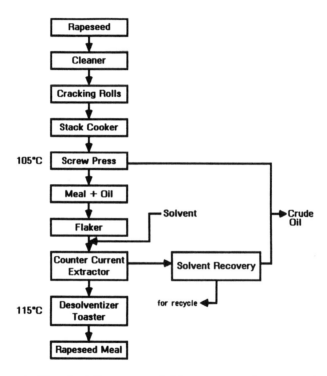

Figure 4. Flowsheet for commercial hexane-extraction processing of rapeseed and canola.

zero" or "double-low" and contain less than 30μmol/g of one or any combinations of the four known aliphatic glucosinolates (gluconapin, progoitrin, glucobrassicanapin, and gluconapoleiferin) in their meal and less than 2% erucic acid in their oil (22). Table II summarizes the total content of glucosinolates in selected *Brassica* oilseeds. All canola seeds contain less than approximately 0.3% glucosinolates in their defatted meals as compared with up to 7% in traditional rapeseed meals. Recently, new varieties of canola have been developed within the Canadian breeding program. Some of these varieties contain as little as 6μmol (approximately 0.06%) aliphatic glucosinolates/g meal (Table II).

The content and chemical nature of individual glucosinolates present in canola varieties depends on the cultivar being studied. As an example, Triton canola contained 23.68 μmol glucosinolate/g defatted meal. Its dominant glucosinolates were progoitrin and gluconapin. Other glucosinolates contributed less than 7.71 μmol/g to this total (Table III). On the other hand, Hu You 9 variety of Chinese rapeseed contained up to 135.70 μmol (approximately 1.36%) glucosinolates/g defatted meal and is rich in progoitrin. Enzyme-assisted chemical-induced breakdown of glucosinolates may be governed by the chemical nature of their side chains. Furthermore, volatiles so produced could vary greatly in both their concentration and flavor intensity. In addition, it is evident that breeding has a pronounced effect on the elimination of individual glucosinolates, particularly progoitrin. However, a slight increase in the content of indolyl glucosinolates paralleled a decrease in the content of aliphatic glucosinolites. Possible beneficial effects of indolyl glucosinolates and their breakdown products have already been discussed.

Detoxification of *Brassica* Oilseeds and Alternative Methods of Processing

Much research has been carried out towards developing detoxification methods for canola, rapeseed and their respective meals. The detoxification methods reported in the literature may be categorized as:

1) genetic improvements (breeding);
2) chemical degradation of glucosinolates and their subsequent removal by oxidation, addition of metal salts; or acids and bases;
3) microbiological degradation of glucosinolates by fungus, mold or bacteria and their subsequent removal;
4) physical extraction of glucosinolates and/or their degradation products;
5) enzymatic methods using endogenous/exogenous myrosinase and subsequent removal of breakdown products by extraction or adsorption on carbon;
6) combination methods such as diffusion extraction;
7) enzyme inactivation methods by heat, steaming, infrared, microwave, etc.;
8) protein isolation.

Table II. The content of four specific canola glucosinolates in some rapeseed and canola varieties [a]

Rapeseed/Canola	Glucosinolates, μmoles/g meal	
	Canola Designation	Total
Midas	137.68	144.05
Hu You 9	131.87	135.79
Triton	17.86	23.68
Altex	17.21	23.03
Regent	22.36	31.65
Westar	11.92	16.24
Profit[b]	9.50	ND
Tobin	6.91	9.00
Vanguard[b]	6.80	ND
Bountry[b]	6.60	ND
Celebra[b]	6.10	ND

[a]Determined according to HPLC method of Shahidi and Gabon (32).
[b]Determined by a GC method. ND, not determined.

Table III. Major glucosinolates of selected *Brassica* Oilseed Meals (μmol/g deoiled, dried meal)[a]

Glucosinolate	Hu You 9	Midas	Triton	Mustard
Sinigrin	0.56	0.70	0.89	236.62
Gluconapin	32.93	23.22	5.05	–
Progoitrin	92.52	101.33	10.92	–
Glucobrassicapapin	4.53	7.28	0.58	–
Gluconapoleiferin	1.89	5.85	1.31	–
Gluconasturtin	1.27	1.23	0.10	1.27
Glucobrassicin	0.30	2.32	2.82	1.35
4-Hydroxyglucobrassicin	1.79	2.12	2.01	0.89
Neoglucobrassicin	–	–	–	0.21

[a]Determined by HPLC Method of Analysis (Ref. 33).

However, only a few breeding programs have so far been used successfully to reduce the level of glucosinolates of rapeseed, and none of the other methods of processing has reached the commercial stage. Disadvantages such as loss of protein, poor functional properties of the resultant meal, or high processing costs are among the main reasons.

The Alkanol/Ammonia-Hexane Extraction Process

The flow diagram for the alkanol/ammonia-hexane extraction system is schematically shown in Figure 5. In this process, crushed seeds are exposed to a two-phase solvent extraction system consisting of a polar alkanol/ammonia phase and a non-polar hexane phase. The use of alkanol/ammonia was first reported by Schlingmann and co-workers (22, 23). These authors indicated that methanol/ammonia was most effective in removing glucosinolates from rapeseed without providing the relevant data. Ammoniation of other *Brassica* seeds has been reported in the literature (24-26).

The effect of the two-phase extraction system with different solvent combinations, on the content of individual glucosinolates in the *Brassica* oilseeds determined by an HPLC method of analysis, is shown in Figures 6 to 9. The concentration of individual glucosinolates in rapeseed, canola and mustard seeds was considerably reduced. The efficacy of different solvent extraction systems in the removal of glucosinolates was primarily due to the presence of ammonia in absolute or 95% methanol; methanol/ammonia/water-hexane extraction system being most efficient. However, no apparent preferential trends were observed for the extraction of individual glucosinolates from the meals.

Chemical Transformation of Glucosinolates

The concentration of glucosinolates in the meals of Triton and Hu You 9 was markedly reduced by the treatment of crushed seeds with methanol/ammonia/water. However, it is necessary to examine whether glucosinolates were extracted out of the seed, as such or were transformed to other products. Due to the complexity of glucosinolates in rapeseed and canola and lack of apparent concern over the fate of indolyl glucosinolates, only the four aliphatic glucosinolates were considered.

In a model system study, possible degradation of pure sinigrin, gluconapin and progoitrin under simulated experimental conditions of extraction with methanol/ammonia/water was tested. While at 25°C only 1-2% loss of these glucosinolates due to their degradation was evident, at 45°C 16-21% degradation occurred (Figure 10). Results in Table IV indicate that the major transformation products were nitriles. Minor quantities of isothiocyanates and/or epithionitriles were also produced from sinigrin, gluconapin and progoitrin. However, no oxazolidinethione was detected in this process.

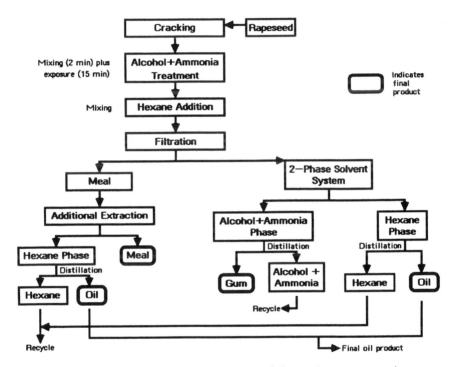

Figure 5. Flowsheet for alkanol/ammonia/water-hexane extraction processing of canola.

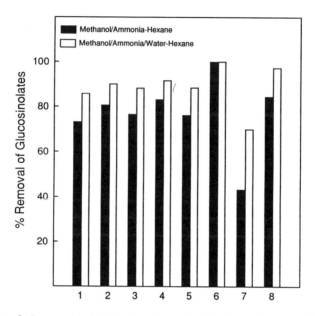

Figure 6. Removal of individual glucosinolates (numbers are defined in Table I) of Hu You 9 rapeseed upon solvent extraction process.

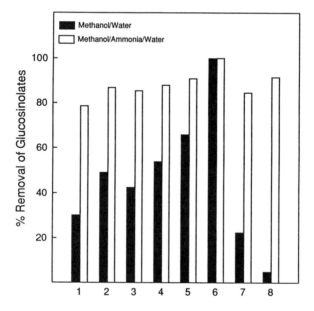

Figure 7. Removal of individual glucosinolates (numbers are defined in Table I) of Midas rapeseed upon solvent extraction process.

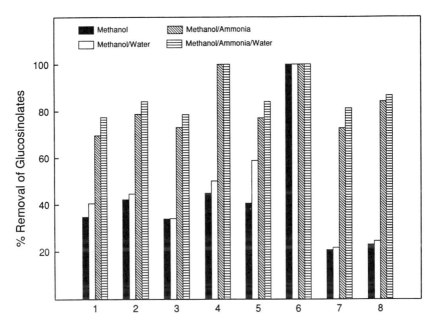

Figure 8. Removal of individual glucosinolates (numbers are defined in Table I) of Triton canola upon solvent extraction process.

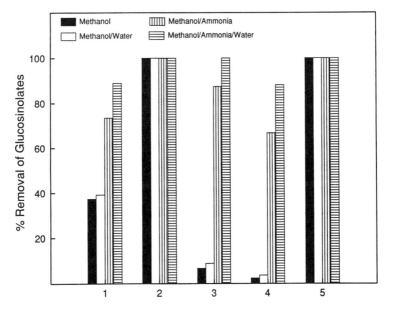

Figure 9. Removal of individual glucosinolates (numbers are defined in Table I) of mustard seed upon solvent extraction process.

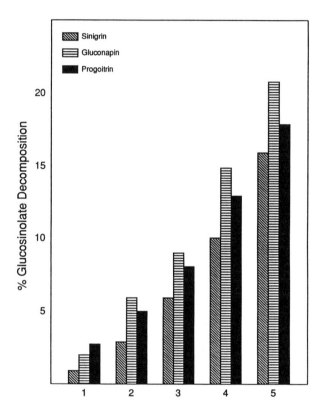

Figure 10. Breakdown of standard sinigrin, gluconapin and progoitrin upon exposure to methanol/ammonia/water over a 24-h period at 1, 25°C; 2, 30°C; 3, 35°C; 4, 40°C and 5, 45°C.

Table IV. Degradation products of pure glucosinolates and their distribution in methanol/ammonia/water

Glucosinolate	Degradation Product	
	Aglucon (%)	Sugar-Related (%)
Sinigrin	But-3-enyl nitrile (90.6)	Thioglucose equivalents (91.7)
	Allyl isothiocyanate (5.7)	Glucose (6.3)
	3,4-Epithiobutane nitrile (3.1)	Furfuryl alcohol (1.1)
	Unidentified (0.1)	Unidentified (0.9)
Gluconapin	Pent-4-enyl nitrile (95.0)	Thioglucose equivalents (95.8)
	But-3-enyl isothiocyanate (3.1)	Glucose (2.9)
	4,5-Epithiopentane nitrile (1.1)	Furfuryl alcohol (0.5)
	Unidentified (0.8)	Unidentified (0.8)
Progoitrin	1-Cyano-2-hydroxybut-3-ene (93.9)	Thioglucose equivalents (95.0)
		Glucose (2.2)
	(β)1-Cyano-2-hydroxybut-3,4-epithiobutane (5.2)	Furfuryl alcohol (1.3)
		Unidentified (1.5)
	Unidentified (0.9)	

Application of the two-phase extraction system to canola meal resulted in a similar pattern for transformation of glucosinolates. Due to the complexity of the glucosinolates in canola, only the four specific aliphatic glucosinolates used for canola designation were examined before and after the treatment. Table V summarizes the mass balance of these glucosinolates, their degradation products, and their whereabouts after the extraction process. A close examination of the results indicates that approximately 75% of the four canola glucosinolates were extracted in the intact form into the polar phase and another 10% of them were retained, as such, in the meal. The balance of glucosinolates was transformed mainly into nitriles and hydroxynitriles. Any oxazolidinethione that might have been formed during the crushing of seeds was extracted into the polar phase.

Based on the results presented in Table V, two distinct patterns for degradation of glucosinolates are evident as the result of methanol/ammonia/water treatment of seeds (Figure 11) (27, 28). The major pathway involves the release of thioglucose from the glucosinolate molecule to yield a stable nitrile or hydroxynitrile. The second pathway, in which isothiocyanates or epithionitriles are produced, may proceed via the cleavage of glucose from the parent molecule to yield an unstable substituted thiohydroximate-O-sulfonate. The latter compound may in turn decompose to yield isothiocyanates, nitriles, thiocyanates and epithionitriles. Furfuryl alcohol may also have been produced from the base-catalyzed dehydration of glucose. The unidentified compounds may have formed from the polymerization of nitriles and isothiocyanates. However, the exact mechanism for these transformations is still subject to speculation.

Conclusions and Recommendations

The two-phase solvent extraction of *Brassica* oilseeds produced meals with reduced levels of glucosinolates. In addition, tannins and phenolic acids were removed, protein content of systems was enhanced and improved (results not shown). However, some breakdown products of glucosinolates were deposited into the meal and the oil. It is imperative to eliminate these toxic breakdown products from the meal and oil. Combination of this process with ultrafiltration has been attempted, however, no detailed studies have been carried out to confirm the removal of the breakdown products from the resultant meal or the extracted oil (29). Similarly, soaking of seeds in citric acid or ammonium carbonate has been shown to remove all oxazolidinethione, nitriles and isothiocyanates formed in the alcohol/ammonia/water-treated samples (20). However, the process was not applied to crushed whole seeds to examine the possible presence of the breakdown products in the extracted meal and oil. It is mandatory to confirm the removal of any traces of glucosinolate degradation products from both the meal and oil in alternate processing of *Brassica* seeds. Enzyme or chemically-assisted water leaching of glucosinolates and their degradation products (31, 32) requires similar confirmatory studies.

Table V. Mass balance of the four specific canola glucosinolates in solvent–extracted canola (μmol/100g seeds, on a dry weight basis)

Class of Components	Meal	Oil	Polar Matters
HEXANE EXTRACTED			
Glucosinolates	422.74±1.36	0	–
Desulfoglucosinolates	1.11±0.14	0	–
Isothiocyanates	1.82±0.46	4.32±0.75	–
Oxazolidinethiones	3.68±0.43	5.25±0.82	–
Nitriles[a]	0.54±0.11	0	–
MEHANOL/AMMONIA/WATER-HEXANE EXTRACTED			
Glucosinolates	41.04±0.29	0	308.42±3.68
Desulfoglucosinolates	0.43±0.07	0	13.56±1.11
Isothiocyanates	0.50±0.14	0.75±0.14	9.92±0.79
Oxazolidinethiones	0	0	8.20±0.75
Nitriles[a]	1.53±0.35	1.25±0.18	67.17±3.03

[a]Includes hydroxy–and epithionitriles.

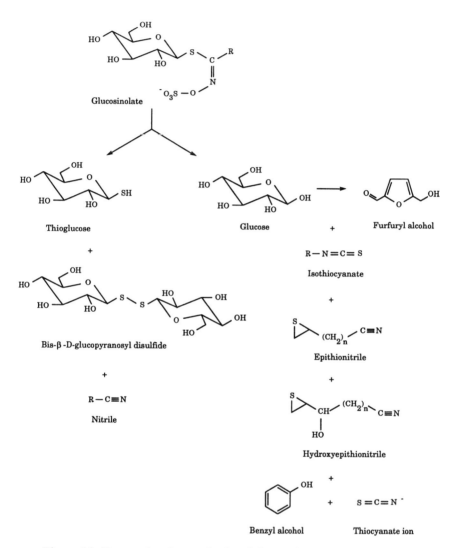

Figure 11. Proposed pathways for breakdown of glucosinolates upon extraction of seeds with methanol/ammonia/water.

Literature Cited

1. Maheshwari, P.N.; Stanley, D.W.; Gray, J.I. *J. Food Prot.* **1981**, *44*, 459.
2. Astwood, B.E. *J. Pharmacol. Exp. Ther.* **1943**, *78*, 78.
3. Clandinin, D.R.; Bayley, I.; Cahllero, A. *Poult. Sci.* **1966**, *45*, 833.
4. deMan, J.M.; Pogorzelska, E.; deMan, L. *J. Am. Oil Chem. Soc.* **1983**, *60*, 558.
5. Ettlinger, M.G.; Lundeen, A.J. *J. Am. Chem. Soc.* **1956**, *78*, 4172.
6. Sørensen, H. In *Canola and Rapeseed in Production, Chemistry, Nutrition and Processing Technology*; Shahidi, F., Ed.; Van Nostrand Reinhold: New York, NY, **1990**, pp 149-172.
7. Fenwick, G.R.; Heaney, R.K.; Mullin, W.J. *CRC Crit. Rev. Food Sci. Nutr.* **1983**, *18*, 123.
8. Bjerg, B.; Sørensen, H. In *World Crops - Production, Utilization, Description, Glucosinolates in Rapeseeds*. Wathelst, J.-P., Ed.; Analytical Aspects; Martinus Nijhoff: Dordrecht, **1987**, 13; pp 59-75.
9. Jensen, S.K.; Michaelsen, S.; Kachlicki, P.; Sørensen, H. In *Rapeseed in a Changing World - Proceedings of the 8th International Rapeseed Congress*, McGregor, D.I., Ed.; Saskatoon, Canada, **1991**, pp 1359-1364.
10. Buchwaldt, L.; Larsen, L.M.; Plöger, A.; Sørensen, H. *J. Chromatogr.* **1986**, *363(1)*, 71.
11. Oginsky, E.L.; Stein, A.E.; Greer, M.A. *Proc. Soc. Exptl. Biol. Med.* **1965**, *119*, 360.
12. Hill, R. *Brit. Vet. J.* **1979**, *135*, 3.
13. Butler, W.J.; Pearson, A.W.; Fenwick, G.R. *J. Sci. Food Agric.* **1982**, *33*, 866.
14. Tookey, H.L.; Van Etten, C.H.; Daxenbichler, M.E. In *Toxic Constituents of Plant Foodstuffs. Second Edition.* Liener, I.E., Ed.; Academic Press: New York, NY, **1980**, pp 103-142.
15. Nishie, K.; Daxenbichler, M.E. *Food Cosmet. Toxicol.* **1980**, *18*, 159.
16. Dietz, H.M.; Panigrahi, S.; Harris, R.V. *J. Agric. Food Chem.* **1991**, *39*, 311.
17. Loft, S.; Otte, J.; Poulsen, H.E.; Sørensen, H. *Fd. Chem. Toxic.* **1992**, *30*, 927.
18. Shahidi, F.; Naczk, M.N.; Rubin, L.J.; Diosady, L.L. *J. Food Prot.* **1988**, *51*, 743.
19. Shahidi, F. *In Engineering and Food*. Spiess, W.E.L.; Schubert, H., Eds.; Advanced Processes, Vol. 3.; Elsevier Applied Science: London and New York, **1989**, pp 50-59.
20. Daun, J.K.; Hougen, F.W. *J. Am. Oil Chem. Soc.* **1977**, *54*, 351.
21. Shahidi, F. In *Canola and Rapeseed - Production, Chemistry, Nutrition and Processing Technology*. Shahidi, F., Ed.; Van Nostrand Reinhold: New York, **1990**, pp 3-14.
22. Schlingmann, M.; Praere, P. *Fette Seifen Anstrichm.* **1978**, *80*, 283.

23. Schlingmann, M.; von Rymon-Lipinski, G.W. *Can. Patent*, 1, 120, 1979, **1982**.
24. Kirk, W.D.; Mustakas, G.C.; Griffin, E.L., Jr. *J. Am. Oil Chem. Soc.* **1966**, *43*, 550.
25. Kirk, W.D.; Mustakas, G.C.; Griffin, E.L., Jr.; Booth, A.N. *J. Am. Oil Chem. Soc.* **1971**, *48*, 845.
26. McGregor, D.I.; Blake, J.A.; Pickard, M.O. In *Proceedings of the 6th International Rapeseed Conference*; Paris, France, **1983**, Vol. 2; pp 1426-1431.
27. Shahidi, F.; Gabon, J.E.; Rubin, L.J.; Naczk, M. *J. Agric. Food Chem.* **1990**, *38*, 251.
28. Shahidi, F.; Gabon, J.E. *Lebensm.-Wiss. U. Technol.* **1990**, *23*, 154.
29. Rubin, L.J.; Diosady, L.L.; Tzeng, Y-M. In *Canola and Rapeseed - Production, Chemistry, Nutrition and Processing Technology*; Shahidi, F., Ed.; Van Nostrand Reinhold: New York, NY, **1992**, pp 307-330.
30. Jensen, S.K.; Olsen, H.S.; Sørensen, H. In *Canola and Rapeseed - Production, Chemistry, Nutrtion and Processing Technology*; Shahidi, F., Ed.; Van Nostrand Reinhold: New York, NY, **1992**, pp 331-341.
31. Schwenke, K.D.; Kroll, J.; Lange, R.; Kujawa, M.; Schnaak, W. *J. Sci. Food Agric.* **1990**, *51*, 391.
32. Kroll, J.; Kujawa, M.; Schnaak, W. *Fat. Sci. Technol.* **1991**, *93(2)*, 61.
33. Shahidi, F.; Gabon, J.E. *J. Food Quality* **1989**, *11*, 421.
34. Shahidi, F.; DeClercq, D.R.; Daun, J.K., unpublished results.

RECEIVED March 23, 1994

Chapter 10

Kinetics of the Formation of Methional, Dimethyl Disulfide, and 2-Acetylthiophene via the Maillard Reaction

F. Chan and G. A. Reineccius

Department of Food Science and Nutrition, University of Minnesota, 1334 Eckles Avenue, St. Paul, MN 55108

Strecker aldehydes are quantitatively the major products of the Maillard reaction. In addition to their intrinsic flavor, they are very reactive and participate in numerous reactions that make additional contributions to flavor development in foods. There is a lack of information on the reaction kinetics of these Strecker aldehydes as well as other flavor compounds. Thus a kinetic study on the formation of methional and two secondary products (dimethyl disulfide and 2-acetylthiophene) from the reaction of amino acids (0.075 mole) and glucose (0.5 mole) in aqueous model systems was conducted. Systems were heated at temperatures from 75 to 115°C at times from 5 min. to 7.5 h and pH's of 6, 7, and 8. Kinetic data are presented and discussed.

Nearly all foods are made up of a complex mixture of components, including carbohydrates, amino acids, and proteins. When these foods are heated, the Maillard reaction occurs resulting in the formation of a large variety of volatile flavor compounds (1-3). The Maillard reaction is responsible for both desirable and undesirable aromas in foods. The aroma of bread, chocolate, coffee, and meat are all examples of desirable aromas resulting from the Maillard reaction. The aromas of burned food, canned products, stale milk powder, cereal, and dehydrated potatoes are typical examples of the undesirable aspects of this reaction.

Nearly all foods examined in our laboratory by aromagram methodology (4) have contained aromas characteristic of Strecker aldehydes (caramel, potato, and floral notes). This raised a question as to whether these aromas were always present because the Strecker aldehydes have exceptionally low sensory thresholds, or are they readily formed in foods. This question prompted the study to be reported here.

The Strecker reaction involves the oxidative deamination and decarboxylation of an α-amino acid in the presence of a dicarbonyl compound. The products formed from this reaction are an aldehyde, containing one less carbon than its original amino acid, and an α-aminoketone. The interaction of Maillard reaction products with those of the Strecker degradation leads to the formation of many additional flavor compounds (e.g., aldehydes, pyrazines, and thiophenes). An immense amount of time and effort have been spent on the isolation/identification of these compounds in foods and model systems (1-7). Many studies have focused on the different pathways and mechanisms while others have been very empirical in nature - simply heating model systems under various conditions and examining the products formed. A literature search revealed that while some kinetic work has been done on pyrazines (8, 9) and oxygen containing heterocyclic compounds (10), there is scant information available on the kinetics of the Strecker degradation.

This study focuses on the reaction kinetics of the Strecker aldehyde, methional and two other secondary products, dimethyl disulfide and 2-acetylthiophene. While additional sulfur containing compounds were detected in this study, this paper will focus only on those that yielded sufficient kinetic data at the different pHs and temperatures.

Materials and Methods

Sample Preparation. Glucose (0.5 mole = 90 g; Aldrich Chemical, Milwaukee, WI), methionine (0.075 mole = 11.19 g; Sigma Chemical, St. Louis, MO), phenylalanine (0.075 mole = 12.39 g; Sigma Chemical, St. Louis, MO), proline (0.075 mole = 8.64 g; Aldrich chemical, Milwaukee, WI), and leucine (0.075 mole = 9.84 g; Aldrich Chemical, Milwaukee, WI) were dissolved in 400 mL of distilled water (Glenwood Inglewood, Minneapolis, MN), and the pH was adjusted to 6, 7, or 8 with a 0.1M phosphate buffer and the appropriate amount of NaOH. A 600 mL pressure reactor with a temperature controller (Parr 4563, Parr 4842; Parr Instrument Co., Moline, IL) was filled with the entire solution, sealed and heating and stirring started. Initial temperature was noted and start time determined when the solutions reached reaction temperatures. Aliquots of 50 mL were withdrawn from the reactor via a sampling port at the following schedule:

75°C	1.5, 3.0, 4.5, 6.0, 7.5 h
85°C	0.5, 1.0, 1.5, 2.0, 2.5 h
95°C	20, 40, 60, 100, 120 min.
105°C	10, 20, 30, 40, 50 min.
115°C	5, 10, 15, 20, 25 min.

Ultra high purity N_2 (ca. 50 mL) was added after each sampling to restore pressure to the reactor. Each sample was cooled to room temperature in an ice bath, the pH measured and the sample extracted with 3 x 5 mL of dichloromethane (EM Science, Gibbstown, NJ) containing 500 ppm of 4-methylthiazole (Aldrich Chemical, Milwaukee, WI) as an internal standard in a 150 mL separatory funnel. The dichloromethane fraction was dried with anhydrous magnesium sulfate (Fisher Scientific, Fairlawn, NJ), filtered and evaporated to 0.5 to 1.0 mL volume under a stream of high purity nitrogen.

Gas Chromatography/Mass Spectrometry. A Hewlett Packard (HP) Model 5890 gas chromatograph (GC) (Hewlett Packard, Avondale, PA) equipped with a mass selective detector (MSD) HP Model 5970 was used to identify compounds in this study. The capillary column was interfaced directly into the MSD operating at 70 eV ionization potential, with an ion source temperature of 220°C, and a scan threshold of 750, scanning from m/z 29 to 400 at 0.86 s/cycle. GC separation was achieved on a 30 m x 0.32 mm i.d. x 1 µm film thickness DB-5 fused silica capillary column (J & W Scientific, Folsom, CA). All injections were performed under the following conditions: injector port temperature 230°C, initial column temperature 40°C, initial time 3 min., final temperature 250°C, final time 10 min., and an oven ramp rate of 5°C/min. Helium was used as a carrier gas at 7 psi of head pressure; a 20:1 inlet split was used and 1.0 µL of the extract was injected. The compounds studied were identified by comparing the mass spectrum obtained with the National Bureau of Standards Mass Spectra library, published retention indices, and cochromatography with authentic compounds.

Gas Chromatography/Atomic Emission Detection. A HP Model 5890 Series II GC equipped with an Atomic Emission Detector (AED) HP Model 5921A (sulfur, carbon and nitrogen signal) were used to collect kinetic data. Compounds were separated on a 30 m x 0.32 mm i.d. x 1 µm film thickness DB-5 fused silica capillary column (J & W Scientific, Folsom, CA). All runs were performed under the following conditions: initial temperature 40°C, initial time 3.0 min., final temperature 250°C, final time 10.0 min., ramp rate of 5°C/min., injection port temperature 250°C, and AED transfer line temperature 275°C. Helium (ca. 2 mL/min.) was used as a carrier gas in the system at 15 psi of head pressure. For the AED detector, the flow rates of the reagent gases are based on (11). For the carbon (193 nm), nitrogen (174 nm), and sulfur signal (181 nm), oxygen and hydrogen were used as scavenger gases, and the makeup flow was 30 mL/min. A split ratio of 45:1 was used, and the injection size was 1.0 µL.

Results And Discussion

General Observations. Generally, the amount of volatile compounds formed increased both in number and quantity as either time or temperature of heating increased. This result is expected based on previous studies too numerous to mention.

There were no volatile compounds that appeared at any given pH and not at another pH. However, the quantity of volatiles produced increased as pH increased from 6 to 8. This resulted in some volatiles being detected only at higher temperatures because their formation at lower pH's may have been slowed. Kinetic studies by previous workers (8, 9) on the formation of pyrazines also found that rate of formation of pyrazine, 2-methylpyrazine and dimethylpyrazine increased with either an increase in temperature or pH.

Individual flavor compounds. While numerous other volatile compounds (sulfur and nonsulfur-containing) were formed in this model system during heating, only the sulfur compounds which were formed in sufficient quantity to be subjected to kinetic analysis are going to be discussed in this paper. The formation of additional nonsulfur-containing volatiles will be discussed in another article (12).

The amount of dimethyl disulfide formed increased as either heating time or pH increased (Figure 1). At pH 8, dimethyl disulfide concentration leveled off at longer heating times. Since analysis is in progress to determine the amount of

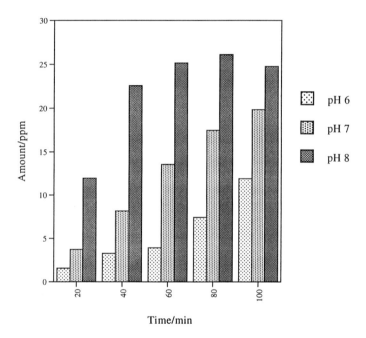

Figure 1. The influence of pH and heating time on the formation of dimethyl disulfide at 95°C.

reactants remaining during heating, we can not comment at this time as to whether this leveling off is due to the exhaustion of reactants or the further reaction of dimethyl disulfide to other end products.

The concentration of methional in the model system (Figure 2) also increased with heating time but initially increased and then decreased with increasing pH. The higher pH's likely favored either degradation (or further reaction) of the methional or a shift in pathways to channel reactants into alternate end products. We have no data to suggest one alternative over the other.

The formation of 2-acetylthiophene (Figure 3) was very interesting as the compound was only formed at pH 6 and pH 8 and none was detected at pH 7. This again could be due to degradation (or further reaction) being favored at this pH.

Kinetic Analysis. Before discussing the kinetic data, it is worthwhile to make the point that the kinetic data obtained in this study were from a complex model system i.e. there was more than one amino acid present in the model system. This is likely less significant for the data on Strecker aldehydes (i.e. methional) than it would be for secondary volatiles (dimethyl disulfide and 2-acetylthiophene). The secondary volatiles are those which are formed from the reaction of two or more Maillard products. In our system, one reactant may come from the degradation of one amino acid while another may come from a different amino acid. Thus including a very reactive amino acid in a model system will rapidly give many reactive fragments which will greatly change the kinetics of the formation of a volatile versus having the system composed of a single amino acid or amino acids which are very slow in reaction. Thus the kinetics observed in any study are very system dependent and cannot be readily compared. This will become evident when the current data are compared to previous studies. An additional consideration is that the data represent the concentration of each volatile in the system at any given time. It is well documented in the literature that few flavor compounds are end products of the Maillard reaction. The flavor compounds formed react with potentially numerous other volatiles to form new compounds. Thus the kinetics observed are the dynamic result of formation versus further reaction. Since we know so little about the reactions occurring in the Maillard reaction, we can do little other than accept an "observed" reaction rate and the resultant kinetics.

The Macintosh program "Water Analyzer Series - Reaction Kinetics Program V. 2.09" (13) was used to do kinetic analysis of the data. The reaction order for the compounds selected (dimethyl disulfide, methional, and 2-acetylthiophene) were determined by plotting concentration versus time and then using linear regression to determine how well the data fit the straight line. The r^2 for all three compounds at zero order ranged form 0.83 to 0.99 and were consistently better than any other order. It was, therefore, concluded that the formation of dimethyl disulfide, methional, and 2-acetylthiophene all followed zero order reaction kinetics.

The rate constants, k, were obtained by calculating the slope of concentration (linear portion of plot) versus time. The rate constant is a characteristic of each particular reaction. It's value depends on the conditions of the reaction, especially the reaction temperature. This temperature dependence is expressed by the Arrhenius equation. Therefore, the Arrhenius equation was used to determine the activation

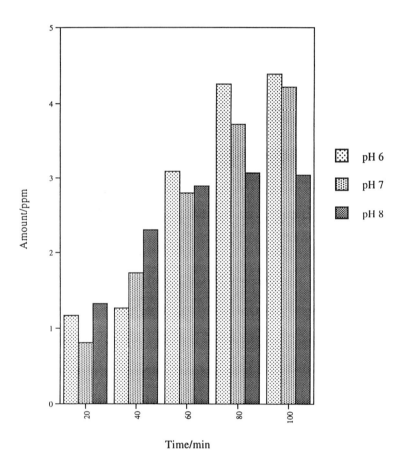

Figure 2. The influence of heating time and pH on the formation of methional at 95°C.

10. CHAN & REINECCIUS *Kinetics of Flavor Compound Formation* 133

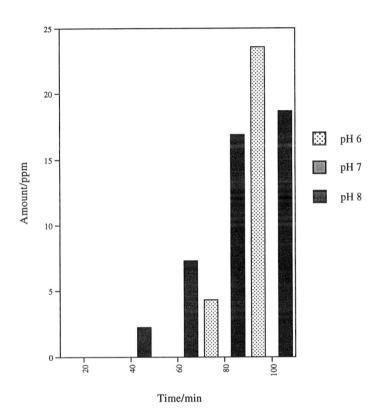

Figure 3. The influence of heating time and pH on the formation of 2-acetylthiophene at 95°C.

energy (Ea). The value of Ea was obtained by doing linear regression on a plot of ln k vs. 1/T.

A plot of reaction rates vs. temperature and pH for methional and dimethyl disulfide are shown in Figures 4 and 5, respectively. The pH had little effect at lower temperatures (up to 115°C for dimethyl disulfide and 105°C for methional). While pH did not follow a trend for dimethyl disulfide at 115°C, higher pH's lowered the reaction rate of methional at higher temperatures.

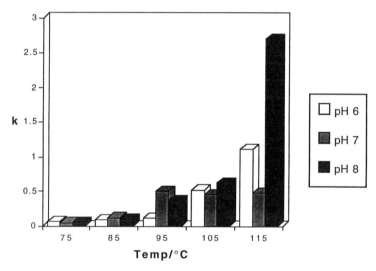

Figure 4. The influence of temperature and pH on the rate constant of dimethyl disulfide.

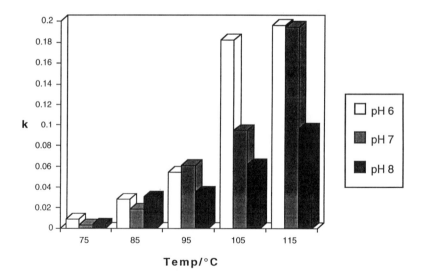

Figure 5. The influence of temperature and pH on the rate constant of methional.

As was noted before, there was a good fit of the data to zero order kinetics (the r^2 ranged from 0.81 to 0.98). A summary of the kinetic data is presented in Table I. It was interesting to see that when Ea was plotted against pH (Figure 6), the expected pattern of a higher pH having a lower Ea wasn't observed. This may be due to the fact that the variation in pH from 6 to 8 was insufficient to discern a trend. When the 95% confidence limits of the calculated Ea's are taken into account, there is essentially no difference in Ea between pH 6, 7, and 8. Earlier studies by Leahy and Reineccius (8, 9) on the effect of pH (5 to 9) on the activation energy of pyrazine found that the Ea was highest at pH 5 and lowest at pH 7, but the study did not have sufficient data to calculate the 95% confidence limits.

Conclusions

While analytical aspects of this study were not discussed in this paper, the usefulness of the AED as a sensitive and selective detector in flavor research was appreciated (14). This greatly simplified the task of quantifying sulfur-containing volatiles in the model system.

In general, the quantity of sulfur containing and other non-sulfur containing compounds increased with time, temperature of heating, and pH (with some exception). The formation of sulfur-containing volatiles was found to have zero order reaction kinetics and activation energies ranged from 15 to 31 kcal/mole. The Ea of methional is moderate (17-19 kcal/mole) and is thus, readily formed with minimal heating. The fact that it is nearly always present in the aromagrams of flavor isolates is likely due both to its moderate Ea and extremely low sensory threshold.

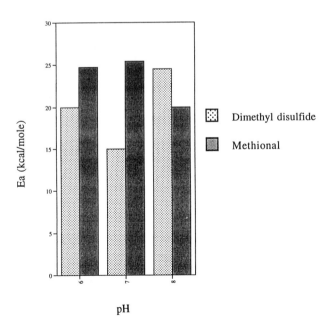

Figure 6. The influence of pH on average activation energy.

Table I. Activation energy for selected flavor compounds

1ST REPLICATE

pH 6

Compound	r^2	Ea (kcal/mole)
Dimethyl disulfide	0.935	20.07
Methional	0.851	27.07
2-Acetylthiophene	Insufficient data	

pH 7

Compound	r^2	Ea (kcal/mole)
Dimethyl disulfide	0.877	15.02
Methional	0.912	24.95
2-Acetylthiophene	Insufficient data	

pH 8

Compound	r^2	Ea (kcal/mole)
Dimethyl disulfide	0.857	22.80
Methional	0.797	19.39
2-Acetylthiophene	0.976	32.71

2ND REPLICATE

pH 6

Compound	r^2	Ea (kcal/mole)
Dimethyl disulfide	0.900	19.56
Methional	0.961	21.74
2-Acetylthiophene	Insufficient data	

pH 7

Compound	r^2	Ea (kcal/mole)
Dimethyl disulfide	0.805	16.20
Methional	0.956	26.02
2-Acetylthiophene	Insufficient data	

pH 8

Compound	r^2	Ea (kcal/mole)
Dimethyl disulfide	0.974	25.63
Methional	0.848	19.50
2-Acetylthiophene	0.972	30.81

CAS Registry No. Dimethyl disulfide, 624-92-0; methional, 3268-49-3; 2-acetylthiophene, 88-15-3; 4-methylthiazole, 693-95-8.

Literature Cited

1. Maarse, H.; Visscher, C.A. *Volatiles Compounds in Foods*; TNO-CIVO: Netherlands, 1989; Vol. 1.
2. Maarse, H.; Visscher, C.A. *Volatiles Compounds in Foods*; TNO-CIVO: Netherlands, 1989; Vol. 2.
3. Maarse, H.; Visscher, C.A. *Volatiles Compounds in Foods*; TNO-CIVO: Netherlands, 1989; Vol. 3.
4. Grosch, W. *Trends in Food Sci. Technol.*, **1993**, *4*, pp 68-73.
5. *Amino-Carbonyl Reactions in Food and Biological Systems*; Fujimaki, M.; Namiki, M.; Kato, H., Eds.; Elsevier:Amsterdam, 1985; pp 1-579.
6. *The Maillard Reaction in Food Processing, Human Nutrition and Physiology*; Finot, P.A.; Aeschbacher, H.U.; Hurrell, R.F.; Liardon, R., Eds.; Birkhäuser Verlag:Berlin, 1989; pp 1-510.
7. *5th International Symposium on the Maillard Reaction*; Reineccius, G.A.; Labuza, T.P., Eds.; Royal Society of Chemistry:London, 1994; in press.
8. Leahy, M.M.; Reineccius, G.A. In *Flavor Chemistry*; Teranishi, R., Buttery, R., Shahidi, F., Ed.; Advances in Chemistry Series No. 388; American Chemical Society: Washington, D.C., 1989, pp 76-91.
9. Leahy, M.M.; Reineccius, G.A. In *Thermal Generation of Aromas*; Parliment, T.H., McGorrin, R.J., Ho, C.T., Ed.; ACS Symposium Series No. 409; American Chemical Society: Washington, D.C., 1989, pp 196-208.
10. Schirlé-Keller, J.P.; Reineccius, G.A. In *Flavor Precursors*; Teranishi, R., Takeoka, G.R., Guntert, M., Ed.; American Chemical Society Symposium Series No. 490; American Chemical Society: Washington, D.C., 1989, pp 244-258.
11. Fox, L.; Wylie, P. *Hewlett-Packard Appl. Note.* **1989**, *228-275*, pp 1-3.
12. Chan, F.; Reineccius, G.A. In *5th International Symposium on the Maillard Reaction*; Reineccius, G.A., Labuza, T.P., Ed.; Royal Society of Chemistry:London, 1994, in press.
13. Labuza, T.P.; Nelson, K.A.; Nelson, G.J. *Water Analyzer Series - Reaction Kinetics Program V.2.09*. University of Minnesota, 1991.
14. Mistry, B.S.; Reineccius, G.A.; Jasper, B. In *Sulfur Compounds in Foods*; Mussinan, C.J., Keelan, M.E., Ed.; ACS Symposium Series; American Chemical Society:Washington, D.C., 1994, in press.

RECEIVED April 27, 1994

Chapter 11

Kinetics of the Release of Hydrogen Sulfide from Cysteine and Glutathione During Thermal Treatment

Yan Zheng and Chi-Tang Ho

Department of Food Science, Cook College, New Jersey Agricultural Experiment Station, Rutgers, The State University of New Jersey, New Brunswick, NJ 08903

> 0.1M of cysteine or glutathione solutions with a pH of 3.0, 5.0, 7.0 or 9.0 were incubated at 80°C, 90°C, 100°C or 110°C for a certain period of time. The amount of hydrogen sulfide released during the thermal treatment was determined using a sulfide/silver electrode. The results showed that both reactions followed first order kinetics. The rate constants of the reactions increased with the pH value. The release of hydrogen sulfide from these two compounds was catalyzed by the hydroxide ion, while the β-elimination was favored at higher pH. The activation energies for both reactions were calculated using the Arrhenius equation; 31.3, 31.8, 32.2 and 29.4 kcal/mol for cysteine and 18.8, 30.8, 22.8 and 19.9 kcal/mol for glutathione at pH 3.0, 5.0, 7.0 and 9.0,and respectively. The lower activation energies for glutathione implies that the molecular structure and micro-environmental condition of the amino compound plays a significant role in the release of hydrogen sulfide.

Sulfur-containing compounds are well-recognized as major contributors to meat flavor. Among the sulfur-containing compounds, hydrogen sulfide was one of the first identified in early meat flavor studies (*1-4*). Apart from its direct contribution to meat flavor, hydrogen sulfide participates in the formation of other sulfur-containing volatiles, thus contributing indirectly to the meat flavor (*5-7*).

Mabrouk *et al.* (*8*) used extraction, dialysis, and gel-permeation chromatography to fractionate an aqueous extract prepared from fat-free lyophilized meat. Out of the twelve resulting fractions, seven had a broiled beef aroma. Those fractions with an intense aroma contained cysteic acids (cysteine + cystine). Two fractions which contained methionine but no cysteic acid did not have any broiled beef flavor. Other reports (*9-10*) demonstrated that both cysteine and glutathione are the major precursors of hydrogen sulfide and other sulfur-containing flavor compounds in meat.

Previous studies have focused on the formation of flavor from cysteine and glutathione under various processing conditions (*12-13*). The kinetics of the formation of hydrogen sulfide and other sulfur-containing compounds have not been systematically studied. The objective of this research was to study the rate of release of hydrogen sulfide from cysteine and glutathione during thermal treatment. We hoped that the information obtained would enable us to manipulate the quality and quantity of the desired meat flavor formed during thermal processing of meat.

Experimental

Chemicals. All chemicals were purchased from Sigma Chemical Co. except as indicated. The water used in this study was prepared from distilled water with a Milli-Q deionized water system. Water was degassed by sonication for 15 minutes prior to use.

Reactions. 0.1M cysteine or glutathione solutions were prepared in 0.1M citrate-phosphate or 0.1M phosphate-sodium hydroxide buffer with pH values of 3.0, 5.0, 7.0, and 9.0. Ten mL of the solution were then transferred into a reaction vessel (Kimax brand glass test tube with a PTFE-coated liner screw). The headspace of the tubes was flushed with nitrogen to expel the oxygen. The tubes were capped and tightened. The tubes were then incubated in a glycerine bath, set at temperatures of 80°C, 90°C, 100°C or 110°C. Two tubes were withdrawn from the bath at 10 minute intervals for the first 30 minutes and at 30 minutes intervals thereafter. The test tubes withdrawn from the bath were immersed immediately in an ice-water bath to quench the reaction. The hydrogen sulfide generated was determined by the sulfide/silver electrode method described below.

Determination of hydrogen sulfide. The determination of hydrogen sulfide produced in the reactions was performed with an Orion Sulfide/Silver ion selective electrode connected to an Orion digital pH/mV meter. The electrode includes a silver/sulfide sensing element which develops an electrode potential in contact with a solution containing either silver or sulfide ions. The measured potential corresponding to the level of silver or sulfide ions in solution is described by the Nernst equation:

$$E = E_o + S \times \log(A)$$

Where:

    E: measured electrode potential.
    E_o: reference potential (a constant)

A: level of silver or sulfide ion in solution
S: electrode slope

A standardized Na_2S solution in sulfur antioxidation buffer (SAOB, consisting of 2N NaOH, 1N EDTA and 4% ascorbic acid) was used to established a standard curve. Samples were diluted with the SAOB to an electrode potential within the range of the standard curve. The H_2S concentration in the samples could then be calculated from the electrode potential and the dilution factor.

Results and Discussion

Experimental data on the release of H_2S from cysteine solution at different temperatures and pHs were fitted with zero-order, first-order and second-order methods. As shown in Figures 1 to 4, the reaction followed first-order kinetics. The R^2s of the first-order regressions as shown in Table I were in the range of 0.955 to 0.999. The results show that the rates of H_2S release from cysteine were influenced by the pH of the solution. With the increase of pH from acidic condition to basic condition, the rate constant increased (Table I). The results indicate that the release of H_2S from cysteine is easier under a basic environment. A similar observation has been reported for the release of H_2S from chicken muscles (10). It is believed that the sulfhydryl group in cysteine undergoes a β-elimination reaction at basic pH conditions (14). The initial step in β-elimination is the abstraction of the proton by the hydroxide ion from the α-carbon atom (the β position to the sulfhydryl group). Dehydroalanine is then formed with the subsequent loss of the sulfhydryl group which could combine with the proton to form hydrogen sulfide. A linear relationship between the rate constants and the corresponding pH or hydroxide ion concentration was observed except for those reactions at pH 3.0. The equations for the relationship are listed in Table II. The nonconformity of reactions at pH 3.0 may mean that a different reaction mechanism is involved in the release of hydrogen sulfide from cysteine at pH 3.

The release of hydrogen sulfide from glutathione during the thermal treatment was similar to that of cysteine. The rate constants derived from first-order kinetics, however, were higher than those for cysteine under the same reaction conditions. The rate constants for glutathione and the corresponding R^2 values are listed in Table III. This result confirmed earlier reports that glutathione evolves H_2S more rapidly than cysteine (11). The molecular environment could be the main reason for the difference. It has been reported that β-elimination would be preferred if the α-amino group of cysteine is acetylated or the carboxyl group is esterified (15). In the case of glutathione, the negative charge of carboxyl group is further away from the α-carbon than it is in cysteine. Therefore, the hydrogen atom at the α-carbon is more readily abstracted. This mechanism is further supported by the activation energies obtained in this study.

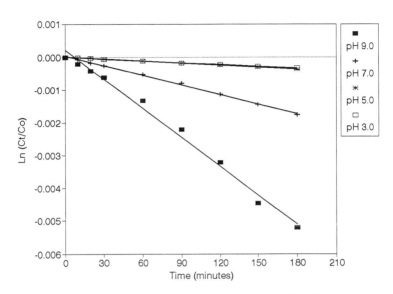

Figure 1. First-order plot of the release of hydrogen sulfide from cysteine at 80°C.

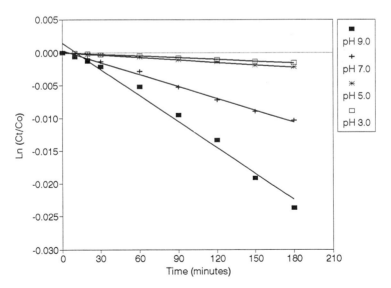

Figure 2. First-order plot of the release of hydrogen sulfide from cysteine at 90°C.

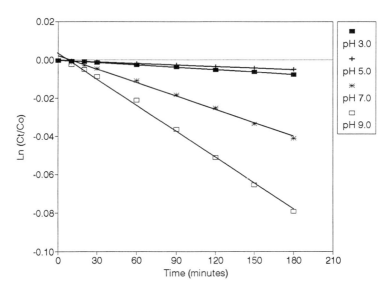

Figure 3. First-order plot of the release of hydrogen sulfide from cysteine at 100°C.

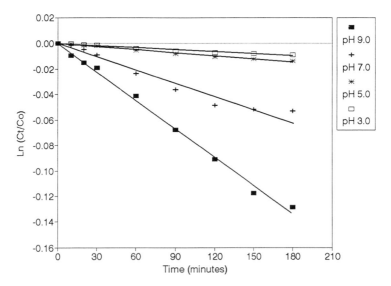

Figure 4. First-order plot of the release of hydrogen sulfide from cysteine at 110°C.

Table I. First-order Rate Constants (K), the Corresponding Regression Coefficients (R^2) and the Half-Life ($T_{1/2}$) of the Release of Hydrogen Sulfide from Cysteine during Thermal Processing at Different pH Value

Temp	pH	K (min^{-1})	$T_{1/2}$ (hrs)	R^2
80°C	pH 3.0	1.87 x 10^{-6}	6171.2	0.999
	pH 5.0	2.02 x 10^{-6}	5713.4	0.998
	pH 7.0	9.71 x 10^{-6}	1188.5	0.998
	pH 9.0	2.95 x 10^{-5}	391.3	0.991
90°C	pH 3.0	8.30 x 10^{-6}	1395.2	0.995
	pH 5.0	1.24 x 10^{-5}	930.1	0.997
	pH 7.0	5.99 x 10^{-5}	192.9	0.996
	pH 9.0	1.33 x 10^{-4}	86.7	0.985
100°C	pH 3.0	4.15 x 10^{-5}	278.2	0.998
	pH 5.0	2.65 x 10^{-5}	435.6	0.998
	pH 7.0	2.30 x 10^{-4}	50.3	0.995
	pH 9.0	4.53 x 10^{-4}	25.5	0.996
110°C	pH 3.0	5.23 x 10^{-5}	220.8	0.973
	pH 5.0	7.97 x 10^{-5}	144.8	0.985
	pH 7.0	3.35 x 10^{-4}	34.5	0.955
	pH 9.0	7.45 x 10^{-4}	15.5	0.995

Table II. Linear Relationship between the Rate Constants of the Release of Hydrogen Sulfide from Cysteine and the pH

Temp.	Relationship	R^2
80°C	K = 6.87 x 10^{-6} x log[OH$^-$] + 6.19 x 10^{-5}	0.939
90°C	K = 3.01 x 10^{-5} x log[OH$^-$] + 2.78 x 10^{-4}	0.985
100°C	K = 1.07 x 10^{-4} x log[OH$^-$] + 9.88 x 10^{-4}	0.999
110°C	K = 1.66 x 10^{-4} x log[OH$^-$] + 1.55 x 10^{-4}	0.982

Table III. First-order Rate Constants (K), the Corresponding Regression Coefficients (R^2) and the Half-Life ($T_{1/2}$) of the Release of Hydrogen Sulfide from Glutathione during Thermal Treatment at Different pH Values

Temp	pH	K (min^{-1})	$T_{1/2}$ (hrs)	R^2
80°C	pH 3.0	4.04 x 10^{-6}	2859.5	0.996
	pH 5.0	4.11 x 10^{-6}	2810.8	0.996
	pH 7.0	3.25 x 10^{-5}	355.5	0.996
	pH 9.0	1.34 x 10^{-4}	86.2	0.989
90°C	pH 3.0	1.17 x 10^{-5}	987.4	0.987
	pH 5.0	1.57 x 10^{-5}	735.8	0.992
	pH 7.0	7.69 x 10^{-5}	150.2	0.997
	pH 9.0	2.44 x 10^{-4}	47.3	0.989
100°C	pH 3.0	2.28 x 10^{-5}	506.7	0.965
	pH 5.0	4.94 x 10^{-5}	233.9	0.991
	pH 7.0	2.26 x 10^{-4}	51.1	0.993
	pH 9.0	7.74 x 10^{-4}	14.9	0.995
110°C	pH 3.0	4.57 x 10^{-5}	252.8	0.979
	pH 5.0	7.46 x 10^{-5}	154.9	0.956
	pH 7.0	3.89 x 10^{-4}	29.7	0.992
	pH 9.0	1.09 x 10^{-3}	10.6	0.990

The linear relationship between rate constants and pH was also observed for the reactions of glutathione in the range from pH 5.0 to pH 9.0 (Table IV). Again, the relationship between the rate constants and the pH at pH 3.0 did not follow those observed in the higher pH conditions.

The activation energies for the hydrogen sulfide released from cysteine and glutathione at pH 3.0, 5.0, 7.0 and 9.0 were calculated using the Arrhenius equation:

$$K = K_o \times \text{Exp}(-E_a/RT)$$

Where:

K_o = pre-exponential (absolute) rate constant
E_a = activation energy in kcal/mol
R = gas constant, 1.987 cal/mol/°K
T = temperature in °K

Table V shows the activation energies for the elimination of hydrogen sulfide from cysteine and glutathione. The fact that the activation energy for the release of H_2S from glutathione is lower than for cysteine again suggested the structural effect on the release of hydrogen sulfide. The result also indicate that pH affected the activation energy for the elimination. As the pH increased from

Table IV. Linear Relationship between the Rate Constants of the Release of Hydrogen Sulfide from Glutathione and the pH

Temp.	Relationship	R^2
80°C	$K = 3.25 \times 10^{-5} \times \log[OH^-] + 2.85 \times 10^{-4}$	0.905
90°C	$K = 5.71 \times 10^{-5} \times \log[OH^-] + 5.12 \times 10^{-4}$	0.933
100°C	$K = 1.81 \times 10^{-4} \times \log[OH^-] + 1.62 \times 10^{-3}$	0.920
110°C	$K = 2.54 \times 10^{-4} \times \log[OH^-] + 4.82 \times 10^{-3}$	0.954

Table V. Activation Energy of Release of Hydrogen Sulfide from Cysteine and Glutathione at Different pH Values

	Cysteine		Glutathione	
	Ea (kcal/mol)	Linearity R^2	Ea (kcal/mol)	Linearity R^2
pH 3.0	31.3	0.940	18.8	0.997
pH 5.0	31.8	0.975	30.8	0.996
pH 7.0	32.3	0.942	22.8	0.990
pH 9.0	29.4	0.967	19.9	0.970

3.0 to 9.0, the activation energy reached a maximum and then declined. For both cysteine and glutathione, the maximum activation energy showed at about 2 pH units above the isoelectric point (pI) of the molecule (i.e. pI of cysteine = 5.07; pI of glutathione = 2.83). Such a phenomenon could be related to the ionization of the molecule. At pH 7.0, cysteine exists mainly as $HSCH_2CH(NH_3^+)COO^-$. These molecules could associate with each other so that the attack at the α-hydrogen by the hydroxyl ion is slowed down. Subsequently, the activation energy is higher than those at the other pHs. On the other hand, the major structure of glutathione at pH 5.0 is

$^+H_3N(COO^-)CHCH_2CH_2CONHCH(CH_2SH)CONHCH_2COO^-$.

In this case, the interaction between the carboxyl group in the glycine residue with the amino group in the glutamate residue will also reduce the attack of hydroxide ion on the α-hydrogen atom. Therefore, the activation energy is higher.

Acknowledgements

New Jersey Agricultural Experiment Station Publication No. D-10205-11-93 supported by State Funds.

Literature Cited

1. Osborne, W.A. *Biochem. J.* **1928**, *22*, 1312-1316.
2. Crocker, E.C. *Food Res.* **1948**, *13*, 179-183.
3. Bouthilet, R.J. *Food Res.* **1951**, *16*, 137-141.
4. Pippen, E.L.; Erying, E.J. *Food Technol.* **1957**, *11*, 53-56.
5. Pippen E.L.; Mecchi, E.P. *J. Food Sci.* **1969**, *34*, 443-446.
6. Schutte, L.; Koenders, E.B. *J. Agric. Food Chem.* **1972**, *20*, 181-184.
7. Boelens, H.; van der Linde, L.M.; de Valois, P.J.; van Dort, J.M.; Takken, H.J. *J. Agric. Food Chem.*, **1974**, *22*, 1071-1076.
8. Mabrouk, A.F.; Jarboe, J.K.; O'Conner, E.M. *J. Agric. Food Chem.* **1969**, *17*, 5-9.
9. Obataka, Y.; Tanaka, H. *Agric. Biol. Chem.* **1965**, *29*, 191-195.
10. Mecchi, E.P.; Pippen, E.L.; Lineweaver, H. *J. Food Sci.* **1964**, *29*, 393-399.
11. Ohloff, G.; Flament, I.; Pickenhagen, W. *Food Rev. Intl.* **1985**, *1*, 99-148.
12. Shu, C.K.; Hagedorn, M.L.; Mookherjee, B.D.; Ho, C.-T. *J. Agric. Food Chem.* **1985**, *33*, 442-446.
13. Zhang, Y.; Chien, M.; Ho, C.-T. *J. Agric. Food Chem.* **1988**, *36*, 992-996.
14. Tarbell, D.S.; Harnish, D.P. *Chem. Rev.* **1951**, *49*, 1-90.
15. Schneider, J.F.; Westley, J. *J. Biol. Chem.* **1969**, *244*, 5735-5744.

RECEIVED March 23, 1994

Chapter 12

Volatile Sulfur Compounds in Yeast Extracts

Jennifer M. Ames

Department of Food Science and Technology, University of Reading, Whiteknights, Reading RG6 2AP, United Kingdom

Yeast extracts represent an important source of volatile sulfur compounds, many of which possess low odor threshold values. They are used as sources of flavor for a range of savory foods, especially when a meaty note is required. In spite of the usefulness of yeast extracts, there are very few reports of their volatile flavor components. The production of yeast extracts is briefly reviewed, and the volatile sulfur compounds which have been identified are discussed. A recent study is presented in which the aroma components of some yeast extracts were analyzed. A total of 268 compounds were identified, including 67 sulfur compounds. The 34 sulfur compounds reported for the first time comprised 3 aliphatic sulfur compounds, one sulfur-substituted benzene derivative, 10 thiophenes, 18 thiazoles and 2 alicyclic sulfur compounds. Their importance as components of flavors and routes to their formation are considered.

Yeast extracts (YEs) are sources of natural flavor compounds, and their exact composition varies according to the type of yeast and the conditions used for its propagation and the production of the final YE. This chapter is devoted to a review of the volatile sulfur compounds of YEs (many of which play a crucial role in determining the overall flavor) and to a report of some sulfur compounds which have recently been identified as components of selected YEs.

The first step in the preparation of a yeast extract (YE) involves autolysis of a yeast slurry or cream at 50°C. A yeast autolysate (YA) results and removal of the unhydrolyzed cell walls from the soluble autolysate yields a clear extract (1). Caramel, vegetable concentrates, spice infusions, monosodium glutamate and 5'-ribonucleotides, all of which are flavor modifiers, may be added to the YE ($1,2$). Products are marketed as liquids, pastes or powders and find uses as flavoring agents for a range of foods, especially when a meaty flavor is demanded (3).

Many of the amino acids, peptides, sugars, lipids, nucleotides and B vitamins present in YEs are important meat flavour precursors ($2,4,5$). Despite their potential as meat flavoring agents, until recently there were few reports of the aroma components of YAs and YEs.

The acidic, basic and neutral flavor components of a YA were examined in the first published study, and 48 compounds were identified (6,7). Three sulfur compounds, i.e., 2-thiophenecarboxylic acid, its 5-methyl derivative and 4-methyl-5(2'-hydroxyethyl)thiazole were mentioned. Later, Golovnya et al. (8) identified 37 sulfur compounds from a simulated meat flavor produced by heating a bakery YA with sugar. They included aliphatic sulfides and thiols, alicyclic sulfur compounds, thiophenes and sulfur-substituted furans.

The analysis of a paste comprising YE, salt, spices and vegetable extract resulted in the identification of 125 compounds. Twenty-four sulfur compounds were reported including some sulfur-substituted furans (9,10). In a study of a yeast/sucrose homogenate, Schieberle (11) evaluated the most important odor compounds in the heated and unheated material using aroma dilution analysis. Methional was one of four compounds possessing the highest flavor dilution (i.e. FD) factor for the heated homogenate.

Extrusion cookers may be used as continuous flavor reactors. Izzo and Ho (12) investigated their use for the production of flavoring materials from YEs, glucose and ammonium bicarbonate, but only pyrazine production was monitored.

Two excellent papers on YE aroma have arisen recently from Werkhoff's group at Haarmann and Reimer in Germany (13,14). Fifty sulfur compounds were reported in the first study (13) while 115 such components were described in the second paper (14). Aliphatic compounds, sulfur-substituted furans, thiophenes, thiazoles, 3-thiazolines, cyclic polysulfides, perhydro-1,3,5-dithiazines, a perhydro-1,3,5-oxathiazine, a perhydro-1,3,5-thiadiazine and aliphatic and heterocyclic dithiohemiacetals were all mentioned, and the structures of most of the newly identified compounds were confirmed by synthesis. These sulfur compounds elicit a wide variety of aromas but are often savory, e.g., meaty, vegetable-like, and YEs are a valuable source of potent components which provide meaty notes.

All the reported sulfur-containing volatile components of YAs and YEs are summarized in Table I. Numerically, different classes of sulfur compounds predominate in the various investigations. None of the compounds reported in YEs has been mentioned in more than two studies, and most have been reported in only one study. This may be due to the investigation of YEs processed under different conditions and/or to different analytical conditions used for the volatiles analysis.

Currently, a selection of YEs with a range of aroma attributes is available from a number of suppliers. The flavour characteristics of YEs vary according to factors such as the nature of the yeast, the presence of additional components which may contribute to flavour development and the temperature of concentration or drying.

The study described here reports the aroma components of three YEs obtained from the same manufacturer but processed in different ways and with different aroma characteristics.

Experimental

The YEs were supplied by the manufacturer and were described as follows:

Table I. Volatile sulfur components of yeast autolysates and yeast extracts

Chemical class	6,7	8	9,10	11	13,14	Total
Aliphatic sulfur compounds						
Monothiaalkanes	-	3	-		-	3
Dithiaalkanes	-	8	2	-	-	8
Trithiaalkanes	-	-	1	-	2	3
Tetrathiaalkanes	-	-	-	-	2	2
Thiols	-	5	1		5	10
1,1-Bis(alkylthio)alkanes	-	3	-	-	1	4
Sulfur-containing carbonyl compounds	-	-	1	1	3	5
Benzene derivatives	-	1	-	-	-	1
Furans						
Furanthiols	-	1	2	-	2	4
Furanalkyl sulfides and disulfides	-	2	3	-	1	5
Bis-alkylfuryl sulfides and disulfides	-	1	1	-	1	3
Thienyl-substituted furans	-	-	-	-	2	2
Other derivatives	-	-	1	-	4	4
Thiophenes						
Alkylthiophenes	-	2	1	-	4	6
Thiophenes with a carbonyl substituent	-	3	7	-	2	9
Thiophene carboxylic acids	2	-	-	-	-	2
Thiophenes with a sulfur-containing substituent	-	2	-	-	2	4
Dihydrothiophenes	-	-	-	-	3	3
Tetrahydrothiophenes	-	-	2	-	1	3
Other derivatives	-	-	1	-	-	1
Thiazoles						
Alkylthiazoles	1	-	1	-	11	13
Thiazolines	-	-	-	-	6	6
Other derivatives	-	-	-	-	2	2
Dithiazines	-	-	-	-	50	50
Thiadiazines	-	-	-	-	1	1
Oxathiazines	-	-	-	-	1	1
Cyclic polysulfides						
Dithiolanes	-	3	-	-		3
Trithiolanes	-	-	-	-	14	14
Dithianes	-	3	-	-	1	4
Trithianes	-	-	-	-	2	2
Tetrathianes	-	-	-	-	1	1

Sample A: Pure, full-bodied, delicate, brothy flavor without an undesirable meaty flavor; clean aftertaste.

Sample B: Pleasant, full-bodied, delicate, savory, brothy-flavored; free of bitter or harsh notes and no astringent aftertaste.

Sample C: Robust, savory, roasted flavor; produced by blending different yeast extracts and further processing with sugars under mild conditions.

All the YEs were prepared from *Saccharomyces cerevisiae*. They were supplied as powders and possessed the following analysis: total solids, 96%; nitrogen, 6-7%; sodium chloride, 40%.

Isolates of the volatile components of the YEs were obtained from 10g portions of powder suspended in 10 mL distilled water held in a flask at 60°C. The volatile components were swept onto a Tenax GC trap using a stream of purified oxygen-free nitrogen. The volatile components were separated by GC, identified by GC-MS and their aromas were described by means of GC-odor port assessment (GC-OPA). Further experimental details have been given previously (*15*).

Results and Discussion

The aromas of the YEs were assessed sensorially after the isolation procedure. Isolates prepared from Sample A possessed the weakest aroma and Sample A was described as "yeast extract, hydrolyzed vegetable protein, slightly brothy, slightly burnt and slightly roasted". Aroma descriptors applied to Sample B were "cereal, savory, slightly hydrolyzed vegetable protein, slightly yeast extract and slightly cooked beef". Isolates from Sample B were 2.5 times stronger than those from Sample A. Isolates from Sample C were stronger than those from Samples A and B by factors of 7.5 and 3, respectively. Sample C was described as "strongly roasted, roasted cereal, burnt, slightly yeast extract, not meaty".

Two hundred and sixty-eight compounds were identified in this study as components of YEs, and they are listed by class in Table II. Sixty-seven of the compounds identified contained sulfur (*15*) and are listed in Table III. The 34 compounds identified for the first time as components of YEs are denoted by (*) in Table III. Twenty-eight of the newly identified components of YE were detected in Sample C while seven were identified in Sample B. None could be identified from Sample A. Eighteen of the newly-identified representatives listed are thiazoles. Some of the thiazoles reported in this study, i.e., thiazole, 4,5-dimethylthiazole and 2,4-dimethyl-5-ethylthiazole possess meaty qualities at certain dilutions (*16*). 2,4-Dimethyl-thiazole, which has not previously been reported as a component of YE aroma, possesses a "meaty, cocoa" or "skunky-oily" odor (*16*). Other thiazoles listed in Table III may be expected to contribute green, nutty and vegetable-like notes (*16*).

The major routes which account for the formation of many of the compounds identified in YEs are (1) interactions involving carbonyl compounds, sulfur and ammonia (often via the degradation of sulfur-containing amino acids, with or without the intervention of reducing sugars) and, (2) the thermal degradation of thiamine.

Table II. Compounds identified in Yeast Extracts A, B and C by chemical class

Chemical class	Number of representatives
Aliphatic hydrocarbons	28
Alicyclic hydrocarbons	8
Terpenes	3
Aliphatic alcohols	2
Aliphatic aldehydes	12
Aliphatic ketones	16
Alicyclic ketones	10
Aliphatic acids	5
Aliphatic esters	3
Aliphatic sulfur compounds	11
Benzene derivatives	36
Furans (including sulfur-substituted furans)	17
Thiophenes	20
Pyrroles	7
Pyridines	1
Pyrazines	45
Oxazoles	9
Thiazoles	32
Cyclic polysulfides	2

Table III. Sulfur compounds identified for the first time as components of yeast extract aroma.

	Relative percentage abundance of total aroma isolate		
	Sample		
Component	A	B	C
Aliphatic sulfur compounds			
Methyl propyl sulfide	0.9	2.8	tr
*Di(3-methylpropyl)sulfide	-	5.9	0.8
A sulfide, M132	-	tr	-
A sulfide, M146	-	4.2	0.4
*Carbon disulfide	-	tr	-
Dimethyl disulfide	0.9	0.7	tr
Dimethyl trisulfide	3.0	1.8	0.3
Dimethyl tetrasulfide	-	tr	-
3-(Methylthio) propanal (methional)	tr	3.6	0.6
2-(2-Thiapropyl)-2-butenal	-	0.8	-
*Ethylthiopropan-2-one	-	4.2	0.6
Benzene derivatives			
*A methylbenzenethiol	-	-	3.3
Furans			
2-Methyl-3-(methyldithio)furan	-	tr	tr
Thiophenes			
2(or 3)-Methylthiophene	tr	-	0.5
3(or 2)-Methylthiophene	-	-	0.2
2-Ethylthiophene	-	-	0.3
2-Propylthiophene	-	-	tr
*?2-Pentylthiophene	-	-	tr
*2,3-Dimethylthiophene	-	-	tr
*2,4-Dimethylthiophene	-	-	tr
*2,5-Dimethylthiophene	-	-	tr
Thiophene-2-carboxaldehyde	-	-	0.9
3-Methylthiophene-2-carboxaldehyde	-	tr	-
5-Methylthiophene-2-carboxaldehyde	-	tr	-
*3-Acetyl-2,5-dimethylthiophene	-	tr	tr
*?2(or 3)-Thiophenethiol	-	0.4	tr
*?2,5-Dimethyltetrahydrothiophene	-	-	tr
2-Methyltetrahydrothiophen-3-one	-	-	tr
*2-Benzylthiophene	-	-	tr
*?4(or 5)-Methylbenzylthiophene	-	-	tr
*?2(or 3)-Phenylthiophene	-	tr	4.1
A thiophene, M 144	-	-	tr
A thiophene, M 146	-	1.1	-
Thiazoles			
Thiazole	-	-	tr
2-Methylthiazole	-	-	0.9
4-Methylthiazole	-	-	-

Continued on next page

Table III (continued)

Component	Relative percentage abundance of total aroma isolate Sample		
	A	B	C
*5-Methylthiazole	-	-	0.2
*2-Ethylthiazole	-	-	0.6
*4-Ethylthiazole	-	-	0.3
5-Ethylthiazole	-	-	tr
*2-Propylthiazole	-	-	tr
*2-Isobutylthiazole	-	-	tr
*2,4-Dimethylthiazole	-	-	tr
*2,5-Dimethylthiazole	-	-	tr
4,5-Dimethylthiazole	-	-	tr
*?4,5-Dimethylisothiazole	-	-	tr
*2-Methyl-5-ethylthiazole	-	-	tr
4-Methyl-5-ethylthiazole	-	tr	tr
*A C_3 alkyl substituted thiazole or isothiazole	-	-	0.7
*A methylpropylthiazole	-	-	tr
*?2-Methyl-4-isopropylthiazole	-	-	0.2
*2-Isopropyl-4-methylthiazole	-	-	tr
*A methylisopropylthiazole	-	-	0.8
2,4-Dimethyl-5-ethylthiazole	-	-	1.1
2,5-Dimethyl-4-ethylthiazole	-	-	tr
?2,4-Diethylthiazole	-	tr	-
*2-Propyl-4,5-dimethylthiazole	-	-	tr
*A C_5 alkyl substituted thiazole	-	-	tr
*?2-Isopropyl-4-ethyl-5-methylthiazole	-	-	0.4
*?2-Propanoyl-4-methylthiazole	-	-	0.1
A thiazole, M 155	-	0.4	-
A thiazole, BP 169	-	tr	-
A thiazole, M 183	-	tr	-
A thiazole, M 169	-	-	0.2
A thiazoline, M 129	-	-	0.2
Alicyclic sulfur compounds			
*cis-3,5-Dimethyl-1,2-dithiolan-4-one	-	tr	-
*trans-3,5-Dimethyl-1,2-dithiolan-4-one	-	tr	-
TOTAL	4.8	25.9	17.7

Cysteine, cystine and methionine all yield many volatile compounds on pyrolysis, including acetaldehyde, ammonia and hydrogen sulfide, and methanethiol is also formed from methionine (17). The intervention of reducing sugars greatly increases the range of reaction products, largely due to the production of additional carbonyl compounds, e.g., propanal, 2-pentanone, butanedione, pyruvaldehyde and 2-furfural (18,19) which are able to act as intermediates in the formation of many sulfur volatiles. Selected volatile sulfur compounds formed in (A), cysteine/cystine-ribose; (B), cystine-glucose; (C), cysteine-glucose; (D), cysteine-ribose; (E), cysteine-pyruvaldehyde; and (F) cystine-pyruvaldehyde model systems, and identified in YEs are listed in Table IV. In particular, large numbers of thiophenes and thiazoles are formed as a result of interactions involving cysteine and cystine. Many thiazoles are formed as a result of the Maillard reaction in the presence of hydrogen sulfide (27).

Acetaldehyde can undergo various interactions with butanedione, hydrogen sulfide, methanethiol and ammonia to give various aliphatic sulfur compounds, thiazoles, thiadiazines and dithiazines as shown in Fig. 1. Compounds identified in YEs are indicated by (*) (29). The involvement of alternative aldehydes, such as formaldehyde in place of acetaldehyde or mixtures of aldehydes, will lead to the corresponding homologues, some of which have also been reported in YEs. Another aldehyde, 3-methylbutanal, reacts with hydrogen sulfide and methanethiol to give the hemidithioacetal, 1-methylthio-3-methyl-1-butanethiol which has been reported in YEs (14). Compounds of this class possess characterisitic meaty aromas with an onion note (14).

Some of the compounds identified in YEs which are formed either by the thermal degradation of thiamine or on the interaction of thiamine degradation products with other components are shown in Fig. 2. They include aliphatic sulfur compounds, furans, thiophenes and thiazoles. 2-Methyl-3-furanthiol and 2-methyl-3-thiophenethiol have been identified in YEs (9,13,14) and are well known thermal degradation products of thiamine (29). As well as possessing meaty aromas and low odor threshold values (34), these compounds are key precursors of several other sulfur-substituted furans and thiophenes, including the derivatives in Fig. 2. Most possess meaty aromas at low concentrations and several have been identified in YEs (see Tables I and III).

A further range of compounds, including dimethyl disulfide, 1,1-ethanedithiol, 1-methylthio-1-ethanethiol, 3,5-dimethyl-1,2,4-trithiolane, 2,4,6-trimethyl-5,6-dihydrodithiazine,2,6-dimethyl-4-ethyl-5,6-dihydrodithiazine,4,6-dimethyl-2-ethyl-5,6-dihydrodithiazine, 2,4,5-trimethylthiazole and 2,4,5-trimethyl-3-thiazoline have recently been identified from thiamine heated at pH 9.5 (29,33) and have also been detected in YEs (2-5). However, it seems unlikely that thiamine is the source of these aroma compounds in YEs since such a high pH is not encountered during manufacture. When thiamine degrades at a pH of 9.5, the amounts of degradation products such as formaldehyde, acetaldehyde, ammonia and hydrogen sulfide are quite high (29,33), and it is these compounds which are the immediate precursors

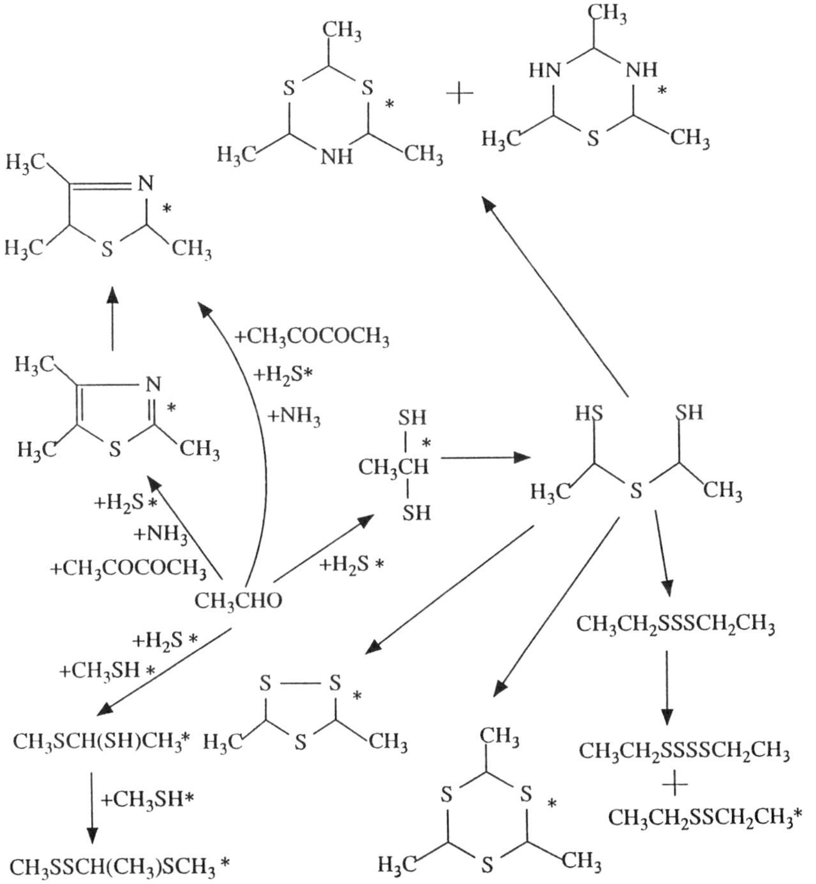

Figure 1 Formation of selected compounds from interactions between acetaldehyde, hydrogen sulfide, methanethiol and ammonia (3,28) (Compounds identified in YE aroma are indicated by (*).)

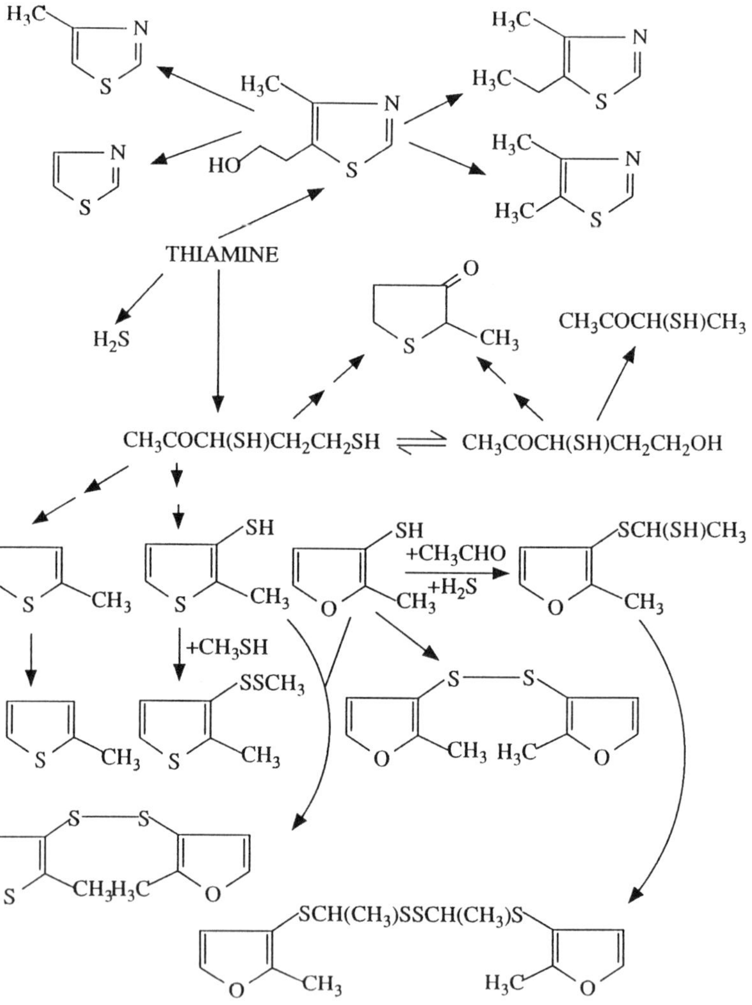

Figure 2 Formation of selected thiamine degradation products identified in YE aroma (*27, 29-33*)

Table IV. Selected volatile sulfur compounds reported in cysteine (or cystine)-reducing sugar (or carbonyl compound) model systems which have been identified in yeast extracts

Compound	Model system (ref)	Compound	Model system (ref)
Ethanethiol	A(20), B(21)	*2-Methyl-3-thiophenethiol	D(23)
Propanethiol	A(20), B(21), C(21)	3-(Methylthio)-thiophene	C(21)
*2-Mercapto-3-butanone	D(22)	2-Methylthiazole	A(20), C(21), D(23,24), F(26)
*2-Methyl-3-furanthiol	D(23,24)	Trimethylthiazole	A(20), B(21), D(24)
*Bis(2-methyl-3-furyl)-disulfide	D(25)	5-Ethyl-2,4-dimethylthiazole	A(20), D(23,24)
Thiophene	E(26)	2-Acetylthiazole	A(20), D(23,24)
2-Propylthiophene	D(24)	2-(2-Furyl)thiazole	D(23)
3-Methylthiophene-2-carboxaldehyde	A(20)	3-Methyl-1,2,4-trithiane	A(20)
2-Propionylthiophene	C(21), D(23,24)	*3-Methyl-1,2-dithian-4-one	D(23)

of compounds such as the dihydrodithiazines. These precursors may derive from various alternative sources, some of which have been mentioned above.

Several of the entries of Table IV may form on thiamine degradation as well as resulting from amino-carbonyl interactions and are indicated by (*). The relative importance of these two reaction pathways for the production of potent sulfur-containing aroma components has been discussed recently (29) and represents a fertile area for further research.

Conclusion

This chapter illustrates the diversity of aroma compounds present in YEs and the variability in volatile composition. While YEs possess aromas which may generally be described as 'savory', it seems that variations in the cultural and processing conditions lead to differences in the profile of volatiles which may be expected to be reflected in modified aromas. For the future, it would be interesting to investigate the range of yeast extracts which may be produced, in terms of aroma properties, by applying extreme cultural and processing conditions, and also by studying the products formed from other yeasts, e.g., *Candida*.

Acknowledgments

The author thanks Dr. J.S. Elmore, AFRC Institute of Food Research, Reading, for the GC-MS analyses.

Literature Cited

1. Cogman, D.G.K.; Sarant, R. *Food Trade Rev.* 1977, 15-16.
2. Acraman, A.R. *Process Biochem.* 1966, 1, 313-317.
3. MacLeod, G.; Seyyedain-Ardebili, M. *CRC Crit. Revs. Food Sci. and Nut.* 1981, 12, 309-437.
4. Hough, J.S.; Maddox, I.S. *Process Biochem.* 1970, 5, 50-52.
5. Goossens, A.E. *The Flav. Ind.* 1974, Nov./Dec., 273-274, 276.
6. Davidek, J.; Hajšlová, J.; Kubelka, V.; Velíšek, J. *Die Nahrung* 1979, 23, 673-680.
7. Hajšlová, J.; Velíšek, J.; Davidek, J.; Kubelka, V. *Die Nahrung* 1980, 24, 875-881.
8. Golovnya, R.V.; Misharina, T.A.; Garbuzov, V.G.; Medvedyev, F.A. *Die Nahrung* 1983, 27, 237-249.
9. Ames, J.M.; MacLeod, G. *J. Food Sci.* 1985, 50, 125-131, 135.
10. MacLeod, G.; Ames, J.M. In *Amino-Carbonyl Reactions in Food and Biological Systems*; Fujimaki, M.; Kato, H.; Namiki, M., Eds.; Elsevier, Amsterdam, 1986; pp 263-272.
11. Schieberle, P. In *The Maillard Reaction in Food Processing, Human Nutrition and Physiology*; Finot, P.A.; Aeschbacher, H.U.; Hurrell, R.F.; Liardon, R., Eds.; Birkhäuser, Basel, 1990; pp 187-196.
12. Izzo, H.V.; Ho, C.T. *J. Food Sci.* 1992, 57, 657-674.
13. Werkhoff, P.; Brüning, J.; Emberger, R.; Güntert, M.; Köpsel, M.; Kuhn, W. *Proceedings of the 11th International Congress of Essential Oils, Fragrances and Flavours* 1989, 4, 215-243.
14. Werkhoff, P.; Bretschneider, R.; Emberger, R.; Güntert, M.; Hopp, R.; Köpsel, M. *Chem. Mikrobiol. Technol. Lebensm.* 1991, 13, 30-57.
15. Ames, J.M.; Elmore, J.S. *Flavour and Fragrance Journal* 1992, 7, 89-103.
16. Fors, S. In *The Maillard Reaction in Foods and Nutrition*; Waller, G.R.; Feather, M.S., Eds.; ACS Symposium Series No. 215; American Chemical Society: Washington, DC, 1983, pp 185-286.
17. Fujimaki, M.; Kato, S.; Kurata, T. *Agric. Biol. Chem.* 1969, 33, 1144-1151.
18. Hodge, J.E. *J. Agric. Food Chem.* 1953, 1, 928-943.
19. Heynes, K.; Stute, R.; Paulsen, H. *Carbohydr. Res.* 1966, 2, 132-149.
20. Mulders, E.J. *Z. Lebensm. Unters. Forsch.* 1973, 152, 193-201.
21. Arroyo, P.T.; Lillard, D.A. *J. Food Sci.* 1970, 35, 769-770.
22. Farmer, L.J.; Mottram, D.S. *J. Sci. Food Agric.* 1990, 53, 505-525.
23. Farmer, L.J.; Mottram, D.S.; Whitfield, F.B. *J. Sci. Food Agric.* 1989, 49, 347-368.
24. Whitfield, F.B.; Mottram, D.S.; Brock, S.; Puckey, D.J.; Salter, L.J. *J. Sci. Food Agric.* 1988, 42, 261-272.

25. Farmer, L.J.; Mottram, D.S. In *Flavour Science and Technology*; Bessière, Y.; Thomas, A.F., Eds; Wiley, Chichester, UK, 1990; pp 113-116.
26. Kato, S.; Kurata, T.; Fujimaki, M. *Agric. Biol. Chem.* 1973, 37, 539-544.
27. Güntert, M.; Brüning, J.; Emberger, R.; Köpsel, M,; Kuhn, W.; Thielman, T.; Werkhoff, P. *J. Agric. Food Chem.* 1990, 38, 2027-2041.
28. Takken, H.J.; van der Linde, L.M.; de Valois, P.J.; van Dort, H.M.; Boelens, M. In *Phenolic, Sulfur and Nitrogen Compounds in Food Flavors*; Charalambous, G.; Katz, I., Eds.; ACS Symposium Series No. 26; American Chemical Society: Washington, DC, 1976; pp 114-121.
29. Güntert, M.; Bertram, H.-J.; Emberger, R.; Hopp, R.; Sommer, H.; Werkhoff, P. In *Progress in Flavour Precursor Studies*; Schreier, P.; Winterhalter, P., Eds.; Allured: Carol Stream, IL, 1993; pp 361-378.
30. Werkhoff, P.; Brüning, J.; Emberger, R.; Güntert, M.; Köpsel, M,; Kuhn, W.; Surburg, H. *J. Agric. Food Chem.* 1990, 38, 777-791.
31. Ames, J.M.; Hincelin, O.; Apriyantono, A. *J. Sci. Food Agric.* 1992, 58, 287-289.
32. Hincelin, O.; Ames, J.M.; Apriyantono, A.; Elmore, J.S. *Food Chem.* 1992, 44, 381-389.
33. Güntert, M.; Brüning, J.; Emberger, R.; Hopp, R.; Köpsel, M,; Surburg, H.; Werkhoff, P. In *Thermal Precursors. Thermal and Enzymatic Conversions*; Teranishi, R.; Takeoka, G.R.; Güntert, M., Eds.; ACS Symposium Series No. 490; American Chemical Society: Washington, DC, 1992; pp 140-163.
34. MacLeod, G. In *Developments in Food Flavours*; Birch, G.G.; Lindley, M.G., Eds.; Elsevier, London, 1986; pp 191-223.

RECEIVED March 23, 1994

Chapter 13

Generation of Furfuryl Mercaptan in Cysteine–Pentose Model Systems in Relation to Roasted Coffee

Thomas H. Parliment and Howard D. Stahl

Kraft General Foods, 250 North Street, White Plains, NY 10625

It has been known for more than 50 years that furfuryl mercaptan is one of the very few critical coffee aroma compounds. Aqueous solutions of ribose and cysteine were reacted in a high temperature/short time reactor and the aromatics generated were separated and identified by GC/MS. This study investigates the effect of pH, time and temperature on furfuryl mercaptan formation. At higher temperatures furfuryl mercaptan is the major compound generated. The kinetics describing its formation are covered.

More than 60 years ago Reichstein and Staudinger (1) reported the presence of furfuryl mercaptan (Fur-SH) or 2-furfurylthiol in coffee. Tressl reported that the level of Fur-SH in Robusta coffee is about double that of Arabica coffee, ranging from about 1-2 ppm in Arabicas to about 2-3.8 ppm in Robustas (2). While not present in many other foods, it is also a flavor constituent of chicken, beef and pork.

The threshold of furfuryl mercaptan in water is reported to be 5 ppt (3), while its odor threshold in air has been reported at 0.01-0.02 ng/L (4). Several workers have reported the organoleptic importance of this compound. Arctander (5) describes it as being very powerful, penetrating and coffee-like at suitable dilution. Tressl and Silwar (6) consider it to be either a positive impact component or off-flavor component depending upon concentration. Holscher and Steinhardt (7) used GC sniffing and aroma dilution analysis techniques (ADA) to study the organoleptic importance of separated GC peaks. They reported that furfuryl mercaptan is one of the more important odor components in roasted coffee. Recently Blank et al, compared the potent odorants of a roasted and ground Arabica coffee to the brew prepared from this coffee using ADA. The sensory impact of furfuryl mercaptan was found to decrease somewhat in the brew (relative to the ground coffee) which they attribute to either lower solubility in the brew or degradation by the hot water in brewing (4).

Model Systems. Model system studies have investigated the reaction of sulfur sources and pentoses and have established that furfuryl mercaptan may be a significant product. The study of the cysteine/xylose model system under aqueous conditions at 180°C and pH 5 led Tressl to propose a mechanism for the production of furfuryl mercaptan via dehydration and reduction of the 3-deoxpentosone which is a known Maillard reaction product (8).

The formation of furfuryl mercaptan in model systems was recently investigated by Grosch and Zeiler-Hilgart. The model systems included various combinations of cysteine, ribose, thiamine, hydrogen sulfide, reduced glutathione, and 5'-inosinic acid in pH 5.7 pyrophosphate buffer. They found that the level of Fur-SH generated in aqueous model systems at boiling temperatures was always low (9).

Farmer, Mottram and Whitfield studied volatile compounds produced in Maillard reactions involving cysteine/ribose with and without an egg phospholipid upon autoclaving at 140° C for 1 hour. The pH was maintained at 5.7 with phosphate buffer. They found that Fur-SH was a major volatile component, and its generation was independent of the presence of phospholipid (10).

Shibamoto investigated the formation of sulfur and nitrogen containing compounds upon heating an aqueous model system consisting of furfural, hydrogen sulfide and ammonia at 100°C for 2 hours. Furfuryl mercaptan was identified as a prominent, but not the major, component in the reaction mixture (11). The pH of the system was not reported.

The effect of pH upon furfuryl mercaptan formation was investigated by Mottram and Leseigneur (12). They reported the level of Fur-SH increased as the pH was decreased in the cysteine/ribose model system.

Coffee Precursors. The primary precursors of furfuryl mercaptan in green coffee are postulated to be the free or polymeric forms of pentose sugars and sulfur amino acids. Hexose sugars may also be a source upon fragmentation.

Tressl et al. reported that there is between 0.05% and 0.07% free ribose in Arabica and Robusta green coffee (13).

Arabinose is the major pentose sugar in coffee and is a component of the highly branched arabinogalactan polysaccharide (14). It was found that the arabinose content of Arabicas and Robustas ranges between 3.4% and 4.1% on a dry basis. The high concentration of the terminal arabinose residues on the branches suggests that it would be susceptible to easy hydrolysis during roasting. Small amounts of xylose (0.2%) were also measured. Thaler and Arneth (15) confirmed that about 60% of the arabinose is lost upon roasting green Arabica coffee (decreasing from 1.7% to about 0.68%).

Tressl et al. reported no free cysteine or methionine in Robusta and Arabica coffees (13). Thaler and Gaigl found that the cysteine content of the protein fraction of green coffee ranged from about 2.9-3.9% in Arabica and Robusta coffee and decreased to about 0.14-0.76% after roasting. The proteinaceous methionine decreased to a smaller degree upon roasting (16,17).

In summary, the literature indicates that potential pentose precursors are small amounts of ribose and a larger pool of polymeric arabinose; the major sulfur source appears to be proteinaceous cysteine although cystine or inorganic sulfur sources may also play a role.

Reaction Kinetics. Data on the kinetics of the Maillard reaction are not extensive. It is well known that cooking food under different conditions can result in a variety of flavors.

Reineccius (18) had attributed these differences to reaction kinetics and specifically to the activation energy for a specific reaction. The activation energy determines how temperature affects the rate of formation of a product. Reactions with low activation energies are generally favored at lower temperatures, while reactions with higher activation energies are generally favored at higher temperatures. Leahy and Reineccius examined the kinetics of formation of some alkyl pyrazines. A high correlation was found between the formation of pyrazines and reaction time indicating pseudo zero order kinetics. Activation energies for the formation of various pyrazines were found to range from 27 to 45 Kcal per mole (19).

These results were consistent with the earlier work of Warbier et al. (20) and Labuza et al. (21) who found that formation of browning products in the Maillard reaction generally followed pseudo zero order kinetics when reactant concentrations were in excess.

High temperature short time kinetics of a Maillard reaction in a proline/glucose model system were recently investigated by Stahl and Parliment at temperatures ranging between 160°C and 220°C and reaction times ranging between 0.25 min and 5 min. It was found that one compound, 5-acetyl-2,3-dihydro-1(H)-pyrrolizine, was the major product and formed by pseudo zero order kinetics. It was determined that this compound had a relatively high energy of activation of 45 Kcal/mole (22).

Roasting Process. The flavor of roasted coffee is highly dependent on roasting time and temperature. The roasting of coffee has been typically carried out at temperatures ranging from about 200°C to 260°C and times ranging from about 90 sec to about 20 min. The change in some volatile constituents of roasted coffee with roasting time has been demonstrated by Gianturco (23). He observed that certain coffee volatiles increase to maximum and then decrease with roast time. Three compounds showing this phenomena were clearly of carbohydrate origin: 5-methyl furfural, furfural, and furfuryl alcohol. Very little is known about the kinetics of flavor reactions under roasting conditions.

It is the purpose of this study to investigate the kinetics of the generation of furfuryl mercaptan under high temperature conditions. As a preliminary experiment, the level of furfuryl mercaptan generated during coffee roasting at various time intervals was determined. Reaction kinetics were determined in a small continuous flow reactor which permitted precise control over time and temperature. The model system we chose to investigate was the ribose/cysteine system because of the greater information on this system in the literature. Arabinose is a stereoisomer of ribose and should have equivalent chemical properties; it would have been equally valid for examination in the model system.

EXPERIMENTAL SECTION

Coffee Bean Roasting. One pound batches of green Colombian coffee beans were roasted for various times ranging from six to sixteen minutes in a Jabez Burns coffee roaster. The coffees were ground, and the degree of roast measured in an Agtron near infrared abridged spectrophotometer model   E-10CP (Sparks, Nev.).

Coffee Bean Analysis. Three grams of roasted and ground coffee beans were combined with 5.0 gm of water in a 50 mL pear shaped flask. The sample was indirectly steam distilled and 8 mL condensate was collected. The aqueous phase was extracted in a Mixxor as described by Parliment (24).

Samples were analyzed by gas chromatography using a Hewlett Packard model 5890 gas chromatograph equipped with a 60m x 0.32 mm id column coated with a 1 micron film thickness of DB5 liquid phase. A Sievers model 355 sulfur chemiluminescent detector was employed to detect the sulfur containing species as described by Cohen (25).

Reaction Conditions in Continuous Flow Reactor. For the thermal reactions, 1.2 gm cysteine and 1.5 gm ribose were combined in 100 mL of deionized water (pH= 3.75). This system was pumped through a 1.0 mL sample loop made of 1/16 inch stainless steel tubing which was immersed in a silicone oil bath. After passing through the heated loop, the product was immediately cooled by immersing the connecting tubing in a cooled water bath. The pump permitted various flow rates ranging from 0.1 to 9.9 mL/min. and thus reaction times ranging from 10 min to about 0.1 min. Connecting tubing was 1/16 inch 316 stainless steel tubing (0.031 inch id). Two experimental methodologies were conducted. In the first, the oil bath was held at 200°C and flow rates were varied to give reaction times ranging from 0.50 min to 5 min. In the second methodology, the reaction time was held at 1.0 min and the oil bath temperature was varied from 180°C to 220°C. The second methodology provided single point rate constants for product formation (relative GC/MS counts/min) over a range of temperatures and proved useful in determining Arrhenius activation energy for compound formation. The overall equipment and procedure has been described in detail previously (22).

Most of the studies were conducted at an initial pH of 3.75, yielding a final pH of 4.3, since this approximates the coffee system. In one set of experiments we reduced the pH to 2.0 with phosphoric acid and in another set raised the pH to 10.2 in order to measure the effect of pH on the reaction.

One series of experiments was conducted in which ribose was replaced by either 0.98 gm furfuryl alcohol or 0.96 gm furfural and the system reacted for 1 min at 200°C.

Analysis of Model System. Eight mL of aqueous reaction product were transferred to a 10 mL Mixxor. One gram sodium chloride was added, and the system was extracted with 0.75 gm diethyl ether (containing 1 mg ethyl nonanoate per 10 mL diethyl ether as internal standard). This procedure has been previously described by Parliment (24).

Ether extracts were analyzed via GC/MS. A Varian Model 3700 gas chromatograph was used with a 0.32 mm id x 15 m fused silica column coated with a 1 micron film of DB-5. The following oven conditions were employed: 5 min at 60°C then 5°C/min to 230°C and a final hold of 10 min. The column effluent was passed through an open split interface into a Finnigan model 705 Ion Trap Mass spectrometer. Identifications were achieved by comparison of the generated spectra to those of the NBS Library Compilation or to published spectra. Relative concentrations of the products were determined using the Ion Trap quantitation program.

RESULTS AND DISCUSSION

Generation of Furfuryl Mercaptan in Roasted Coffee. There is convincing evidence that furfuryl mercaptan is an important component of coffee. In the first portion of this study we demonstrated that furfuryl mercaptan is a product of roasting. Figure 1 shows that furfuryl mercaptan increased with degree of roast. A value of 92 represents about six minutes of roast (severely under-roasted) while a value of 32.8 is an over-roasted (espresso-like) coffee roasted for 16 minutes. It is known that little reaction occurs in the early part of the roast cycle (the drying phase) because moisture is leaving the green coffee, and the internal temperature remains low.

Determination of Rate Constants in Model Systems. Figure 2 shows the generation of Fur-SH versus time at temperatures of 190°C, 200°C and 210°C. The generation of Fur-SH is linear with time. This data is consistent with a zero order reaction kinetics model described earlier for most Maillard reaction products (19-21). The rate constants and correlation coefficients were generated from the slope of the regression equation for the data using the Lotus Freelance program and are given in Table I. At the longer time/ higher temperature conditions of these reactions, furfuryl mercaptan becomes the major volatile component produced.

Figure 1. Generation of Furfuryl Mercaptan During Coffee Roasting

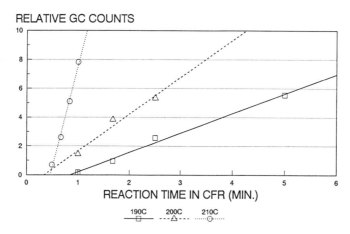

Figure 2. Generation of Furfuryl Mercaptan in Model System at 190°C, 200°C, 210°C.

Table I. Average rate Constants for Formation of Furfuryl Mercaptan at 190°C, 200°C and 210°C

Temp (°C)	Rate Constant (k)	Corr. Coeff. (r^2)
190	1.35	0.992
200	2.55	0.962
210	14.29	0.994

Figure 3 presents the generation of Fur-SH at 220°C. The generation of Fur-SH increases to a maximum between 1 and 2 minutes, then decreases. This suggests that Fur-SH either decomposes or reacts further to produce other products. Thus, there is an optimum time/temperature range for the formation of Fur-SH in roasting. These results are consistent with the work of Schirle-Keller and Reineccius (26) describing the reaction kinetics for the formation of oxygen containing heterocyclic compounds in a glucose/cysteine model system. They reported that some compounds need a certain energy level to form, but too much will result in their degradation and disappearance from the chromatogram.

Effect of pH on Furfuryl Mercaptan Formation. Figure 4 is a plot of the rate of Fur-SH formation versus pH at 200°C using a logarithmic regression curve fitting program (Lotus Freelance TM). The correlation coefficient for this relationship was found to be 0.9996 indicating a definitive relationship between pH and rate of Fur-SH formation. The strong pH dependence demonstrated in our study expands on the results of Mottram and Leseigneur who showed a 6x increase in furfuryl mercaptan by reducing the pH from 6.5 to 4.5 (12). The aforementioned results were obtained at 140°C.

Determination of Activation Energy (E_a). The determination of the activation energy (E_a) of formation of furfuryl mercaptan was done by two procedures.

Methodology 1 involved plotting the log of the rate constants determined at pH 3.75 at 190°C, 200°C and 210°C (from Table I) versus 1/Temperature (Kelvin). This Arrhenius plot is shown in Figure 5. The E_a calculated was 52.6 Kcal/mole (r^2= 0.931) and was found from the slope of the line generated by regression analysis from the following relationship:

$$E_a = (-2.303)(1.987 \text{ cal/degree mole})(\text{slope of line}). \quad \text{(Eqn 1)}$$

A second methodology was also evaluated for calculating the E_a. In this case the relative area counts at one minute reaction time were measured at five temperatures and the activation energy determined from the slope of the plot using Eqn 1.

The activation energies determined by the two methodologies are summarized in the Table II below:

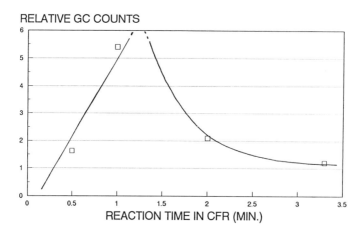

Figure 3. Generation of Furfuryl Mercaptan in Model System at 220°C

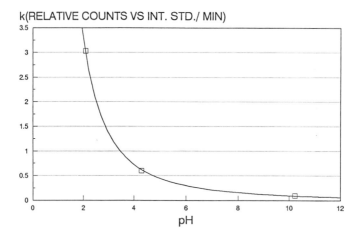

Figure 4. Generation of Furfuryl Mercaptan in Model System at Various pH Values

Table II. Activation Energies for Formation of Fur-SH at pH 3.75 (Kcal/mole)

Methodology	E_a Fur-SH	Corr. Coef. r^2
1	52.6	0.931
2	48.6	0.965

The values of the E_a for Fur-SH formation determined at pH 3.75 by the two methods compare very well. The average for the two methods is 50.6 Kcal/mole.

Furfuryl alcohol and furfural were also detected in the reaction mixture. This suggests that these carbohydrate decomposition products may be intermediates in the formation of Fur-SH. Figure 6 is a plot of the rate of formation at 210°C of furfural (Fur-AL) and furfuryl alcohol (Fur-OH) in our model system. The reaction kinetics are again pseudo zero order. The rates of formation are summarized in Table III below. The rate of formation of Fur-SH is significantly faster than Fur-AL and Fur-OH.

Table III. Kinetic Data for Compounds formed from Cysteine/Ribose at 210°C

	Fur-SH	Fur-AL	Fur-OL
k (rel cts/min)	14.29 (r^2=0.994)	0.422 (r^2=0.938)	0.178 (r^2=0.959)
Rel. Rate of Form.	80.2	2.37	1.00

Figure 7 is the Arrhenius plot (log rate versus 1/Temperature Kelvin) for the activation energy of formation of Fur-SH at pH 2 and pH 3.75 using methodology 2. The activation energy (E_a) was calculated from the slope of the line from Eqn 1. The activation energy for formation of Fur-AL was also determined by the Arrhenius plot of the relative area counts at one min reaction for 4 temperatures at pH 2 and 3.75. The activation energy data for both experiments are summarized in Table IV below. The activation energies of formation of both compounds decrease about 30% as the pH is lowered indicating that the formation of both are favored at acid pH conditions and suggests a common mechanism for their formation.

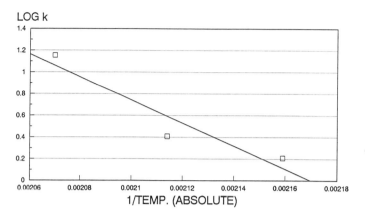

Figure 5. Arrhenius Plot for Formation of Furfuryl Mercaptan Based on Reaction Temperatures of 190°C, 200°C, 210°C

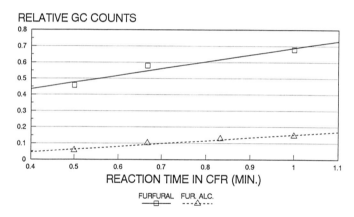

Figure 6. Generation of Furfural and Furfuryl Alcohol in Model System at 210°C

Table IV. Activation Energies for Formation of Fur-SH and Fur-AL at pH 3.75 and pH 2.0 (Kcal/mole)

	Fur-SH	Fur-AL
E_a at pH 3.75	48.6 (r^2=0.965)	25.0 (r^2=0.942)
E_a at pH 2.0	31.0 (r^2=0.986)	17.5 (r^2=0.962)

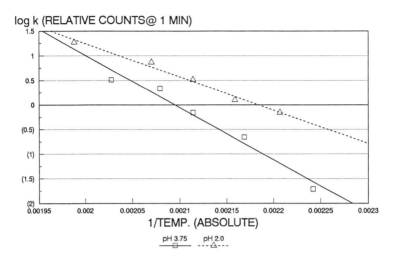

Figure 7. Arrhenius Plot for Formation of Furfuryl Mercaptan at pH 2.0 and 3.75

CONCLUSIONS

A study of roasted coffee demonstrated that the level of furfuryl mercaptan increased with roast time.

A pentose/cysteine model systems study, using a continuous flow reactor, demonstrated that furfuryl mercaptan forms under pseudo zero order kinetics. The energy of activation for the formation of furfuryl mercaptan was determined to be 48.6 Kcal/mole at about pH 3.75. The relatively high energy of activation for Fur-SH, relative to other compounds, suggests that its formation would be favored under high temperature regimes such as occur during coffee roasting. The presence of furfuryl alcohol and furfural in the reaction mixture suggests that these may be intermediates in the formation of Fur-SH.

LITERATURE CITED

1. Reichstein, T.; Staudinger, H., Brit. Patent 246,454 and 260,960 (**1926**), Improvements in a Method For Isolating the Aromatic Principle Contained in Roasted Coffee.
2. Tressl, R., In: Thermal Generation of Aromas; Parliment, T.H.; McGorrin, R.J.; Ho, C-T, Eds. ACS Symposium Series #409; American Chemical Society, Washington, DC, **1989**; pp 285-301.
3. Flament, I.; Chevallier, C., Chem. Ind. (London), **1988**, 18, 592.
4. Blank, I.; Sen, A.; Grosch W., Z. Lebensm. Unters. Forsch., **1992**, 195, 239.
5. Arctander, S., Perfume and Flavor Chemicals Monograph No. 1417, S. Arctander; Montclair, N. J.; **1969**.
6. Tressl,R.; Silwar J., J. Agric. Food Chem., **1981**, 29, 1078.
7. Holscher, W.; Steinhardt, H., Z. Lebensm. Unters. Forsch.,**1992**, 195, 33.
8. Tressl, R., In: Thermal Generation of Aromas; Parliment, T.H.; McGorrin, R.J.; Ho, C-T, Eds. ACS Symposium Series #409; American Chemical Society, Washington, DC, **1989** ; pp 156-171.

9. Grosch, W.; Zeiler-Hilgart In: Flavor Precursors: Thermal and Enzymatic Conversions; Teranishi, R., Takeoka, G.R, and Guentert, M., ACS Symposium Series #490; American Chemical Society, Washington, DC, **1992**; pp 183-192.
10. Farmer, L. J.; Mottram, D. S.; Whitfield, F.B., J. Sci. Food Agric., **1989**, 49, 347.
11. Shibamoto, T., J. Agric. Food Chem., **1977**, 25, 206.
12. Mottram, D.; Leseigneur, A., In: Flavor Science and Technology; Bessiere, Y., Thomas, A. Eds. J. Wiley, New York, **1990** ; pp 121-124.
13. Tressl, R.; Holzer, M.; Kamperschroer, H., ASIC 10th Colloquium, Salvidor, **1982**, 279-292.
14. Bradbury, A.W.; Halliday, D.J., ASIC 12th Colloquium, Montreux, **1987**, 265-269.
15. Thaler, H.; Arneth, W., Z. Lebensm. Unters. Forsch., **1969**, 140, 101.
16. Thaler H.; Gaigl, R., Z. Lebensm. Unters. Forsch., **1963**, 119, 10. 17.
17. Thaler H.; Gaigl, R., Z. Lebensm. Unters. Forsch., **1963**, 120, 357.
18. Reineccius, G.A. In: The Maillard Reaction in Food Processing, Human Nutrition and Physiology; Pinot, P.A.; Aeschbacher, H.U.; Hurrell, R.F.; Liardon, R., Eds.; Birkhauser Verlag: Boston, **1990**; pp 157-170.
19. Leahy, M.M.; Reineccius, G.A. In: Flavor Chemistry: Trends and Developments, Teranishi, R.; Buttery, R.G.; Shahidi, F. Eds.; ACS Symposium Series No. 388; American Chemical Society: Washington, DC, **1989**; pp 76-91
20. Warbier, H.C.; Schnickels, R.A.; Labuza, T.P., J. Food Sci. **1976**, 41, 981.
21. Labuza, T.P.; Tannenbaum, S.R.; Karel, M., Food Technol. **1970** 24, 35.
22. Stahl, H.D.; Parliment, T.H.; In: Thermally Generated Flavors: Maillard, Microwave, and Extrusion Processes, Parliment, T.H. ; Morello, M.J. ; McGorrin, R.J., Eds. ACS Symposium series #543 ; American Chemical Society, Washington ; DC, **1994** ; pp 251-262.
23. Gianturco, M. In Symposium on Foods: The Chemistry and Physiology of Flavors, Schultz, H.W.; Day, E.A.; Libbey, L.M. Eds.; AVI Publishing Company, Inc.; Westport, Conn. **1967**; pp 431-449.
24. Parliment, T.H., Perfum. Favorist **1986**, 11, 1.
25. Cohen, G. ASIC 15th Colloquium, Montpellier, France, **1993**, 528-536.
26. Schirle-Keller, J.P.; Reineccius, G.A., In: Flavor Precursors: Thermal and Enzymatic Conversions; Teranishi, R.; Takeoka, G.R.; Guentert, M. Eds.; ACS Symposium Series No. 490; American Chemical Society; Washington, DC, **1992**; pp 244-258.

RECEIVED March 23, 1994

Chapter 14

Heat-Induced Changes of Sulfhydryl Groups of Muscle Foods

Fereidoon Shahidi[1,2], Akhile C. Onodenalore[1], and Jozef Synowiecki[1]

Departments of [1]Biochemistry and [2]Chemistry, Memorial University of Newfoundland, St. John's, Newfoundland A1B 3X7, Canada

> The effects of heat processing temperature and time schedules on changes in sulfhydryl groups and disulfide bonds of mechanically separated seal meat (MSSM) and mechanically separated chicken meat (MSCM) were studied. The content of free sulfhydryl groups in samples decreased by approximately 60% as a result of heat processing at 99°C for 40 min. A corresponding increase in their content of disulfide bonds was observed. Aqueous washing of samples resulted in a decrease in their content of sulfhydryl groups. The amount of disulfide bonds formed upon heat processing in samples corresponded well with the solubility of proteins in 0.035M sodium dodecyl sulfate or the degree of their thermal coagulation.

Heating of meat is accompanied by changes in appearance, taste, texture, smell and nutritional value (*1*). The sulfhydryl groups and disulfide bonds which are moieties of cysteine and cystine contribute to these qualities in meat. Sulfur-containing compounds including disulfides are often found in foodstuffs and usually contribute to the overall aroma of such food because of their low threshold values (*2*). Sulfur compounds are major contributors to the flavor of cooked products (*3*). Furthermore, destruction of labile amino acids during heat processing operations may have detrimental effects on nutritional value of such products.

The sulfhydryl groups and disulfide bonds play important roles in the functional and structural properties of heat-processed meat. During thermal processing, disulfide cross-linking of protein molecules may occur. This has been reported in fish, meat, squid, and seal (*4-7*). Sulfhydryl-dependent gelation of hen egg lysozyme at 80°C has been demonstrated. These proteins had several intramolecular disulfide bonds with a good correlation between concentration of sulfhydryl groups and strength of heat-induced gels so produced (8). Furthermore, sulfhydryl groups have been implicated to enhance gel formation in fish proteins during heating. Minced and washed fish muscles produced high quality kamaboko gels due to their high content of sulfhydryl groups (9). Changes in surface hydrophobicity in proteins of frozen cod have been suggested as a possible cause of disulfide bonding which might have contributed to the muscle fibrousness observed during extended frozen storage at elevated (-5°C) or fluctuating temperatures (10). The formation of intermolecular disulfide bonds during gel formation may occur, and this has been demonstrated in carp actomyosin, thus implicating sulfhydryl groups in the formation of polymeric protein molecules in the resultant gel (11).

The effect of alkali on the degradation of disulfide bonds in proteins has been studied. It has been reported that alkali treatment of disulfide-containing proteins was accompanied by degradation of disulfide bonds through a β-elimination mechanism that induced the formation of dehydroalanine, elemental sulfur and H_2S (12). Addition of alkali to protein caused an increase in sulfhydryl groups and viscosity of native protein solutions of bovine serum albumin, hemoglobin and rapeseed, which was attributed to unfolding of proteins at high pH (13). However, reactivity of sulfhydryl groups, which enhances both oxidation of sulfhydryl groups into disulfide bonds and sulfhydryl/disulfide interchange reactions, has been reported to decrease significantly under acidic conditions (14).

The presence of sulfhydryl groups in organisms and foods has been reported to be important for detoxification mechanism because they can bind toxic elements and molecules as well as carcinogens (1). Another important function of sulfhydryl groups in food is their preferential reaction with nitrite as compared with competing reactions with amines and amino acids in the stomach. Thus, the presence of large amounts of sulfhydryl groups in foods prevents/retards the formation of N-nitrosamines in the stomach (15).

In the present study, we assessed the effect of heating and aqueous washing of mechanically separated seal meat (MSSM) and mechanically separated chicken meat (MSCM) on the content of sulfhydryl groups and disulfide bonds in their proteins. An attempt was made to correlate the content of sulfhydryl groups with changes in the solubility and thermal coagulation of mechanically separated meat proteins.

Experimental

Materials. Beaters (3 weeks to 1 year) and bedlamer (1-4 years) harp seals (*Phoca groenlandica*) hunted in the coastal regions of Newfoundland/Labrador during May-July were bled and skinned, the blubber fat was removed, and the carcasses were eviscerated. Whole seal carcasses, without head and flippers, were placed inside plastic bags and stored in containers with ice for up to 3 days. Each carcass was then washed with a stream of cold water (10°C) for about 15 s to remove most of their subcutaneous fat. Samples of mechanically separated seal meat (MSSM) were prepared from these seal parts and mechanically separated chicken meat (MSCM) was prepared after removing skins from necks and backs of 40 day-old birds (*Arbor acre*). Mechanical separation of seal and chicken meats was carried out using a Poss deboner (Model PDE 500, Poss Limited, Toronto, ON). Small portions of mechanically separated meats were vacuum packed in polythene pouches and kept frozen at -20°C for 2-3 wk until use. Samples were thawed at 4°C for 12 h before use.

The MSSM was washed one to three times with water (pH 5.9-6.0) using a water to meat ratio of 3:1 (v/w). Each washing was carried out at 2-5°C for 10 min while stirring manually. The washed meat was then filtered through two layers of cheese cloth with 1 mm holes. Samples of MSCM were washed two times. First washing used water and the pH of the mixture was 6.9; second washing used water (pH of mixture 5.2 or 7.2) or a 0.5% solution of NaCl (pH 6.9) or $NaHCO_3$ (pH 7.8). In another set of experiments, the pH of the meat-water mixture was adjusted to 5.2 using a 5% acetic acid solution. Each washing was carried out at 2-5°C for 10 min while stirring manually. The ratio of the extraction solution to meat was 3:1 (v/w). Washed meats were filtered through three layers of cheese cloth with 1 mm holes. Each washing was replicated three times using samples obtained on the same day.

For determination of disulfide bond formation and changes in protein solubility of MSSM and MSCM in sodium dodecyl sulfate (SDS) (Sigma Chemical Company, St. Louis, MO), samples were heated in sealed polythene pouches in a controlled water bath over 40 min at temperatures ranging from 40 to 99°C or at 80°C over a 60 min period.

Analyses. For determination of protein solubility, 5 g of meat sample was homogenized in ice for 60 s, using a Polytron homogenizer (Brinkmann Instruments, Rexdale, ON) with 100 mL of 5% NaCl solution in 0.003M $NaHCO_3$, 0.035M SDS from Sigma Chemical Company in 0.003M $NaHCO_3$, or 0.035M SDS + 0.15M 2-mercaptoethanol (Sigma Chemical Company) in 0.003M $NaHCO_3$ at pH 7.0. After 30 min of solubilization of meat sample with intermittent mixing, and 10 min of centrifugation at 10,000 x g, total proteins in the supernatant were determined by Kjeldahl method (*16*). The degree of thermal coagulation as loss of solubility was determined by heating the MSSM or MSCM extracts with 5% NaCl solution in 0.003M $NaHCO_3$ for 40 min at 40, 50, 60, 75 and 99°C. The degree of thermal coagulation was expressed as $c_1-c_2/c_1 \times 100$, where c_1 and c_2 are the concentrations of proteins before and after heat treatment, respectively.

Free sulfhydryl (SH) groups were determined in 0.10 to 0.15 g of meat sample dissolved in 8 mL of 0.75% ethylenediaminetetraacetic acid disodium salt (Na_2EDTA) and 0.035M SDS solution in a Tris buffer pH 8.2 (*4*). After standing for 2 h, 0.5 mL of 0.016M solution of 5,5'-dithiobis(2-nitrobenzoic acid) (DTNB) (Sigma Chemical Company) in methanol and 31.5 mL of methanol were added with mixing. The solution was allowed to stand at 20°C for 15 min, centrifuged at 3,000 x g for 15 min, and the absorbance was read at 412 nm (*17*). The calibration curve for sulfhydryl groups was prepared using reduced glutathione (Sigma Chemical Company) in concentrations ranging from 0 to 0.1 mg/mL of sample. A linear correlation in the absorbance range of 0 to 1.5 was noticed. From this, the molar extinction coefficient at 412 nm was calculated as 13620 which compares well with that of 13600 previously reported (*17*).

Disulfide bonds in meat samples containing approximately 35 mg of proteins were determined. These samples were then reduced with 4 mL 0.6M $NaBH_4$ in 8M urea (*4*). The disulfide bonds were calculated from the difference between the content of SH in the reduced samples and those in the original meats.

Statistical analysis. Analysis of variance and Tukey's studentized range test (*18*) were used to determine differences in mean values based on the data collected from 4 to 6 replications of each measurement. Significance was determined at 95% probability.

Results and Discussion

Effect of Aqueous Washing. The content of sulfhydryl groups and disulfide bonds in MSSM and MSCM and as affected by aqueous washing is given in Table I. Washing decreased the content of free sulfhydryl groups of meats as compared with their unwashed counterparts. Single washing of MSSM with water resulted in a 12.7% decrease in the content of sulfhydryl groups and a second and third washing with water brought about a corresponding decrease of 16.6 and 17.7% in the samples, respectively. However, there was a proportionate increase in the disulfide bonds in the washed MSSM. The latter increase may be due to the existing differences in the content of free sulfhydryl groups and disulfide bonds between proteins removed during washings from the samples and those present in the resultant washed meat. A similar trend in the content of these components was observed for unwashed and washed MSCM, except for the sample washed with water at pH 5.2. The second washing with water removed 34.8 and 28.1% of the proteins, mostly sarcoplasmic, from MSSM and MSCM, respectively, thus concentrating the myofibrillar and connective tissue proteins. The use of urea exposes sulfhydryl groups embedded within the protein molecule to DTNB, hence the estimation of initially masked and unmasked sulfhydryl groups was possible in this study. The use of urea has been reported to increase the total sulfhydryl group estimation by 8-10% as compared with determination of sulfhydryl groups in the absence of urea (*4*).

Effect of Heating at 20-99°C. The effect of heating at various temperatures (20-99°C) for 40 min on sulfhydryl groups and disulfide bonds in MSSM and MSCM is given in Figure 1. The content of sulfhydryl groups was inversely related to heating temperatures, showing a maximum at 20°C and a minimum at 99°C. However, the content of disulfide bonds increased progressively, as the temperature of heat processing increased. Increasing the temperature from 20 to 99°C increased the content of disulfide bonds by 150 and 230% in MSSM

Table I. Sulfhydryl groups and disulfide bonds in MSSM and MSCM proteins[1]

Treatment	Sulfhydryl groups (μmol/g protein)		Disulfide bonds (μmol/g protein)
	Initial sample	After reduction of S-S bonds	
Unwashed MSSM	63.80±1.24[a]	84.68±0.27[a]	10.12±0.63[c]
Washed MSSM 1 x H$_2$O	55.70±0.73[b]	78.96±0.35[c]	11.28±0.35[bc]
2 x H$_2$O	53.23±0.59[bc]	78.85±0.30[c]	12.42±0.44[ab]
3 x H$_2$O	52.48±2.59[c]	81.92±0.38[b]	14.27±1.23[a]
Unwashed MSCM	57.95±0.81[d]	76.23±0.51[d]	8.86±0.44[g]
Washed MSCM 1 x H$_2$O	49.79±0.59[e]	68.75±0.41[f]	9.77±0.51[f]
2 x H$_2$O	46.85±0.26[f]	69.50±0.35[f]	10.98±0.63[ef]
1 x H$_2$O, then 0.5% NaCl	47.83±0.25[f]	70.95±0.32[e]	11.21±0.71[e]
1 x H$_2$O, then 0.5% NaHCO$_3$	39.93±0.17[g]	67.19±0.34[g]	13.22±0.92[d]
1 x H$_2$O, then H$_2$O at pH 5.2	49.14±0.73[e]	66.77±0.36[g]	8.54±0.61[g]

[1]Results are mean values of 6 determinations ± standard deviation. Values in the same column with same superscript [a-c] (MSSM) and [d-g] (MSCM) are not significantly ($p > 0.05$) different from each other.

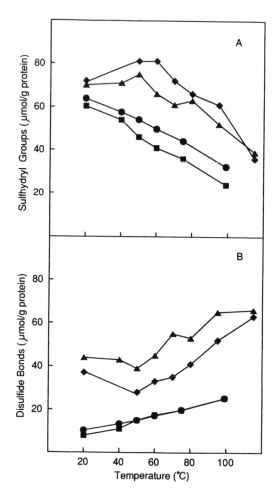

Figure 1. Sulfhydryl groups (A) and disulfide bonds (B) contents of mechanically separated seal meat (MSSM, ●), mechanically separated chicken meat (MSCM, ■) pollock, ▲; and mackerel, ♦; as a function of heat processing temperatures (20-115°C) for 40 min (MSSM and MSCM) and 20 min (pollock and mackerel). Values for pollock and mackerel were adapted from ref. 4 and a factor of 6.25 was used to convert the content of nitrogen to protein.

and MSCM, respectively. Conversely, the content of free sulfhydryl groups decreased by a factor of 2.0 and 2.5 in MSSM and MSCM, respectively, during the same heat processing temperatures. Oxidation of sulfhydryl groups to disulfide bonds may be a major cause for these changes. A decrease of about 50% in total sulfhydryl groups of croaker actomyosin at temperatures at 35°C has been reported (19). The observed increase in disulfide bonds of meat proteins in this study is expected to play an important role in gel formation properties of MSSM and MSCM in processed meat products. Effect of heating on sulfhydryl groups and disulfide bonds of pollock and mackerel proteins has been reported (4). The proteins from both sources followed a similar trend of changes in the content of the sulfhydryl groups and disulfide bonds due to heating. A decrease in the content of sulfhydryl groups was reported from 50 and 60°C for pollock and mackerel, respectively, and disulfide bonds increased linearly with temperature ranging from 50 to 95 and 115°C for pollock and mackerel, respectively (Figure 1). The observed increase in disulfide bonds at these temperatures was attributed to oxidation of sulfhydryl groups and formation of disulfide bonds. The pattern of changes in sulfhydryl groups and disulfide bonds in these fish samples conforms with that observed in this study.

Thermal Denaturation of Proteins. The change in solubility of MSSM and MSCM proteins in 5% NaCl solution was used to assess the extent of thermal denaturation of these proteins during heating at 40-99°C as presented in Figure 2. A marked increase in the degree of thermal coagulation of MSSM proteins at temperatures ranging from 50 to 75°C was apparent, as a decrease in the solubility of these proteins was noted. However, at 75 to 99°C there was only a slight change in these parameters. The MSCM proteins followed a similar trend showing a large decrease in their solubility in 5% NaCl solution at 40 to 60°C. A sharp rise in protein coagulation of croaker actomyosin at 35-60°C has been reported when a solution of this protein was heated for 30 min (19). Although oxidation of sulfhydryl groups may have contributed to coagulation of MSSM and MSCM proteins, hydrophobic interactions may have also contributed to protein aggregation and coagulation. Different functional groups of proteins are involved in the process of thermal denaturation and aggregation of protein molecules (20).

The use of 2-mercaptoethanol, which reduces the disulfide bonds of proteins, is sometimes practised during the determination of disulfide bonds content (1). The influence of disulfide bonds formation on solubility of proteins was evidenced by comparing the protein solubility in sodium dodecyl sulfate (SDS) with or without the addition of 2-mercaptoethanol. Solubilities of meat proteins expressed as percent of their amount in the meat after heating at 80°C in SDS and in the presence or absence of 2-mercaptoethanol were 76 and 66%, respectively, for MSSM and 81 and 83%, respectively, for MSCM.

Formation of disulfide bonds during heating at different temperatures correlated well with the degree of thermal coagulation of MSSM and MSCM proteins with correlation coefficients of 0.897 and 0.917, respectively. The content of sulfhydryl groups also depended on the length of the heating period (Figure 3). The decrease in the content of sulfhydryl groups in MSSM and

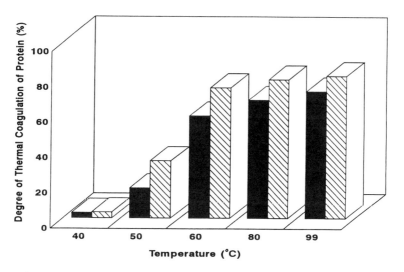

Figure 2. Degree of thermal coagulation of proteins extracted from mechanically separated seal meat (MSSM, ■); and mechanically separated chicken meat (MSCM, ▨) by 5% NaCl in 0.003M NaHCO$_3$ solution. Results are mean values of 4 determinations and standard deviations did not exceed 3.7% of the recorded mean values.

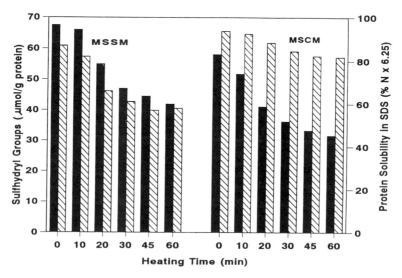

Figure 3. Effect of heat processing time of mechanically separated seal meat (MSSM) and mechanically separated chicken meat (MSCM) at 80°C on free sulfhydryl groups, ■; and protein solubility in 0.035M SDS, ▨. Results are mean values of 4 determinations and standard deviations did not exceed 3.0% of the recorded mean values.

MSCM heated at 80°C for up to 60 min also correlated well with a decrease in solubility of proteins in 0.035M SDS. Correlation coefficients of 0.982 and 0.980 were obtained for MSSM and MSCM, respectively. The decrease in free sulfhydryl groups paralleled the formation of disulfide bonds in the samples.

The role of sulfhydryl groups and disulfide bonds in the gel formation of proteins has been established and already reported. Washing of MSSM and MSCM may favor gelation when washed meats are used as ingredients in formulated products due to the observed increase of disulfide bonds upon washing. The sulfhydryl groups also contribute to the solubility characteristics of mechanically separated meat proteins in salt solution, thus affecting their gelling and emulsification properties.

Literature Cited

1. Hoffman, K.; Hamm, R. *Adv. Food Res.* **1978**, *24*, 1-111.
2. Buttery, R.G.; Guadagni, D.G.; Ling, L.C.; Seifert, R.M.; Lipton, W. *J. Agric. Food Chem.* **1976**, *24*, 829-832.
3. Shahidi, F. In *Flavour Chemistry: Trend and Developments*; Teranishi, R; Buttery, R.G.; Shahidi, F., Eds.; ACS Symposium Series 388, American Chemical Society, Washington, DC, **1989**, pp 188-201.
4. Opstvedt, J.; Miller, R.; Hardy, R.W.; Spinelli, J. *J. Agric. Food Chem.* **1984**, *32*, 929-935.
5. Hamm, R.; Hoffman, K. *Nature* **1965**, *18*, 1269-1271.
6. Synowiecki, J.; Sikorski, Z.E. *J. Food Biochem.* **1988**, *12*, 127-135.
7. Synowiecki, J.; Shahidi, F. *J. Agric. Food Chem.* **1991**, *39*, 2006-2009.
8. Margoshes, B.A. *J. Food Sci.* **1990**, *55*, 1753, 1756.
9. Jiang, S.T.; Lan, C.C.; Tsao, C.Y. *J. Food Sci.* **1986**, *51*, 310-312,351.
10. LeBlanc, E.L.; LeBlanc, C.C. *Food Chem.* **1992**, *43*, 3-11.
11. Itoh, Y.; Yoshinaka, R.; Ikeda, S. *Bull. Jap. Soc. Sci. Fish.* **1980**, *46*, 621-624.
12. Nashef, A.S.; Osuga, D.T.; Lee, H.S.; Ahmed, A.I.; Whitaker, J.R.; Feeney, R.E. *J. Agric. Food Chem.* **1977**, *25*, 245-251.
13. Gaucher-Choquette, J.; Boulet, M. *Cereal Chem.* **1982**, *59*, 435-439.
14. Shimada, K; Cheftel, J.C. *J. Agric. Food Chem.* **1988**, *36*, 147-153.
15. Cantoni, C; Cattaneo, P. *Ind. Aliment.* **1974**, *13*, 63-68.
16. AOAC. *Methods of Analysis*, 15th ed. Association of Official Analytical Chemists, Washington DC, **1990**.
17. Ellman, G.L. *Arch. Biochem. Biophys.* **1959**, *82*, 70-77.
18. Snedecor, G.W.; Cochran, W.G. *Statistical Methods*, 7th ed.; The Iowa State University Press: Ames, IA, **1980**.
19. Liu, Y.M.; Lin, T.S.; Lanier, T.C. *J. Food Sci.* **1982**, *47*, 1916-1920.
20. Madovi, P.B. *J. Food Technol.* **1980**, *15*, 311-318.

RECEIVED March 23, 1994

Chapter 15

Important Sulfur-Containing Aroma Volatiles in Meat

Donald S. Mottram and Marta S. Madruga

University of Reading, Department of Food Science and Technology, Whiteknights, Reading RG6 2AP, United Kingdom

Sulfur-containing furans and thiophenes and related disulfides are known to possess strong meat-like aromas and exceptionally low odor threshold values. Such compounds have been found in model Maillard reaction systems and in cooked meat where they are considered to contribute to the characteristic aroma. Possible precursors of these compounds in meat are pentose sugars and cysteine. One of the main sources of pentoses in meat is the ribonucleotide, inosine-5'-monophosphate (IMP), which accumulates in meat during post-mortem glycolysis. In this study the role of IMP as a precursor of meat flavor has been examined by heating muscle with and without added IMP. A number of thiols and novel disulfides containing furan groups were isolated from the meat systems, with much larger amounts formed in the meat containing IMP. The amounts of these sulfur compounds were also higher in meat systems in which the pH had been reduced by the addition of acid.

The investigation of characteristic flavors associated with cooked meats has been the subject of much research over the past four decades but, although compounds with "meaty" aromas had been synthesized, compounds with such characteristics were not found in cooked meats until recently (1). In the search for compounds with characteristic aromas it was found that furans and thiophenes with a thiol group in the 3-position possessed meat-like aromas (2). The corresponding disulfides formed by oxidation of furan and thiophene thiols were also found to have meat-like characteristics, and exceptionally low odor threshold values (3). A number of such compounds are formed in heated model systems containing hydrogen sulfide or cysteine and pentoses or other sources of carbonyl compounds (4,5). The thermal degradation of thiamine also produces 2-methyl-3-furanthiol and a number of sulfides and disulfides (6,7).

Despite the acknowledged importance of these furan- and thiophenethiols and their disulfides in meat-like flavors, it is only recently that compounds of this type have been reported in meat itself *(8-10)*. Grosch and co-workers *(11)* have suggested that the 2-methyl-3-furanthiol and its disulfide are more important in the aroma of boiled meat than in roast beef where alkylpyrazines, 2-acetyl-2-thiazoline and 2,5-dimethyl-4-hydroxy-3(2*H*)-furanone make greater contributions.

The Maillard reaction between reducing sugars and amino acids is primarily responsible for the formation of these meaty aroma compounds during the cooking of meat, although the thermal degradation of thiamine may also be important. In the Maillard reaction the precursors for 2-methyl-3-furanthiol are likely to be pentose sugars and cysteine. Indeed early work on meat flavor showed that meaty aromas were generated by heating cysteine with pentoses. In meat the principal sources of pentose sugars are the ribonucleotides, in particular inosine-5'-monophosphate (inosinic acid) which accumulates in post-slaughter muscle from the hydrolysis of adenosine triphosphate, the ribonucleotide associated with muscle function. IMP has been used as a flavor enhancer in savory foods and is believed to contribute to the "umami" taste which is considered to be an important part of the taste characteristics in meat. However, its thermal degradation and reaction with cysteine and hydrogen sulfide during cooking may be one of the main routes to the characteristic aromas of cooked meat.

The formation of 2-methyl-3-furanthiol and related compounds in model systems is influenced by factors such as pH, temperature and buffer, as well as the nature and concentration of reactants *(5,11,12)*. The Maillard reaction is known to be pH dependent, alkaline conditions favoring the formation of colored compounds and nitrogen-containing volatiles, while other volatiles are only formed at lower pHs *(13-15)*. Recently the effect of small changes in pH between 4.5 and 6.5 have been examined in meat-like model systems containing cysteine and ribose or 4-hydroxy-5-methyl-3(2*H*)-furanone *(5,12,16)*. Small changes in pH had a major effect on certain classes of volatiles; nitrogen heterocyclics such as pyrazines were only produced at pH above 5.5, while the formation of 2-methyl-3 furanthiol and its disulfide were favored by lower pH. In unbuffered model systems, large changes in pH occur during heating; therefore, buffers such as phosphate or pyrophosphate are generally used. It is known that phosphate can catalyze the Maillard reaction *(17)* and alternative means of maintaining a constant pH during the reaction would be preferable. Meat normally has a pH in the range 5.6 - 6.0 and has a high buffering capacity which results in little pH change during cooking. It would therefore provide the most appropriate medium in which to study the effect of pH on the Maillard derived volatiles associated with meat flavor.

In the work reported in this paper, meat has been used as the substrate in which to study the effect of pH on the formation of thiols and disulfides, using acid or alkali to adjust the pH. The importance of IMP concentration in meat on the formation of sulfur-containing furans during cooking was also examined in meat containing added IMP.

Experimental

Samples (100g) from a beef fillet (*M. psoas major*), obtained from a local meat supplier, were chopped in a laboratory blender and mixed with 12 ml water containing sufficient hydrochloric acid (1 M) to adjust the pH to 5.0, 4.5 or 4.0. The pH of the untreated meat was 5.6 and blanks were prepared without pH adjustment. Other samples were prepared with water containing 2.7 g inosine-5'-monophosphate (free acid obtained from yeast, Sigma Chemical Co). The addition of IMP resulted in a drop in pH of approximately 1.1 pH units, therefore the pH of some of these samples was adjusted to 5.6 by the addition of sodium hydroxide (1 M). The meat samples were then heated in glass bottles in an autoclave at 140 °C for 30 min. All the reactions were carried out in triplicate.

Headspace volatiles from the cooked meat were analyzed by headspace concentration followed by GC or GC-MS. The meat was placed in a 250 ml conical flask, fitted with a Drechsel head, and oxygen-free nitrogen was passed over the sample, for 2 h at a rate of 40 ml/min, to sweep the volatiles onto a trap packed with Tenax GC (SGE Ltd) *(18)*. The sample was maintained at 60 °C throughout collection. After collection, the volatiles were thermally desorbed, using a modified injector port, directly onto the front of a DB-5 fused silica column (30 m x 0.32 mm i.d.) in the oven of a Hewlett Packard HP5890 GC. The oven was held initially at 0 °C for 5 min while the volatiles were desorbed from the Tenax trap (held at 250 °C in the modified injector). The temperature was increased to 60 °C over 1 min, and then held for 5 min at 60 °C before programming to 200 °C at a rate of 4 °C/min. The column effluent was split equally between an FID detector and an odor port. GC-MS analyses were performed under similar GC conditions using a Hewlett Packard HP5988A mass spectrometer. Quantitation was based on peak area integration of the GC-MS chromatograms using 2-dichlorobenzene as internal standard. This compound, in ethanolic solution (1μl containing 65 ng/μl), was added to the Tenax trap just before collection.

Results and Discussion

The heated meat systems contained several thiols, namely 2-methyl-3-furanthiol, 2-furylmethanethiol, 2-mercapto-3-pentanone and 3-mercapto-2-butanone, and di- and trisulfides containing 2-methyl-3-furyl and/or 2-furylmethyl groups (Figure 1). Dimethyldisulfide, dimethyltrisulfide and dimethyltetrasulfide were also produced presumably from methanethiol, which was not isolated by the headspace entrainment method. Methanethiol also gave rise to methyl disulfides containing 2-methyl-3-furyl and 2-furylmethyl moieties. The concentrations of many of the compounds were low, and some were only detected in systems containing added IMP. No thiol-substituted thiophenes were found, although significant quantities of formyl- and acetylthiophenes, thiophenones and dithiolanones were present in all the cooked meat preparations. Many of these thiols and disulfides have not been previously reported in meat systems (Table I). Their identifications were based on mass spectra and comparison of linear retention indices (LRI) with those of authentic compounds. Authentic samples of the mercaptoketones and most of the di- and trisulfides were not available

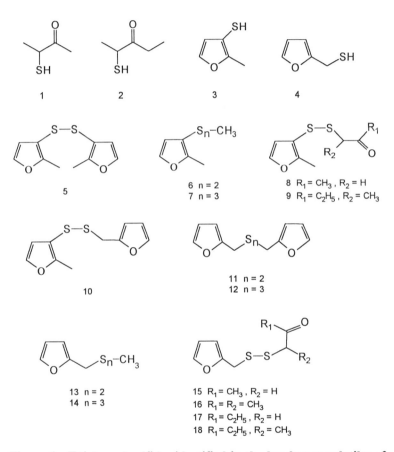

Figure 1. Thiols and sulfides identified in the headspace volatiles of heated meat systems.

Table I. Approximate concentrations (ng/100g meat) of some sulfur compounds in the headspace volatiles of heated meat systems

	Compound	LRI	Meat Blank pH 5.6	Meat pH 5.0	Meat pH 4.5	Meat pH 4.0	Meat + IMP pH 5.6	Meat + IMP pH 4.5
1	3-mercapto-2-butanone [a]	813	tr	2	7	5	17	25
2	2-mercapto-3-pentanone [a]	908	tr	3	1	7	tr	7
3	2-methyl-3-furanthiol	873	tr	3	6	13	13	19
4	2-furylmethanethiol	913	142	72	58	310	250	605
5	2-methyl-3-furyl methyl disulfide	1175	17	1	1	1	23	65
6	2-methyl-3-furyl methyl trisulfide [a]	1392	1	1	1	2	3	1
7	2-furylmethyl methyl disulfide	1220	21	31	46	81	74	124
8	2-furylmethyl methyl trisulfide [a]	1451	30	33	37	46	55	75
9	bis(2-methyl-3-furyl) disulfide	1535	nd	nd	nd	nd	tr	tr
10	2-methyl-3-furyl 2-furylmethyl disulfide	1635	4	1	1	14	8	19
11	2-methyl-3-furyl 2-oxopropyl disulfide [a]	1466	nd	nd	nd	nd	tr	tr
12	2-methyl-3-furyl 1-methyl-2-oxobutyl disulfide [a]	1584	nd	nd	nd	nd	tr	2
13	bis(2-furylmethyl) disulfide	1687	30	tr	tr	tr	20	22
14	bis(2-furylmethyl) trisulfide [a]	1932	7	tr	tr	tr	13	29
15	2-furylmethyl 2-oxopropyl disulfide [a]	1552	1	1	4	3	5	7
16	2-furylmethyl 1-methyl-2-oxopropyl disulfide [a]	1584	5	9	16	32	15	61
17	2-furylmethyl 2-oxobutyl disulfide [a]	1649	1	3	6	5	5	5
18	2-furylmethyl 1-methyl-2-oxobutyl disulfide [a]	1659	nd	2	5	6	4	22

tr, trace (<0.2 ng/100g meat); nd, not detected
[a] compound not found previously in meat

from commercial sources and, therefore, these compounds were prepared by reacting 2,3-butanedione or 2,3-pentanedione with hydrogen sulfide to give mercaptoketones and mixing appropriate furanthiols and mercaptoketones to give mixtures of symmetrical and unsymmetrical disulfides (Mottram, Whitfield and Madruga, unpublished results). The aromas of some of the compounds from the meat systems were evaluated by odor-port sniffing (GC olfactometry) of the chromatographed volatiles (Table II). In general, compounds containing the 2-methyl-3-furyl group possessed meaty aromas, but were more sulfurous at higher concentrations, while compounds with the 2-furylmethyl group were more roast and nut-like. This agrees with previous observations on the aromas of sulfur-substituted furans *(2)*. In the meat systems these sulfur compounds were only present at low concentrations and none were major peaks in the chromatogram. However, these compounds have extremely low odor threshold values and consequently they exhibited strong, characteristic odors in the reaction systems.

Table II. Odors of some sulfur-compounds formed in heated meat systems evaluated by GC-olfactometry

Compound	Odor description
3-mercapto-2-butanone	fried onion, sulfury, cooked meat
2-mercapto-3-pentanone	brothy, mashed potatoes
2-methyl-3-furanthiol	meaty, roast meat, boiled meat, fresh onion, bovril
bis(2-methyl-3-furyl) disulfide	meaty, boiled meat
2-methyl-3-furyl methyl disulfide	meaty, sulfury, fatty
2-methyl-3-furyl 2-oxopropyl disulfide	roast meat pork, onion
2-methyl-3-furyl 1-methyl-2-oxobutyl disulfide	meaty
bis(2-furylmethyl) disulfide	roast, brazil nuts, baked mushroom
2-furylmethyl methyl disulfide	brothy, spices, roast, fatty
2-furylmethyl 1-methyl-2-oxopropyl disulfide	onion, burnt rubber, burnt wood
2-furylmethyl 1-methyl-2-oxobutyl disulfide	sweet, onion, roast nuts

In the meat blank at pH 5.6, a number of the sulfur compounds were either not detected or were only found at very low concentrations (Table I). These were 2-methyl-3-furanthiol, the two mercaptoketones and the disulfides containing these moieties. A reduction in pH to 5.0 resulted in an increase in the amounts of the thiols and mercaptoketones formed and further increases were found as the pH decreased further. The lower pH also gave increased amounts of some of the di- and trisulfides containing the 2-methyl-3-furyl group. The important meaty compound, bis(2-methyl-3-furyl) disulfide, was not found in these meat systems even at lower pH, although the corresponding thiol was readily detected. The disulfide has been found previously in cooked meat *(9-11)*, but in the present work relatively small quantities of meat were used together with headspace collection, which would be relatively inefficient for the concentration of high molecular weight volatiles. 2-Furylmethanethiol was found in the meat blank in higher concentrations than the 2-methyl-3-furanthiol or the mercaptoketones, as were its di- and trisulfides. In general pH had little effect on the concentrations of these compounds.

In order to evaluate the role of inosine-5'-monophosphate as a precursor of sulfur compounds, meat systems were prepared containing added IMP. The concentration of IMP in beef varies considerably between animals and between muscles but a typical value for *M. psoas major* is 300 mg/100g (Madruga and Mottram, unpublished results). IMP was added to the meat to give a concentration approximately 10 times higher than this level. The systems containing added IMP showed increased levels of all the thiols, mercaptoketones, di- and trisulfides compared with the meat blank, but the increases were less pronounced for compounds containing the 2-furylmethyl group. When the pH decreased, the concentration of nearly all the compounds increased.

Previous studies which have shown an effect of pH on aroma volatiles have been carried out in model systems with pH controlled by the addition of phosphate buffers or in the absence of buffer. The present work has demonstrated that in a food system with strong buffering capacity (meat) the formation of furan thiols and sulfides is strongly influenced by pH, thus confirming the earlier work in buffered model systems.

The work also demonstrates that IMP in meat is a precursor for 2-methyl-3-furanthiol and mercaptoketones, although it does not seem to be as important in the formation of 2-furylmethanethiol. The roles of IMP and ribose as sources of these thiols have been discussed previously *(12,19)*. The mechanism involves the Maillard reaction and could require the intermediate formation of 4-hydroxy-5-methyl-3(2H)-furanone and dicarbonyls, such as butanedione and pentanedione, which is then followed by their reaction with hydrogen sulfide or cysteine. The concentrations of IMP in meat vary considerably between different animals and different muscles, and are affected by production conditions both pre- and post-slaughter. The present results indicate that the amount of IMP in the meat at the time of cooking may be an important factor in determining the amount of meaty flavor.

Acknowledgment

The authors are grateful to Dr. S. Elmore for his advice and assistance with the GC-MS analyses.

Literature Cited

1. Mottram, D. S. In *Volatile Compounds in Foods and Beverages*; Maarse, H., Ed.; Marcel Dekker: New York, 1991; pp 107-177.
2. Evers, W. J.; Heinsohn, H. H.; Mayers, B. J.; Sanderson, A. In *Phenolic, Sulfur and Nitrogen Compounds in Food Flavors*; Charalambous, G. and Katz, I., Ed.; American Chemical Society: Washington DC, 1976; pp 184-193.
3. Buttery, R. G.; Haddon, W. F.; Seifert, R. M.; Turnbaugh, J. G. *J. Agric. Food Chem.* **1984**, *32*, 674-676.
4. Farmer, L. J.; Mottram, D. S.; Whitfield, F. B. *J. Sci. Food Agric.* **1989**, *49*, 347-368.
5. Farmer, L. J.; Mottram, D. S. In *Flavour Science and Technology*; Bessiere, Y. and Thomas, A. F., Ed.; Wiley: Chichester, 1990; pp 113-116.
6. van der Linde, L. M.; van Dort, J. M.; de Valois, P.; Boelens, B.; de Rijke, D. In *Progress in Flavor Research*; Land, D. G. and Nursten, H. E., Ed.; Applied Science: London, 1979; pp 219-224.
7. Werkhoff, P.; Bruning, J.; Emberger, R.; Guntert, M.; Kopsel, M.; Kuhn, W.; Surburg, H. *J. Agric. Food Chem.* **1990**, *38*, 777-791.
8. MacLeod, G.; Ames, J. M. *Chem. Ind. (London)* **1986**, 175-176.
9. Gasser, U.; Grosch, W. *Z. Lebensm. Unters. Forsch.* **1988**, *186*, 489-494.
10. Farmer, L. J.; Patterson, R. L. S. *Food Chem.* **1991**, *40*, 201-205.
11. Grosch, W.; Zeiler-Higart, G.; Cerny, C.; Guth, H. In *Progress in Flavour Precursor Studies*; Schreier, P. and Winterhalter, P., Ed.; Allured Publ. Co.: Carol Stream IL, 1993; pp 329-342.
12. Mottram, D. S.; Whitfield, F. B. In *Microwave, Extruded and Maillard Generated Aromas*; McGorrin, R. J., Parliment, T. H. and Morello, M. J., Ed.; American Chemical Society: Washington DC, 1993; in press.
13. Mauron, J. In *Maillard Reactions in Food*; Eriksson, C., Ed.; Pergamon Press: Oxford, 1981; pp 3-35.
14. Leahy, M. M.; Reineccius, G. A. In *Thermal Generation of Aromas*; Parliment, T. H., McCorrin, R. J. and Ho, C. T., Ed.; American Chemical Society: Washington DC, 1989; pp 196-208.
15. Tressl, R.; Helak, B.; Martin, N.; Kersten, E. In *Thermal Generation of Aromas*; Parliment, T. H., McGorrin, R. J. and Ho, C. T., Ed.; American Chemical Society: Washington DC, 1989; pp 156-171.
16. Mottram, D. S.; Leseigneur, A. In *Flavour Science and Technology* ; Bessiere, Y. and Thomas, A. F., Ed.; Wiley: Chichester, 1990; pp 121-124.
17. Potman, R. P.; van Wijk, T. A. In *Thermal Generation of Aromas*; Parliment, T. H., McGorrin, R. J. and Ho, C. T., Ed.; American Chemical Society: Washington DC, 1989; pp 182-195.
18. Whitfield, F. B.; Mottram, D. S.; Brock, S.; Puckey, D. J.; Salter, L. J. *J. Sci. Food Agric.* **1988**, *42*, 261-272.
19. van den Ouweland, G. A. M.; Peer, H. G. *J. Agric. Food Chem.* **1975**, *23*, 501-505.

RECEIVED March 23, 1994

Chapter 16

Volatile Compounds Generated from Thermal Interactions of Inosine-5′-monophosphate and Alliin or Deoxyalliin

Tung-Hsi Yu[1], Chung-May Wu[2], and Chi-Tang Ho[1]

[1]Department of Food Science, Cook College, New Jersey Agricultural Experiment Station, Rutgers, The State University of New Jersey, New Brunswick, NJ 08903
[2]Food Industry Research and Development Institute, P.O. Box 246, Hsinchu, Taiwan, Republic of China

> Alliin and deoxyalliin, two important nonvolatile flavor precursors of garlic, were reacted separately with inosine-5′-monophosphate (IMP) in an aqueous solution at pH 7.5 in a closed sample cylinder at 180 °C for one hour. The volatile compounds generated were isolated by using a modified Likens-Nickerson (L-N) distillation-solvent extraction apparatus, and analyzed by GC and GC-MS. The isolate from the interaction of alliin and IMP possessed slightly fried garlic-like flavor with a roasted meaty character; the isolate from the interaction of deoxyalliin and IMP possessed a pungent garlic flavor with roasted notes. Several pyrazines were identified from both alliin-IMP and deoxyalliin-IMP model systems. Pyrazines, especially methylpyrazine, ethylpyrazine, and 2,5-dimethylpyrazine; thiazoles, especially 2-propylthiazole and 2-ethyl-4-propylthiazole; and pyrrole were found to be the predominant volatile interaction products of alliin and IMP. Pyrazines, especially 2,5-dimethylpyrazine and trimethylpyrazine were found to be the predominant volatile interaction products of deoxyalliin and IMP. Some thiazoles together with some pyrazines, thiophenes, ketones, furans and cyclic sulfur-containing compounds could contribute to the roasted meat-like flavor in the model systems.

γ-Glutamyl alk(en)yl cysteine dipeptides and alk(en)yl cysteine sulfoxides are two important groups of nonvolatile flavor precursors of garlic in the intact garlic cloves. During cold storage or the sprouting of the garlic cloves, γ-glutamyl alk(en)yl cysteine dipeptides can be transformed to respective alk(en)yl cysteines by the action of γ-glutamyl transpeptidase. By the action of hydrogen peroxidase, γ-glutamyl alk(en)yl cysteines can then be oxidized to the corresponding γ-glutamyl alk(en)yl cysteine sulfoxides. During the physical breakdown of the garlic cells, γ-glutamyl cysteine sulfoxides can be transformed to alk(en)yl thiosulfinates, the primary flavor compounds of garlic, by the action of alliinase through the dehydration process and accompanied by the formation of ammonia and pyruvic acid (*1-8*).

Three γ-glutamyl alk(en)ylcysteines, i.e. γ-glutamyl allyl-, γ-glutamyl (*E*)-1-propenyl-, and γ-glutamyl methyl cysteine, have been found in intact garlic cloves with the first two compounds being the predominant compounds. Three alk(en)ylcysteine sulfoxides, i.e. allyl-, (*E*)-1-propenyl-, and methyl cysteine sulfoxide, have also been found in intact garlic cloves with the first one being the predominant compound. After heating in boiling water, γ-glutamyl allylcysteine was converted to S-allylcysteine (deoxyalliin); γ-glutamyl (*E*)-1-propenylcysteine was converted to (*E*)-1- and (*Z*)-1-propenylcysteines, and alliin was completely lost in 8 hours to unknown compounds (*5,8,9*).

Most research on garlic flavor has been focused on the enzymatic generation of flavor compounds from the nonvolatile flavor precursors of garlic and the stability of these flavor compounds. In the study of fried garlic flavor, we (*10*), however, found that the flavor precursors of garlic were probably important contributors to the fried garlic character. In a model system study, we (*11-13*) also observed that the synthesized alliin and deoxyalliin could self-degrade or interact with 2,4-decadienal at 180°C and generate volatile flavor compounds. It is therefore interesting to note that the flavor precursors of garlic, which are derivatives of amino acids, especially cysteine, can participate in the formation of Maillard-type flavor compounds. By reacting glucose with alliin or deoxyalliin, we identified some Maillard-type reaction products which had a roasted meaty flavor (*14*).

Inosine-5'-monophosphate (IMP) and guanosine-5'-monophosphate (GMP) have been found in a wide range of natural foods, i.e. seafood, meats, and vegetables, and can synergistically act with monosodium glutamate (MSG) as flavor enhancers. Due to low dosage and high efficiency, IMP has been used to partially replace MSG in some foodstuffs (*15*). Due to their thermal instability and their ability to release the reducing sugar, ribose during thermal treatment, IMP and/or GMP have been used as one of the ingredients to generate Maillard-type, or meat-like flavors (*16-20*). Garlic has been widely used in food preparation. It would, therefore, be interesting to determine if the flavor precursors of garlic will interact with the nucleotides in food. Alliin, the predominant amino acid derivative of garlic, and deoxyalliin, one of the thermal degradation products of γ-glutamylallyl cysteine, and the predominant glutamyl dipeptides (*4,5,21*) were therefore synthesized and then reacted with IMP to study the potential contribution of the nonvolatile flavor precursors of garlic to the formation of Maillard-type (fried or baked garlic, and roasted meaty) flavor compounds.

Experimental

Synthesis and Purification of Deoxyalliin and Alliin. Deoxyalliin and alliin were synthesized according to the procedures of Iberl et al. (*3*) with a slight modification as shown in our previous report (*11-13*).

Thermal Interaction of IMP and Alliin or Deoxyalliin. 0.005 Mole of synthetic alliin or deoxyalliin was mixed with 0.005 mole of inosine-5'-monophosphate (Sigma, Grade V. from yeast) in 100 ml of distilled water. The solution was adjusted to pH 7.5 using 2N NaOH and then was added to a 0.3 liter Hoke SS-DOT sample cylinder (Hoke Inc., Clifton, NJ) and sealed. This cylinder was heated at 180 °C in a GC oven for 1 hour and then cooled to room temperature. A control sample was also prepared by treating a 0.005 Mole aqueous solution of IMP (adjusted to pH 7.5) exactly as described above.

Isolation of the Volatile Compounds. The total reaction mass was simultaneously distilled and extracted into diethyl ether using a Likens-Nickerson (L-N) apparatus. After distillation, 5 ml of heptadecane stock solution (0.0770 g in 200 ml diethyl ether) was added to the isolate as the internal standard. After drying over anhydrous sodium sulfate and filtering, the distillate was concentrated to about 5 ml using a Kuderna-Danish apparatus fitted with a Vigreaux distillation column. It was further concentrated under a stream of nitrogen in a small sample vial to a final volume of 0.2 ml.

Gas Chromatographic Analysis. A Varian 3400 gas chromatograph equipped with a fused silica capillary column (60 m x 0.25 mm i.d.; 1 μm thickness, DB-1, J & W Inc.) and a flame ionization detector was used to analyze the volatile compounds. The operating conditions were as follows: injector temperature, 270°C, detector temperature, 300°C; helium carrier flow rate, 1 ml/min; temperature program, 40°C (5 min), 2°C/min, 260°C (60 min). A split ratio of 50:1 was used.

Gas Chromatography-Mass Spectrometry (GC-MS) Analysis. The concentrated isolate was analyzed by GC-MS using a Hewlett-Packard 5840A gas chromatograph coupled to a Hewlett-Packard 5985B mass spectrometer equipped with a direct split interface and the same column used for the gas chromatography. The GC operating conditions were the same as described above. Mass spectra were obtained by electron ionization at 70 eV and an ion source temperature of 250°C.

Identification of the Volatile Compounds. Identification of the volatile compounds was accomplished by comparing the mass spectral data with those of authentic compounds available from the Browser-Wiley computer library, NBS computer library or previously published literature (*10-14,22*). The retention indices (using a C_5-C_{25} mixture as a reference standard) were used for the confirmation of structural assignments.

Results and Discussion

The final pH, final appearance and flavor description of the thermal reaction products of IMP and alliin, IMP and deoxyalliin, as well as thermal decomposition products of IMP, are listed in Table I. The flavor of the model system of IMP and alliin can be described as slightly fried garlic-like with a roasted meaty character. On the other hand, the flavor of the model system of IMP and deoxyalliin had a more pungent garlic note with roasted character.

Table I. Final pH, Final Appearance and Flavor Description of IMP + Alliin, IMP + Deoxyalliin, and IMP Model Reaction Systems

Model System	Final pH	Final Appearance	Flavor Description
IMP + alliin	6.7	slightly dark brown	slightly fried garlic like with roasted meaty undertone
IMP + deoxyalliin	7.2	earth-yellow	pungent garlic with roasted odor
IMP	6.8	slightly dark brown	oily, fatty, slightly meaty

The gas chromatographic profiles of the volatile compounds generated from the model reaction systems are shown in Fig. 1. The identification and quantification of the volatile compounds generated from the model systems of IMP and alliin as well as IMP and deoxyalliin are listed in Tables II and III, respectively. As shown in Fig. 1 (C), in the absence of alliin or deoxyalliin, thermal degradation of IMP produced only a few trace components.

Volatile Compounds Generated in the IMP and Alliin Model System

As shown in Table II, some volatile compounds identified from the thermal interaction of IMP and alliin were derived from the thermal degradation of alliin (11, 13), and the others were generated from the interactions of IMP and alliin.

Allyl alcohol was the predominant volatile compound found in the thermal degraded solution of alliin. 2-Formylthiophene, 3-formylthiophene, acetaldehyde, and 4-ethyl-6-methyl-1,2,3,5-tetrathiane were the other major volatile compounds derived from the thermal degradation of alliin (11). The formation of allyl alcohol from alliin could be explained by the [2,3]-sigmatropic rearrangement of alliin followed by the reduction process (11-14). In the proposed mechanisms, cysteine was also generated. Further decomposition of cysteine will lead to the formation of many compounds such as acetaldehyde, 2-formylthiophene, 3-formylthiophene, 3,5-dimethyl-1,2,4-trithiolane and 4-ethyl-6-methyl-1,2,3,5-tetrathiane identified (11-14). The presence of IMP may also contribute to the formation of 2-formylthiophene and 3-formylthiophene.

Among the interaction products of alliin and IMP, acetol and acetoin were obviously derived from ribose, one of the degradation products of IMP. As an amino acid in nature, alliin would certainly catalyze the formation of these sugar degradation products. Other interaction products of IMP and alliin were heterocyclic compounds containing nitrogen and/or sulfur atoms such as pyrazines, thiazoles, pyrroles and thiophenes.

Volatile Compounds Generated in the IMP and Deoxyalliin Model System

As shown in Table III, volatile compounds identified from the thermal interaction of IMP and deoxyalliin can be separated into two groups: those generated from the decomposition of deoxyalliin (13), and those generated from the interactions of IMP and deoxyalliin.

Diallyl sulfide, diallyl disulfide, 3,6-dimethyl-1,4-dithiacyclohexane, 3-methylthiane, and 2-ethyl-1,3-dithiane were found to be the predominant volatile compounds in the degraded solution of deoxyalliin (13). Allyl mercaptan was proposed to be generated from deoxyalliin through the hydrolysis process or free radical rearrangement. Diallyl sulfide and diallyl disulfide could then be formed from allyl mercaptan (11-14). Just like other sulfur-containing amino acids, deoxyalliin is capable of producing ammonia, hydrogen sulfide and acetaldehyde (11-14). 3,6-Dimethyl-1,4-dithiacyclohexane, 3-methylthiane, and most of the volatile compounds which were thought to derive from deoxyalliin could be the interaction products of allyl mercaptan, ammonia, aldehydes and hydrogen sulfide. The formation mechanisms of these compounds have also been reported (11-14,23).

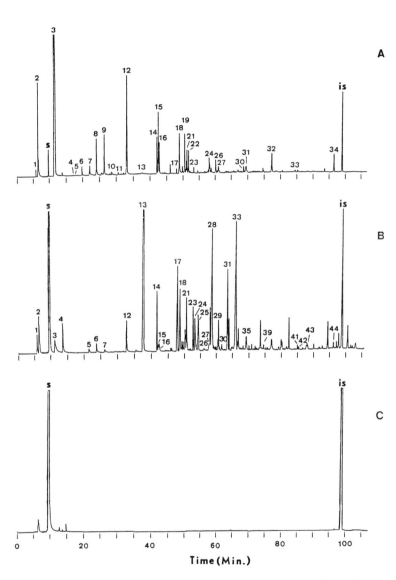

Figure 1. Gas chromatograms of volatile compounds isolated from (A) IMP + alliin, (B) IMP + deoxyalliin, and (C) IMP model systems.

Table II. Volatile Compounds Identified from the Thermal Reactions of Alliin and IMP

Peak No.[a]	Compound	MW	RI[b]	mg/mole of alliin
Compound generated from thermal degradation of alliin				
1	1-propene	42	<500	7.7
2	acetaldehyde	44	<500	162.3
3	allyl alcohol	58	549	3671.6
4	acetic acid	60	611	2.0
6	2-pentenal	84	665	33.0
10	1-mercapto-2-propanol	92	761	14.0
11	1-mercapto-3-propanol	92	780	14.4
18	2-formylthiophene	112	967	208.0
19	3-formylthiophene	112	983	188.1
23	1,4-dithiacyclohept-5-ene	132	1013	24.8
28	3,5-dimethyl-1,2,4-trithiolane	152	1125	10.0
29	3,5-dimethyl-1,2,4-trithiolane	152	1150	7.0
31	5,6-dihydro-2,4,6-trimethyl-4H-1,3,5-dithiazine	163	1199	71.7
34	4-ethyl-6-methyl-1,2,3,5-tetrathiane	198	1570	107.3
Compounds generated from thermal interactions of alliin and IMP				
5	acetol	74	640	2.2
7	acetoin	88	691	43.4
8	pyrazine	80	712	117.6
9	pyrrole	67	738	174.0
12	methylpyrazine	94	804	499.4
14	2,5-dimethylpyrazine	108	894	150.0
15	ethylpyrazine	108	898	276.4
16	2,3-dimethylpyrazine	108	901	111.7
20	trimethylpyrazine	122	988	54.7
21	2-ethyl-3-methylpyrazine	122	990	93.3
22	2-propylthiazole	127	995	104.5
24	2-ethyl-3,5-dimethylpyrazine	136	1067	65.0
25	3-ethyl-2,5-dimethylpyrazine	136	1074	25.0
26	2-ethyl-4-propylthiazole	141	1089	72.4
27	2-formyl-5-methylthiophene	126	1095	8.2
30	acetylmethylthiophene	140	1190	29.3
32	unknown [119(100), 75(17), 45(16), 43(15), 59(12), 85(10), 162(5), 166(1)][c]	-	1298	121.3
33.	3-acetyl-1,2-dithiolane	148	1405	6.6
	Total			6476.9

[a] Peak number refers to that shown in Figure 1(A); [b]Calculated Kovat's retention indices; [c]Mass spectral data, m/z (relative intensity)

Table III. Volatile Compounds Identified from the Thermal Reactions of Deoxyalliin and IMP

Peak No.[a]	Compound	MW	RI[b]	mg/mole of deoxyalliin
Compound generated from thermal degradation of deoxyalliin				
1	1-propene	42	<500	45.4
2	acetaldehyde	44	<500	113.7
3	allyl alcohol	58	540	55.2
4	allyl mercaptan	74	580	72.3
6	ethyl acetate	88	601	8.5
9	1-mercapto-2-propanol	92	761	2.0
10	1-mercapto-3-propanol	92	780	1.6
13	allyl sulfide	114	853	1394.0
17	3-methylthiane	116	957	335.9
18	2-formylthiophene	112	965	197.5
19	3-formylthiophene	112	974	20.9
23	1,3-dithiane	120	1006	105.5
24	(allythio)acetic acid	132	1012	137.6
25	2-methyl-1,3-dithiane	134	1024	104.1
26	(allylthio)propanol	132	1036	10.0
28	allyl disulfide	146	1073	563.8
29	1,4-dithiepane	134	1094	66.0
30	2-(mercaptoethyl)tetrahydrothiophene	148	1107	19.7
31	2-ethyl-1,3-dithiane	148	1129	238.3
32	3,6-dimethyl-1,4-dithiacyclohexane	148	1034	62.0
33	3,6-dimethyl-1,4-dithiacyclohexane	148	1060	573.6
35	5,6-dihydro-2,4,6-trimethyl-4H-1,3,5-dithiazine	163	1195	57.4
36	1,5-dithiacyclooctane	148	1216	13.9
37	5,6-dihydro-2,4,6-trimethyl-4H-1,3,5-dithiazine	163	1234	5.2
38	3,5-dimethyl-1,2,4-trithiane	166	1261	2.6
40	1,2,5-trithiacyclooctane	166	1273	4.8
41	4,6-dimethyl-1,2,5-trithiepane	180	1401	10.3
43	4,6-dimethyl-1,2,5-trithiepane	180	1444	16.2
44	4-ethyl-6-methyl-1,2,3,5-tetrathiane	198	1564	16.2
Compounds generated from thermal interactions of deoxyalliin and IMP				
6	pyrazine	80	708	17.4
7	pyrrole	67	738	7.0

Continued on next page

Table III. continued

Peak No.[a]	Compound	MW	RI[b]	mg/mole of deoxyalliin
11	2,4,5-trimethyloxazoline	113	791	1.2
12	methylpyrazine	94	804	62.5
14	2,5-dimethylpyrazine	108	890	121.5
15	ethylpyrazine	108	895	21.6
16	2,3-dimethylpyrazine	108	897	14.1
20	2-ethyl-6-methylpyrazine	122	980	47.5
21	trimethylpyrazine	122	985	129.7
22	2-ethyl-3-methylpyrazine	122	987	16.8
27	3-formyl-2-methylthiophene	126	1062	16.1
34	acetylmethylthiophene	140	1185	7.4
39	2-ethyl-3-formyldihydrothiophene	142	1264	9.3
42	2-isobutyl-5-propylthiophene	182	1407	5.1
	Total			4732.1

[a] Peak number refers to that shown in Figure 1(B); [b] Calculated Kovat's retention indices

The interaction products of IMP and deoxyalliin shown in Table III were mainly the typical heterocyclic Maillard reaction products such as pyrazines and thiophenes.

Comparison of Volatile Compounds Generated from IMP and Alliin As Well As IMP and Deoxyalliin Model Systems.

The comparison of the yields of volatile compounds generated from the model systems is shown in Table IV. The major differences between these two systems were that the formation of thiazoles was favorable in the IMP and alliin system and the IMP and deoxyalliin system favored the formation of allylthio-containing compounds.

The abundance of thiazoles, especially 2-propylthiazole and 2-ethyl-4-propylthiazole, in the IMP and alliin systems was thought to be the reason that the roasted meaty character in the IMP and alliin model system is stronger than that in IMP and deoxyalliin systems. 2-Propylthiazole was reported to possess green, herbal and nutty odors (24). These thiazole compounds, together with other volatile compounds generated in IMP and deoxyalliin, could then contribute to the roasted meaty character of the reaction products.

On the other hand, the abundance of allylthio-containing volatile compounds

Table IV. Comparison of the Yields of Volatile Compounds Generated from IMP + Alliin, and IMP + Deoxyalliin Model Reaction Systems

Compounds	Concentration (mg/mole of alliin or deoxyalliin)	
	IMP + alliin	IMP + deoxyalliin
Compounds degraded from alliin or deoxyalliin	4521.9	4254.2
Compounds generated from the interactions of IMP and alliin or deoxyalliin		
Nitrogen-containing compounds	1567.1	438.8
Nitrogen and sulfur-containing compounds	176.9	0.0
Sulfur-containing compounds	44.1	37.9
Others	166.9	1.2
Total	6476.9	4732.1

and the lack of thiazole compounds in IMP and deoxyalliin systems may contribute to the strong pungent garlic odor of the reaction products. Some of the allylthio-containing volatile compounds, especially diallyl sulfide and diallyl disulfide, have been identified in the volatile flavor of garlic products and were believed to be responsible for the characteristic heated garlic flavor of thermally-treated garlic products(22,23,25).

As shown in Table IV, many pyrazine compounds were generated in IMP and deoxyalliin as well as in the IMP and alliin model systems. Pyrazines are widely distributed in thermally processed foods, such as roasted beef, roasted peanut, and roasted barley. The Maillard reaction, which involves the interactions of reducing carbonyl compounds and amino-containing compounds, has been shown to be the major mechanism for the formation of pyrazines (26).

Acknowledgements

New Jersey Agricultural Experiment Station Publication No. D-10205-4-93 supported by State Funds. We thanks Mrs. Joan Shumsky for her secretarial aid.

Literature Cited

1. Fenwick, G.R.; Hanley, A.B. *CRC Crit. Rev. Food Sci. Nutr.* **1985**, *22*, 273-340.

2. Carson, J.F. *Food Rev. Internat.* **1987**, *3*, 71-103.
3. Iberl, B.; Winkler, G.; Muller, B.; Knobloch, K. *Planta Med.* **1990**, *56*, 320-326.
4. Lawson, L.D.; Wood, S.G.; Hughes, B.G. *Planta Med.* **1991**, *57*, 263-270.
5. Lawson, L.D.; Wang Z-Y J.; Hughes, B.G. *J. Nat. Prod.* **1991**, *54*, 436-444.
6. Block, E.; Naganathan, S.; Putman, D.; Zhao, S.-H. *J. Agric. Food Chem.*, **1992**, *40*, 2418-2430.
7. Block, E.; Putman, D.; Zhao, S.-H. *J. Agric. Food Chem.*, **1992**, *40*, 2431-2438.
8. Block, E. *Angew. Chem.* **1992**, *31*, 1135-1178.
9. Block E.; Naganathan S.; Putman D.; Zhao S.-H. *Pure & Appl. Chem.* **1993**, 65, 625-632.
10. Yu, T.H.; Wu, C.M.; Ho, C.T. *J. Agric. Food Chem.* **1993**, 41, 800-805.
11. Yu, T.H.; Shu, C.K.; Ho, C.T. In *Food Phytochemicals for Cancer Prevention I: Fruits and Vegetables*; Huang, M. T.; Osawa, T.; Ho, C.-T.; Rosen, R. T., Eds.; ACS Symp. Ser. No. 546, American Chemical Society: Washington, DC, 1994, pp. 144-152.
12. Yu, T.H.; Lee, M.H.; Wu, C.M.; Ho, C.T. In *Lipids in Food Flavors*, Ho, C.-T.; Hartman, T. G., Eds.; ACS Symp. Ser.; American Chemical Society: Washington, DC, 1994, in press.
13. Yu, T.H.; Wu, C.M.; Rosen, R.T.; Hartman T.G.; Ho, C.T. *J. Agric. Food Chem.* **1994**, in press.
14. Yu, T.H.; Wu, C.M.; Ho, C.T. *J. Agric. Food Chem.* **1994**, in press.
15. Duff, M. *FFIPP.* **1980**, 2 -23.
16. Shaoul, O.; Sporns, P. *J. Food Sci.* **1987**, *52*, 810-812.
17. Nguyen, T.T.; Sporns, P. *J. Food Sci.* **1985**, *50*, 812-814.
18. Matoba, T.; Kuichiba, M.; Kimura, M.; Hasegawa, K. *J. Food Sci.* **1988**, *53*, 1156-1160.
19. MacLeod, G. In *Food Flavours*; Birch, G. G.; Lindley, M. G., Eds.; Elsevier Applied Science: London, 1986, pp 191-223.
20. May, C.G. In *Food Flavourings*; Ashurst, P.R., Ed.; AVI: New York, 1991. pp 257-286.
21. Ueda, Y.; Kawajird, H.; Muyamura, N. *Nippon Shokuhin Kogyo Gakkaishi*, **1991**, *38*, 429-434.
22. Yu, T.H.; Wu, C.M.; Liou, Y.C. . *J. Agric. Food Chem.* **1989**, *37*, 725-730.
23. Block, E.; Iyer, R.; Grisoni, S.; Saha, C.; Belman, S.; Lossing F.P. *J. Amer. Chem. Soc.* **1988**, *110*, 7813-7827.
24. Pittet, A.O.; Hruza, D.E. *J. Agric. Food Chem.* **1974**, 22, 264-269.
25. Lawson L.D.; Wang, Z.-Y.J.; Hughes, G. *Planta Med.* **1991**, *57*, 363-370.
26. Mega, J.A. In *Food Flavours-Part A. Introduction*; Morton, I.D.; Macleod, A.J., Eds.; Elsevier Scientific Publ. Co.: New York, USA. **1982**, P. 283-318.

RECEIVED March 23, 1994

Chapter 17

Thermal Degradation of Thiamin (Vitamin B_1)

A Comprehensive Survey of the Latest Studies

Matthias Güntert[1], H.-J. Bertram[1], R. Emberger[2], R. Hopp[1], H. Sommer[1], and P. Werkhoff[1]

[1]Corporate Research and [2]Flavor Division, Haarmann & Reimer GmbH, 37603 Holzminden, Germany

The thermal degradation of thiamin (vitamin B_1) is a very complex reaction consisting of various degradation pathways. This results in the formation of many organoleptically interesting flavor compounds. Most of them contain one or more sulfur and/or nitrogen atoms, and many of them are heterocyclic structures.
Aqueous solutions of pure thiamin hydrochloride as well as mixtures of thiamin hydrochloride with cysteine hydrochloride and of thiamin hydrochloride with methionine were heated in an autoclave. The resulting flavor compounds were obtained by the simultaneous distillation/extraction procedure according to Likens and Nickerson. The concentrates were preseparated by medium-pressure liquid chromatography on silica gel using a pentane-diethyl ether gradient. The different fractions were subsequently analyzed by capillary gas chromatography (HRGC) and capillary gas chromatography-mass spectrometry (HRGC/MS).
Various unknown compounds were isolated by preparative capillary gas chromatography in microgram-quantities in order to elucidate their structures by IR and NMR spectroscopy, and to check their olfactory properties. The spectroscopic data and sensory impressions are given. Many of the analyzed flavor compounds were synthesized. In particular, a series of new thiophenes is presented. Their formation led to the explanation of one of the main hitherto unknown degradation pathways of thermally treated thiamin.

The thermal generation of flavor is a very essential process for the "taste" of many different foodstuffs, e.g. cocoa, coffee, bread, meat. The resulting aromas are formed through non-enzymatic reactions mainly with carbohydrates, lipids, amino acids (proteins), and vitamins under the influence of heat. Thiamin (vitamin B_1) and the amino acids, cysteine and methionine, belong to those food constituents which act as flavor precursors in thermal reactions. The role of thiamin as a potent flavor precursor is related to its chemical structure which consists of a thiazole as well as a pyrimidine moiety. The thermal degradation of this heterocyclic constituent leads to very reactive intermediates which are able to react directly to highly odoriferous flavor compounds or with degradation products of amino acids or carbohydrates.

The thermal degradation of thiamin has long been a matter of analytical investigations. A comprehensive literature survey was given by Güntert et al. [1]. In addition, Clydesdale et al. [2] published a review article about "The effects of postharvest treatment and chemical interactions on the bioavailability of ascorbic acid, thiamin, vitamin A, carotenoids, and minerals". The part therein about thiamin gives an overview of the known literature. In the recent past, Grosch and co-workers began to investigate the flavor of thermally treated foods. In this context they studied the influence of thiamin as a possible precursor for the formation of the important meat flavor compound, 2-methyl-3-furanthiol, and its oxidized form bis-(2-methyl-3-furyl)disulfide [3,4]. The newest results concerning thermally degraded thiamin were published by Ames et al. [5,6]. Other recent articles deal with the flavor compounds of yeast [7,8]. Since thiamin is an important constituent of yeast (1.5mg/100g), its spectrum of flavor compounds can be significantly influenced by the thiamin chemistry.

Thiamin plays an essential role in different foods as a water soluble vitamin. Additionally, its function as a flavor precursor in heated foods, e.g. meat, should not be neglected. But certainly, this aspect depends very much on its amount and the specific conditions in the food system. Another important field in which thiamin plays a remarkable role is the application of flavorings. Along with carbohydrates, amino acids, ribonucleotides, and other constituents, thiamin is widely used as a flavor precursor. This fact is clearly demonstrated by many patented reaction or processed flavors.

It has been of great interest for us to study the very complex reaction pathways and to identify the resulting chemical compounds. Moreover, we were interested to study the interactions between thiamin and cysteine as well as thiamin and methionine when thermally treated. Our recent investigations on the thermal degradation of thiamin itself [1,9,10] and in reaction with the amino acids, cysteine and methionine [11], led to the identification of numerous new flavor compounds and the explanation of several degradation pathways. In the present study, one of the hitherto missing and unknown main pathways is presented along with the resulting compounds.

Experimental Section
Materials.
1) Thiamin hydrochloride (1.05 mol) in 3.5 L distilled water (pH 2.7) was heated in an autoclave to 130°C for 6h.
   Sensory impression: meaty, pungent, sour, rubber, sulfury.
2) A mixture of thiamin hydrochloride (2.77 mol) and cysteine hydrochloride (2.77 mol) in 14 L distilled water (pH 1.3) was heated in an autoclave to 130°C for 6h.
   Sensory impression: pungent, meaty, smoky, burnt.
3) A mixture of thiamin hydrochloride (1.75 mol) and methionine (3.5 mol) in 14 L distilled water (pH 3.3) was heated in an autoclave to 130°C for 6h.
   Sensory impression: cabbage, roasty, burnt, sour, pungent, sulfury.

The subsequent isolation of flavor volatiles by simultaneous distillation/extraction and the preseparation into 6 fractions by medium pressure liquid chromatography as well as the analytical and preparative conditions (capillary gas chromatography, spectroscopy) were previously published [1,9].

Infrared (IR) and nuclear magnetic resonance (NMR) analysis.
Infrared spectra of isolated samples were obtained in CCl_4 using a Perkin Elmer 983G type instrument.
NMR spectra of collected and of synthesized samples were obtained at 400 MHz in $CDCl_3$, C_6D_6 or C_6D_{12} on a Varian VXR-400 instrument with $Si(CH_3)_4$ as internal standard.

The Nuclear Overhauser Difference Spectroscopy (NOED) spectra were recorded with an average of more than 512 scans, with 64k data points each. Irradiation duration was always 10 seconds and the temperature was 25°C. The solvents used were $CDCl_3$ and C_6D_6, with $Si(CH_3)_4$ as the internal standard. Residual oxygen was purged from the samples before measurement using the standard freeze-thaw cycle method. For calculating the structures a force field program was used [12].

Component identification.
Sample components were identified by comparison of the compound's mass spectrum and Kovats index with those of a reference standard. In many cases reference compounds were synthesized in our laboratory. In all cases the respective structures were confirmed by NMR, MS and IR spectroscopy.

Sensory evaluation.
The olfactory evaluation of selected compounds was performed by an expert panel of flavorists. The synthesized compounds were evaluated in water at certain concentrations. The isolated samples were dissolved in ethanol and tested on a smelling blotter.
In addition, the various extracts were judged for their smell. They were also investigated organoleptically by carrying out GC-sniffing.

Results and Discussion
Different aspects of the thermal degradation of thiamin (thiamin hydrochloride) have already been investigated by our research group [1,9-11]. In Figure 1 the primary degradation products are shown. In particular, 5-hydroxy-3-mercapto-2-pentanone is a very reactive and important intermediate. It serves as a precursor mainly for numerous S-containing flavor compounds some of which possess remarkable olfactory properties. The role of this constituent was shown by us in many different examples [1,9-11]. Another important degradation product is 4-methyl-5-(2-hxydroxyethyl)thiazole, the so-called sulfurol. Other main degradation products are 3-mercaptopropanol and 4-amino-5-(aminomethyl)-2-methylpyrimidine as well as the basic chemicals hydrogen sulfide, ammonia, formic acid, acetic acid, formaldehyde, and acetaldehyde.
An interesting new degradation compound whose existence we strongly suggest is 3-mercaptopropanal. This very reactive and volatile aldehyde seems to exist only as monomeric species in the gas phase [13]. Therefore, we were not able to prove its existence by means of a reference compound. But since we very often get a mass spectrum of an unknown sulfur compound with a moleculer weight of 90 at very low k'-values in our extracts, the evidence became fairly strong. The formation of many new compounds identified as degradation compounds of thermally treated thiamin is explainable by the reaction of 3-mercaptopropanal with other degradation products, e.g. 5-hydroxy-3-mercapto-2-pentanone. A summary of these new flavor compounds is given in Figure 2. All of these S-containing heterocycles were identified as thermal degradation compounds of pure thiamin hydrochloride and none of them - with the exception of **9** - is known in the scientific literature [14]. Some of these are major constituents, i.e. 2-ethyl-4-formyl-3-methylthiophene **3** (**29**)*, 1-methyl-3-(2-ethyl-3-methyl-4-thienyl)-2,8-dioxa-4-thiabicyclo[3.3.0]octane **4a/b** (**44**)*, and 1-acetyl-6-oxa-2-thiabicyclo[3.3.0]octane **7** (**31**)*. The numbers in parentheses marked with an asterisk were those used in our previous paper as still unknown structures [11, Figure 6A]. In the meantime we finally succeeded in elucidating the structures of almost all of the main degradation compounds of thermally treated thiamin in our studies. Only one of the main compounds which was numbered (**18**)* in our paper [11, Figure 6A] still remains unknown. Our attempts to isolate this

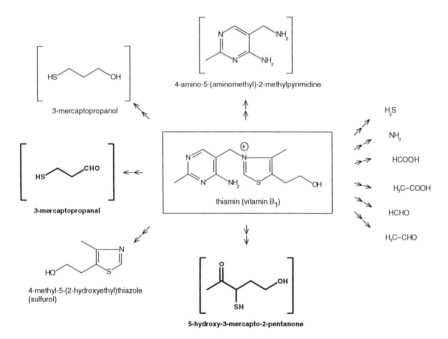

Figure 1: Primary degradation compounds of thermally treated thiamin.

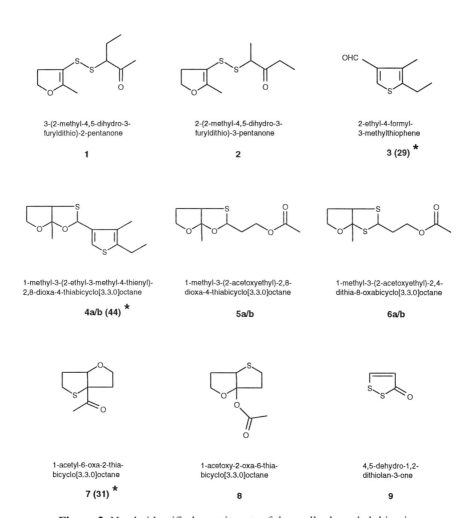

Figure 2: Newly identified constituents of thermally degraded thiamin.

S-containing volatile (molecular weight 114) by preparative GC failed since the substance decomposed. It is very interesting to note that almost any of the shown compounds in Figure 2, with the exception of 1-acetoxy-2-oxa-6-thiabicyclo[3.3.0]-octane **8** and 4,5-dehydro-1,2-dithiolan-3-one **9**, are formed via at least one part of 5-hydroxy-3-mercapto-2-pentanone. This is demonstrated in Figures 3 - 5 where the probable degradation pathways for the formation of the structures in Figure 2 are postulated. Figure 3 depicts the probable degradation pathways for the formation of 3-(2-methyl-4,5-dihydro-3-furyldithio)-2-pentanone **1** and 2-(2-methyl-4,5-dihydro-3-furyldithio)-3-pentanone **2**. Interestingly, these two dihydrofurans fit very well into a series of new heterocyclic S-containing flavor compounds which were described by us in two previous papers [10,11]. One remarkable reaction mechanism (Figure 4) leads to one of the main compounds, 2-ethyl-4-formyl-3-methylthiophene **3** (approximately 5% relative GC area). Apparently, this thiophene is formed by the reaction of 5-hydroxy-3-mercapto-2-pentanone and 3-mercaptopropanal, probably in an intermolecular two step mechanism consisting of an aldol condensation and a condensation between the two SH groups. This interesting reaction type between two carbonyl compounds plays an important role in our studies and is discussed for further in the next section. The substitution pattern of 2-ethyl-4-formyl-3-methylthiophene **3** was elucidated by application of Nuclear Overhauser Difference Spectroscopy (NOED) in the isolated sample. The results are shown in Figure 6. Irradiation of the signal of the formyl group produced Nuclear Overhauser effects at the aromatic methine proton and the methyl group (Figure 6, a, •A and •B), while saturation of the aromatic methine proton resulted in a signal intensification only of the carbonyl proton (Figure 6, b, •C). Irradiation of the signal of the methylene protons produced an NOE at both methyl groups (Figure 6, c, •D and •E), and saturation of the methyl group generated signal intensifications of the carbonyl proton, the methylene protons, and the methyl protons (Figure 6, d, •F, •G, and •H). Using this information, this thiophene could unequivocally be identified as 2-ethyl-4-formyl-3-methylthiophene **3**. The final confirmation was obtained by its synthesis. The ^{1}H-NMR and mass spectra of **3** are depicted in Figure 7. A second heterocycle shown in Figure 4 which is formed by the reaction of 5-hydroxy-3-mercapto-2-pentanone and 3-mercaptopropanal is 1-acetyl-6-oxa-2-thiabicyclo[3.3.0]octane **7**. It also belongs to the hitherto unknown main degradation compounds of pure thiamin hydrochloride (approximately 4% relative GC area). Its ^{1}H-NMR and mass spectra are shown in Figure 7. The third new structure of Figure 4 is the bicyclo[3.3.0]octane compound **4a/b**. Its formation can be easily understood by the reaction of 5-hydroxy-3-mercapto-2-pentanone with the above described thiophene **3**. It shows up as two well separated diastereomers and accounts for approximately 1% (relative GC area). We were able to distinguish for the first time between the two diastereomeric forms of this bicyclo[3.3.0]octane and also of many other compounds of this type using Nuclear Overhauser Difference Spectroscopy (NOED). The ^{1}H-NMR and mass spectra of **4a/b** are shown in Figure 8. The NMR spectrum presents a mixture of the two diastereoisomers of **4** (approximately 28% of 1R*,3S*,5R* and 64% of 1R*,3R*,5R*) and the thiophene **3** (approximately 8%). Apparently, 2-ethyl-4-formyl-3-methylthiophene **3** was regenerated as an artifact in small amounts in the CDCl$_3$ solution in the NMR measuring tube. It can be easily seen in the ^{1}H-NMR spectrum (Figure 8) by means of its carbonyl proton at 9.909ppm. The ratio of the two diastereomers of compounds of this class can be very quickly recognized in the ^{1}H-NMR spectra by the signals of the protons at C-3. Depending upon whether one deals with the 1R*,3S*,5R* or with the 1R*,3R*,5R* isomer the respective proton signals at C-3 are significantly separated whereby the signal belonging to the 1R*,3R*,5R* isomer is shifted down-field. Figure 5 shows further examples of new bicyclo[3.3.0]octanes, 1-methyl-3-(2-acetoxyethyl)-2,8-dioxa-4-thiabicyclo[3.3.0]-octane **5a/b** and 1-methyl-3-(2-acetoxyethyl)-2,4-dithia-8-oxabicyclo[3.3.0]octane **6a/b** each of which also shows up in two diastereomeric forms. Apparently, the

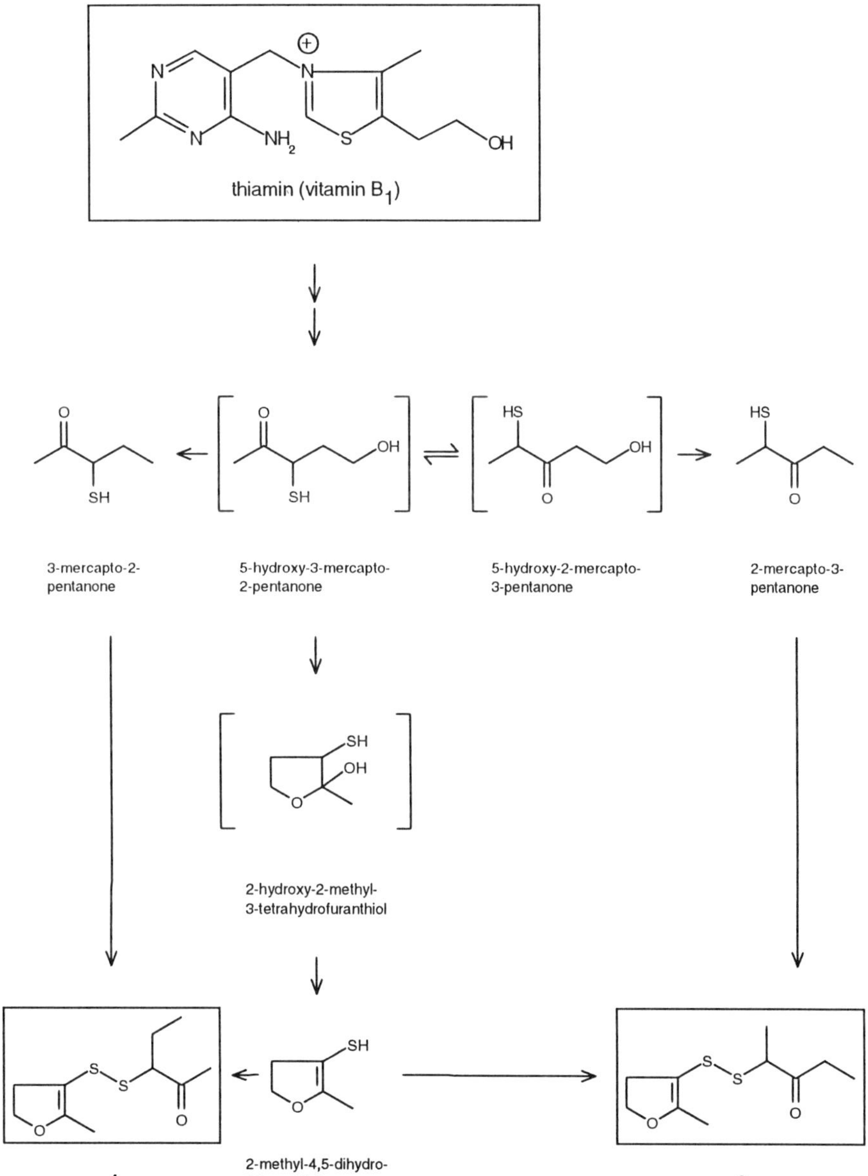

Figure 3: Probable formation pathways of the newly identified compounds **1** and **2**.

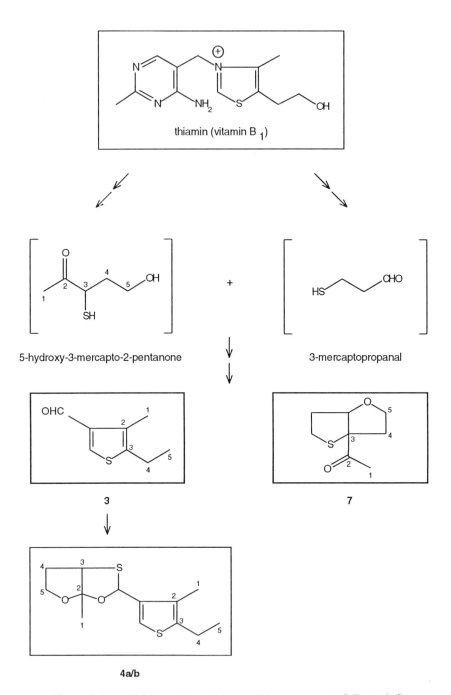

Figure 4: Probable formation pathways of the compounds **3**, **7** and **4a/b**.

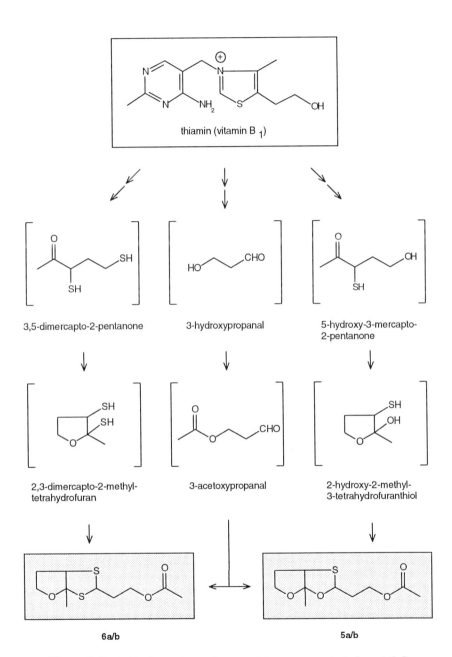

Figure 5: Probable formation pathways of the compounds **5a/b** and **6a/b**.

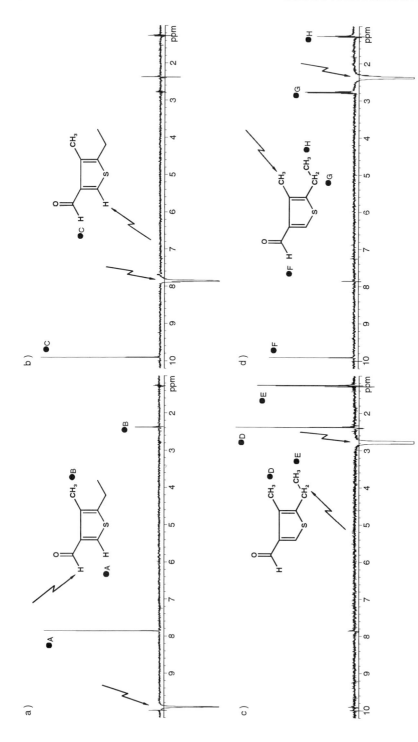

Figure 6: NOED spectra for structure elucidation of 2-ethyl-4-formyl-3-methylthiophene **3** showing signal intensifications from irradiation of **a)** formyl group, **b)** aromatic methine proton, **c)** methylene protons, and **d)** methyl group.

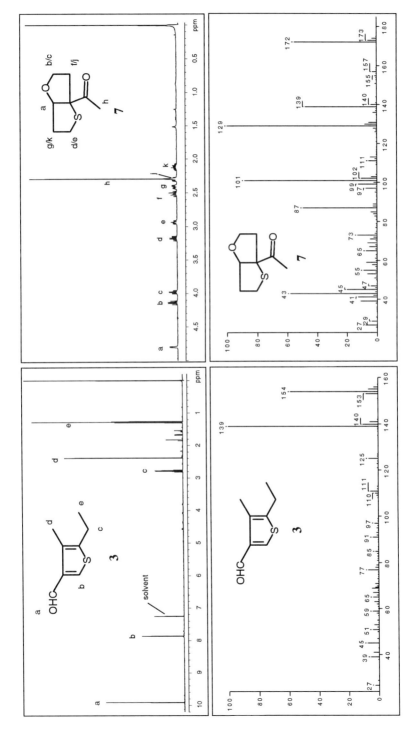

Figure 7: ^{1}H-NMR (CDCl$_3$) and mass spectra of 2-ethyl-4-formyl-3-methylthiophene **3** and 1-acetyl-6-oxa-2-thiabicyclo[3.3.0]octane **7**.

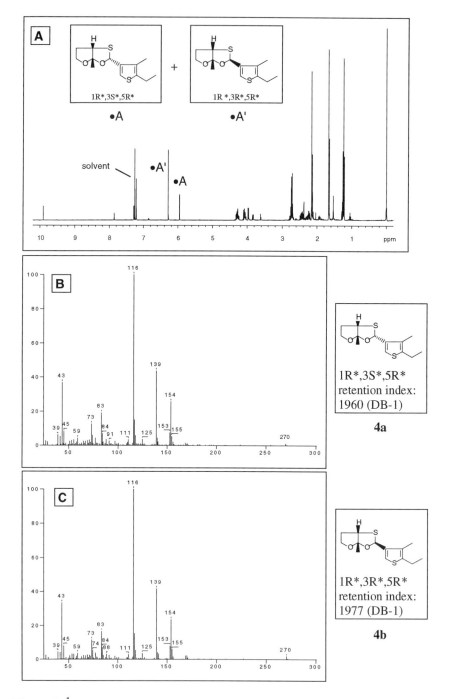

Figure 8: ^{1}H-NMR (A) and mass spectra (B,C) of the diastereoisomers of 1-methyl-3-(2-ethyl-3-methyl-4-thienyl)-2,8-dioxa-4-thiabicyclo[3.3.0]octane **4a/b**.

formation of these compounds requires the existence of 3-hydroxypropanal, an O-analog of the above described 3-mercaptopropanal. Figure 9 summarizes all of the bicyclo[3.3.0]octane compounds identified to date in our reactions. The left side of Figure 9 shows the aldehydes which react with 5-hydroxy-3-mercapto-2-pentanone to yield the identified bicyclics. Most of these were presented by us in previous papers [1,9-11]. All of them are formed during the thermal degradation of thiamin in unspecific mixtures of their diastereomeric forms.

In all cases, these newly identified flavor compounds of thermally degraded pure thiamin (see Figures 2-9) were structurally confirmed by nuclear magnetic resonance (NMR), mass, and infrared (IR) spectra. The sensory properties of many of these compounds are presented in Table I.

The thermal reactions of thiamin and cysteine as well as of thiamin and methionine, were described for the first time in our previous paper [11]. It was demonstrated that in particular the reaction of thiamin with cysteine changed the qualitative and quantitative pattern of the resulting flavor compounds drastically. Hydrogen sulfide, mercaptoacetaldehyde, and cysteamine (from cysteine) as well as methanethiol and methional (from methionine) are the main volatile degradation compounds of these two amino acids. They react further either with themselves or with the degradation compounds of thiamin. Figure 10 depicts a few compounds from the thermal reaction of thiamin with cysteine. Very remarkable is the formation of the three new thiophenes, 3-(2-hydroxyethyl)-2-methylthiophene **10**, 2-(2-hydroxyethyl)-3-methylthiophene **11** (**27**)*, and 2-(3-hydroxypropyl)thiophene **12**. The probable formation pathway of these thiophenes is shown in Figure 11. Apparently, these thiophenes are further examples of the above proposed two step mechanism consisting of an aldol condensation and a condensation between the two *SH* groups. Not surprisingly, mercaptoacetaldehyde, as one of the key degradation products of cysteine, serves as one precursor while the other should be 5-hydroxy-2-pentanone and/or 1-hydroxy-3-pentanone. In our previous paper [11] several reaction products of thiamin and cysteine remained unidentified. The unknown structure of number (**18**)* was already discussed while number (**27**)* belongs to 2-(2-hydroxyethyl)-3-methylthiophene **11**. Thiophene **11** is not known in the literature. Thiophenes **10** and **12** are listed in Chemical Abstracts Service (CAS), but not in a flavor chemistry context. Other new degradation products of the thermal reaction of thiamin and cysteine shown in Figure 10 are 2,3-dimethyl-2,3-dehydrothian-4-one **13**, the two diastereomers of bis-2-[2-methyl-3(2H)]thiophenone **14a/b**, and the two diastereoisomers of 3-mercapto-5-methyl-4,5-dihydro-2(3H)thiophenone **15a/b**. The possible formation pathways of **13** and **15a/b** are depicted in Figure 12. Obviously, both compounds are formed under the influence of acetaldehyde. While **13** seems to be formed from thiamin, i.e. from acetaldehyde and 1,4-dimercapto-3-pentanone, the formation of **15a/b** may proceed via 2-mercaptopropanoic acid which is a known degradation product of cysteine [11]. The sensory properties of some of these newly identified degradation compounds of thiamin and cysteine are presented in Table I.

In Figure 13 various degradation compounds of the thermal reaction of thiamin and methionine are presented. Fortunately, we were able to identify most of the main volatile compounds. The then unknown structures in our previous paper [11, Figure 6A] with numbers (**29**)* and (**31**)* correspond to numbers **3** and **7** above. Two other main compounds, 2-(2-hydroxyethyl)-3-methyl-4-(methylthiomethyl)thiophene **19** (**42**)* and 2-(2-methylthioethyl)-3-methyl-4-formylthiophene **20** (**41**)*, which account for approximately 1% and 0.6% (relative GC area) were numbered in [11, Figure 6A] as (**42**)* and (**41**)*. Their ^{1}H-NMR and mass spectra are depicted in

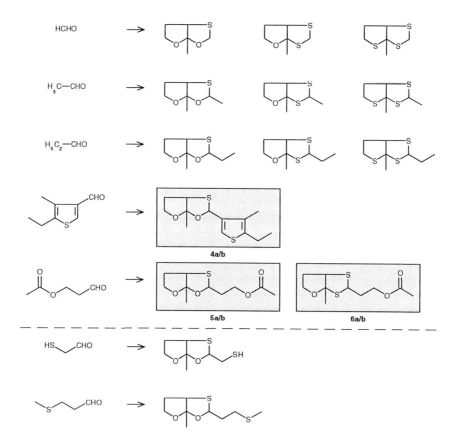

Figure 9: Survey of the bicyclo[3.3.0]octanes identified up to now within our work.

Table I: Newly identified flavor compounds with their sensory properties

#	Sensory properties	#	Sensory properties
3	green, rubbery, naphthalin, coumarin	15a/b	roasty, coffee, rubbery
4a/b	coumarin, phenolic	16	garlic, rubbery, phenolic
5a/b	potato, thiamin, milk	17	garlic, phenolic, sulfury
6a/b	sulfury, meaty, peanut, peas	18	cabbage, phenolic, rubbery
7	vegetables, onion, roasty	19	potato, methional, cabbage, fishy, fatty
8	potato, sulfury, meaty, cabbage	20	sulfury, metallic, hazelnut, mushroom
12	earthy, dusty, potato, sulfury	22	cabbage, potato, sulfury
17	chemical, onion, pungent, coumarin	23	sulfury, rubbery, onion, roasty
14a/b	chemical, phenolic, smoky, tar	24	sulfury, cabbage, meaty, rubbery

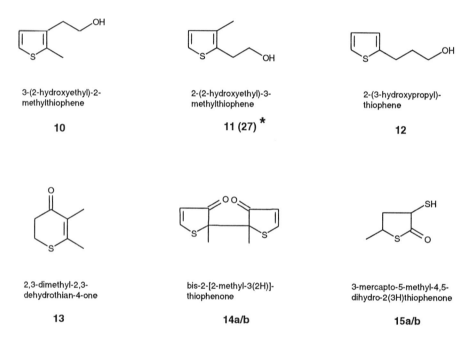

Figure 10: Newly identified constituents in the thermal reaction of thiamin with cysteine.

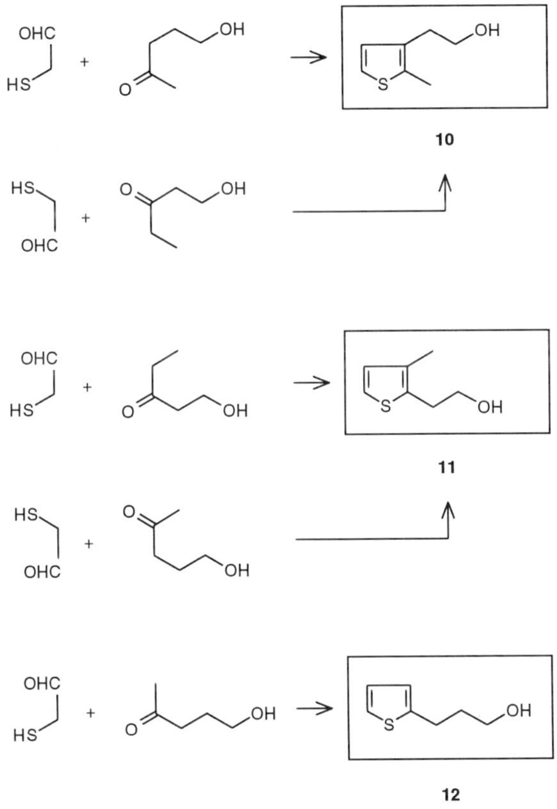

Figure 11: Probable formation pathways of the thiophenes **10**, **11** and **12**.

216

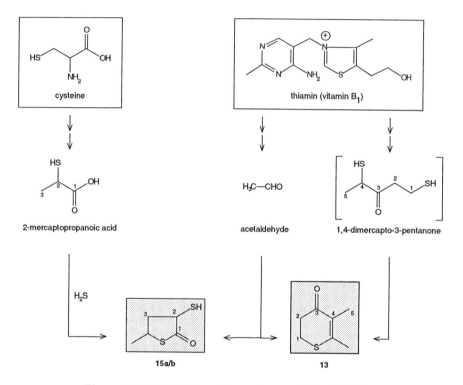

Figure 12: Probable formation pathways of **13** and **15a/b**.

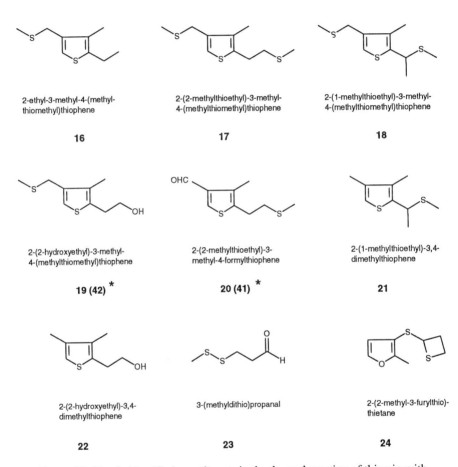

Figure 13: Newly identified constituents in the thermal reaction of thiamin with methionine.

Figure 14. It is very interesting to note that in the reaction of thiamin with methionine another new series of thiophenes was identified. These are 2-ethyl-3-methyl-4-(methylthiomethyl)thiophene **16**, 2-(2-methylthioethyl)-3-methyl-4-(methylthiomethyl)thiophene **17**, 2-(1-methylthioethyl)-3-methyl-4-(methylthiomethyl)thiophene **18**, 2-(1-methylthioethyl)-3,4-dimethylthiophene **21**, and 2-(2-hydroxyethyl)-3,4-dimethylthiophene **22**. Moreover, we were successful in elucidating the structures of 3-(methyldithio)propanal **23** and 2-(2-methyl-3-furylthio)thietane **24**. None of these flavor compounds is mentioned in the literature except 3-(methyldithio)propanal **23** which was described once in a flavor patent [*15*]. The formation of these new thiophenes is depicted in Figure 15. Those with a methylthiomethyl side chain contain methional, a major degradation product of methionine, in their structures. Other precursor aldehydes are propanal and 3-mercaptopropanal. The second part for the formation of these thiophenes originates from thiamin through the postulated C5-compounds. The formation of **23** and **24** (Figure 13) is directly correlated with and supports the existence of 3-mercaptopropanal. The reaction of sulfides with methanethiol and the direct methylation of sulfides are among the main reactions occurring in the thermal reaction of thiamin and methionine [*11*]. Therefore, the identification of 3-(methyldithio)propanal **23** in this reaction mixture is not surprising. Likewise, the formation of 2-(2-methyl-3-furylthio)thietane **24** can be easily explained by an acetalization reaction between 2-methyl-3-furanthiol and 3-mercaptopropanal. The sensory properties of some of these newly identified thermal degradation compounds of thiamin and methionine are presented in Table I.

The last topic regards the occurrence of the structures shown in Figure 16. These heterocycles are among the flavor compounds formed during the thermal reaction of thiamin and methionine. Apparently, they are formed from the well-known thermal degradation products of degraded thiamin, 2-methyl-3-furanthiol and 2-methyl-3-thiophenethiol, by reaction with methanethiol and/or direct radical methylation. We were interested in comparing the chemical and organoleptic properties of these furans and thiophenes. While the two methyltrithio compounds, 2-methyl-3-(methyltrithio)furan **29** and 2-methyl-3-(methyltrithio)thiophene **30**, were never mentioned in the literature so far, their methyldithio and methylthio counterparts were described and discussed several times. The two furans, 2-methyl-3-(methylthio)furan **25** and 2-methyl-3-(methyldithio)furan **27**, have been the subject of controversial statements regarding their sensory properties. This was discussed at length by Werkhoff et al. [*16*]. The two thiophenes, 2-methyl-3-(methylthio)thiophene **26** and 2-methyl-3-(methyldithio)thiophene **28**, were described in our previous paper [*11*]. The syntheses of these six heterocycles (Figure 17) were principally described already by Bertram et al. [*17*]. Purification was by distillation and/or preparative chromatographic methods. It is worthwhile mentioning that the stability of these compounds is reduced with the increasing number of S-atoms in the side chain. We were unable to obtain pure 2-methyl-3-(methyltrithio)furan **29** or 2-methyl-3-(methyltrithio)thiophene **30**. These compounds decomposed during the chromatographic purification. The main decomposition products were formed by disproportionation reactions. Therefore, the sensory impressions of **29** and **30** in Figure 16 were not determined from pure compounds but by GC sniffing (marked by an *). Our results indicate that the methylthio compounds have more interesting organoleptic impact than their methyldithio and methyltrithio counterparts. But interestingly, the taste threshold of 2-methyl-3-(methyldithio)thiophene **28** is extremely low (1ppt in water). In particular, the comparison of the thresholds of the two respective furan and thiophene derivatives, 2-methyl-3-(methyldithio)furan **27** and 2-methyl-3-(methyldithio)thiophene **28**, seems to be very surprising and remarkable.

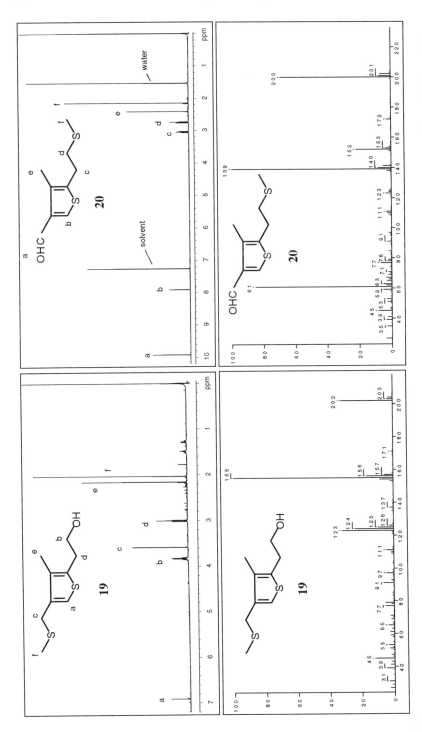

Figure 14: ^{1}H-NMR (CDCl$_3$) and mass spectra of 2-(2-hydroxyethyl)-3-methyl-4-(methylthiomethyl)thiophene **19** and 2-(2-methylthioethyl)-3-methyl-4-formylthiophene **20**.

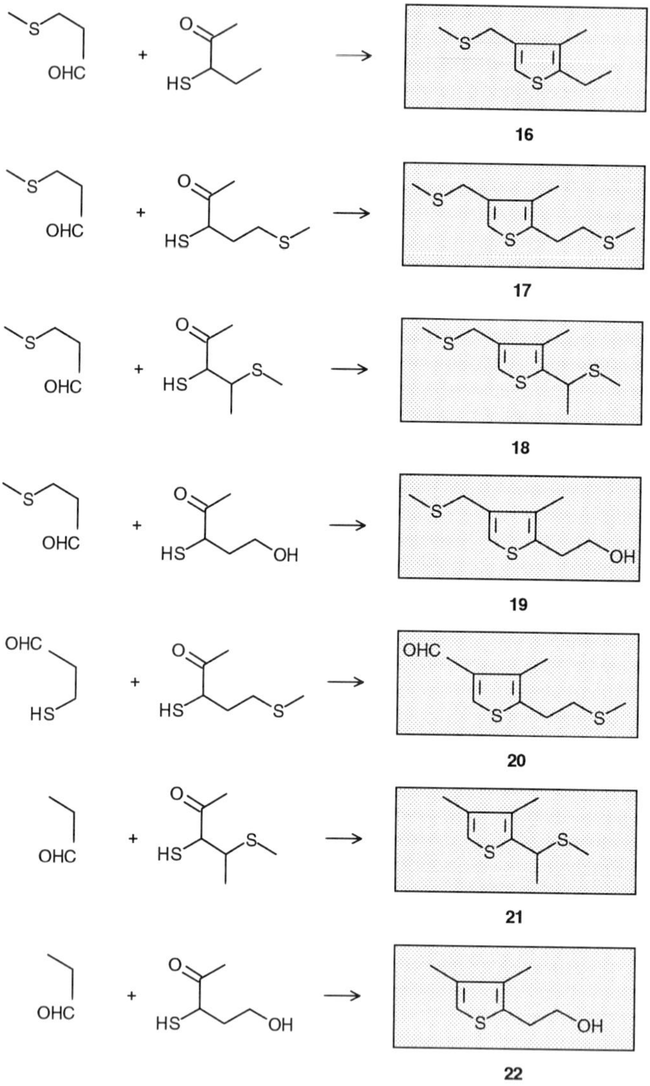

Figure 15: Probable formation pathways of the newly identified thiophenes **16-22**.

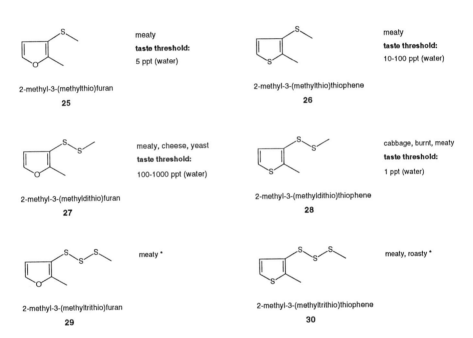

Figure 16: Sensory properties of the furans and thiophenes **25 - 30**.

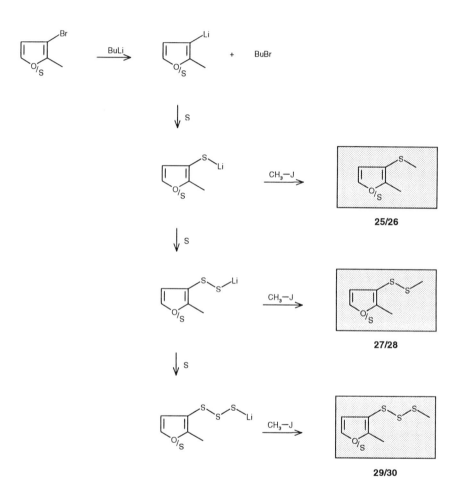

Figure 17: Synthetic pathway to the furans and thiophenes **25 - 30**.

Acknowledgments
The authors would like to thank the entire groups of organic synthesis, chromatography, spectrometry, and flavor research of this corporate for their valuable and skillful work. A special gratitude to A. Sagebiel for drawing the chemical structures.

Literature Cited
1. Güntert, M.; Brüning, J.; Emberger, R.; Köpsel, M.; Kuhn, W.; Thielmann, T.; Werkhoff, P. *J. Agric. Food Chem.* **1990**, *38*, 2027; and literature cited herein.
2. Clydesdale, F.M.; Ho, C.-T.; Lee, C.Y.; Mondy, N.I.; Shewfelt, R.L. *CRC Crit. Rev. Food Sci. Nutr.* **1991**, *30*, 599.
3. Grosch, W.; Zeiler-Hilgart, G. "Formation of Meatlike Flavor Compounds". In *Flavor Precursors - Thermal and Enzymatic Conversions*; Teranishi, R.; Takeoka, G.R.; Güntert, M., Eds.; ACS Symposium Series 490; American Chemical Society: Washington D.C., 1992, pp 183-192.
4. Grosch, W.; Zeiler-Hilgart, G.; Cerny, C.; Guth, H. "Studies on the Formation of Odorants Contributing to Meat Flavours". In *Progress in Flavour Precursor Studies: Analysis - Generation - Biotechnology*; Schreier, P.; Winterhalter, P., Eds.; Allured Publishing Corporation: Carol Stream, Illinois, 1993, pp 329-342.
5. Ames, J.M.; Hincelin, O.; Apriyantano, A. *J. Sci. Food Agric.* **1992**, *58*, 287.
6. Ames, J.M.; Hincelin, O. "Novel Sulphur Compounds from Heated Thiamine and Xylose/Thiamine Model Systems". In *Progress in Flavour Precursor Studies: Analysis - Generation - Biotechnology*; Schreier, P.; Winterhalter, P., Eds.; Allured Publishing Corporation: Carol Stream, Illinois, 1993, pp 379-382.
7. Werkhoff, P.; Bretschneider, W.; Emberger, R.; Güntert, M.; Hopp, R.; Köpsel, M. *Chem. Mikrobiol. Technol. Lebensm.* **1991**, *13*, 30.
8. Ames J.M.; Elmore, J.S. *Flavour and Fragrance Journal* **1992**, *7*, 89.
9. Güntert, M.; Brüning, J.; Emberger, R.; Hopp, R.; Köpsel, M.; Surburg, H.; Werkhoff, P. "Thermally Degraded Thiamin - A Potent Source of Interesting Flavor Compounds". In *Flavor Precursors - Thermal and Enzymatic Conversions*; Teranishi, R.; Takeoka, G.R.; Güntert, M., Eds.; ACS Symposium Series 490; American Chemical Society: Washington D.C., 1992, pp 140-163.
10. Güntert, M.; Bertram, H.-J.; Emberger, R.; Hopp, R.; Sommer, H.; Werkhoff, P. "New Aspects of the Thermal Generation of Flavour Compounds from Thiamin". In *Progress in Flavour Precursor Studies: Analysis - Generation - Biotechnology*; Schreier, P.; Winterhalter, P., Eds.; Allured Publishing Corporation: Carol Stream, Illinois, 1993, pp 361-378.
11. Güntert, M.; Bertram, H.-J.; Hopp, R.; Silberzahn, W.; Sommer, H.; Werkhoff, P. "Thermal Generation of Flavor Compounds from Thiamin and Various Amino Acids". In *Recent Developments in Flavor and Fragrance Chemistry*; Hopp, R.; Mori, K., Eds.; VCH Verlagsgesellschaft: Weinheim, Germany, 1993, pp 215-240.
12. Sommer, H.; Güntert, M. *H&R Contact* **1993**, *59*, 8.
13. Carlsen, L.; Egsgaard, H.; Jorgensen, F.S.; Nicolaisen, F.M. *J. Chem. Soc. Perkin Trans. II* **1984**, 609.
14. *Chemical Abstracts*, searched up to **1993**, *118* (18).
15. Polak's Frutal Works, Neth. Appl. 6812899, 1967.
16. Werkhoff, P.; Brüning, J.; Emberger, R.; Güntert, M.; Hopp, R. "Flavor Chemistry of Meat Volatiles: New Results on Flavor Components from Beef, Pork, and Chicken". In *Recent Developments in Flavor and Fragrance Chemistry*; Hopp, R.; Mori, K., Eds.; VCH Verlagsgesellschaft: Weinheim, Germany, 1993, pp 183-213.
17. Bertram, H.-J.; Emberger, R.; Güntert, M.; Sommer, H.; Werkhoff, P. "Synthesis of New Flavor Constituents". In *Recent Developments in Flavor and Fragrance Chemistry*; Hopp, R.; Mori, K., Eds.; VCH Verlagsgesellschaft: Weinheim, Germany, 1993, pp 241-259.

RECEIVED March 23, 1994

Chapter 18

Formation of Sulfur-Containing Flavor Compounds from [^{13}C]-Labeled Sugars, Cysteine, and Methionine

R. Tressl[1], E. Kersten[1], C. Nittka[1], and D. Rewicki[2]

Institut für Biotechnologie, Technische Universität Berlin, Seestrasse 13, 13353 Berlin, Germany
Institut für Organische Chemie, Freie Universität Berlin, Takustrasse 3, 14195 Berlin, Germany

> The formation of furans, thiophenes, furanones, thiophenones etc. was investigated in a series of [1(or 6)-^{13}C]-glucose and [1-^{13}C]-arabinose/ cysteine and methionine model experiments. The labeled compounds were analyzed by capillary GC/MS and NMR-spectroscopy. From their structures the degradation pathways via different reactive intermediates (e.g. 3-deoxyaldoketose, 1-deoxydiketose) and fragmentations were evaluated. Besides the transformations to flavor compounds via identical labeled precursors, major differences in the flavor compounds result from specific Strecker reaction sequences. Major unlabeled compounds e.g. 3-mercaptopropionic acid from cysteine and 4-methylthiobutyric acid from methionine demonstrate transamination/reduction, and the formation of pyruvate and 2-mercaptopropionic acid from [1-^{13}C]-glucose/cysteine indicates ß-elimination.

Cysteine- and methionine-specific Maillard products are important sulfur containing flavor compounds in meat and roasted coffee. Their formation pathways were postulated from the results of model experiments of cysteine, methionine, sugars and the corresponding degradation products (*1*). According to the specific Strecker degradation of cysteine (*2*) many compounds like 3,5-dimethyl-1,2,4-trithiolane, 3-methyl-1,2,4-trithiane and 1,2,3-trithia-5-cycloheptene result from cysteine without the participation of a fragment from carbohydrates (*3*). During Strecker degradation of methionine, 3-methylthiopropanal is generated and undergoes a cleavage into propenal, propanal and methylmercaptan or a transformation into sulfur compounds (*4,5*). In the initial phase of the Maillard reaction, D-glucose (or D-arabinose) and cysteine form Schiff bases, which undergo keto-enol tautomerism, allylic dehydration and deamination into α-dicarbonyls (Scheme 1). To some extent the Schiff bases are cleaved into C2-, C3- and C4-fragments, which may react as precursors. Therefore, the formation pathways of individual sulfur compounds can not be elucidated without labeling experiments. For this purpose [1-^{13}C]-labeling experiments are most suitable (*6,7*).

In this paper we demonstrate [1(or 6)-^{13}C]-D-glucose/L-cysteine, [1-^{13}C]-D-arabinose/L-cysteine and [1-^{13}C]-D-glucose/L-methionine model experiments, which were carried out comparable to the corresponding L-proline and hydroxyproline experi-

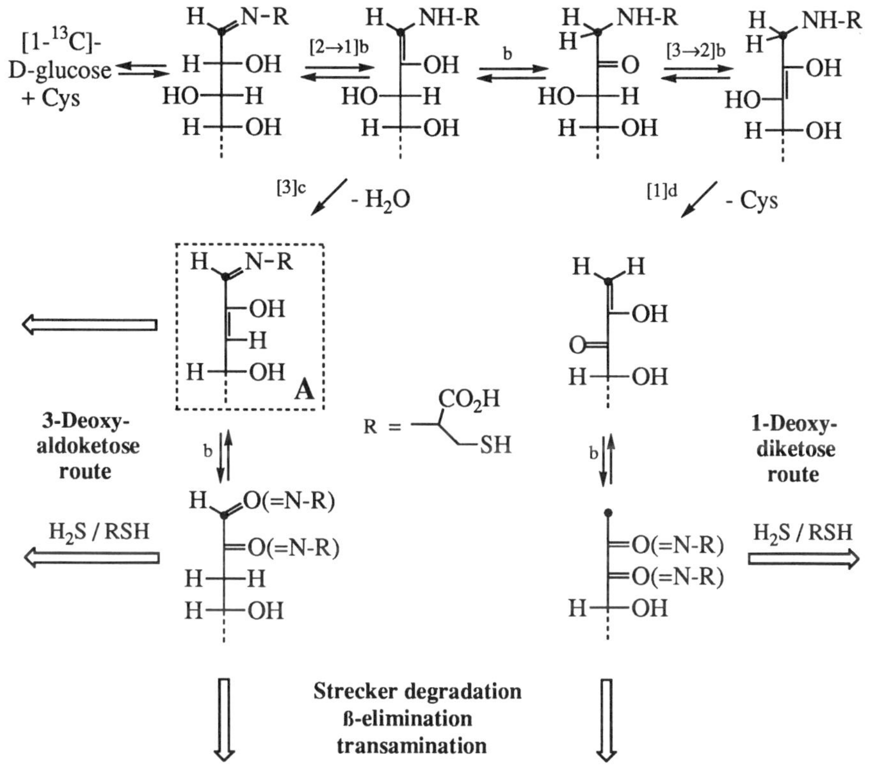

Scheme 1. Nonspecific initial phase in the D-glucose/L-cysteine Maillard system
(b = keto/enol tautomerism, c = allylic dehydration, d= allylic deamination)

ments (7). The described model reactions lead to [^{13}C]-labeled products from which substantial conclusions on their formation routes can be drawn. For this determination the distribution, position and extent of the [^{13}C]-labeling of selected compounds were analyzed by mass spectrometry. Thermal degradation of cysteine and methionine in water (3), in addition with thiamin (8), and with α-dicarbonyls (5) were carried out and numerous volatiles were identified.

Strecker Degradation, ß-Elimination and Transamination of Cysteine/Methionine during Heating with [1(or 6)-^{13}C]-D-Glucose

During Strecker degradation of [1-^{13}C]-D-glucose with primary α-amino acids, pyrroles and pyridinols are formed as major products (6). 4-Aminobutyric acid and peptide bound lysine are transformed into [^{13}CHO]-2-formyl-5-hydroxymethylpyrroles (9). Amino acids like Val, Ile, Leu, Phe and Met are transformed into 2-[^{13}CHO]-pyrrole lactones (10). Equimolar amounts of cysteine (methionine) and [1(or 6)-^{13}C]-D-glucose were heated for 1,5 h at 160°C in aqueous solution at pH 5. The volatiles were extracted with pentane/ether and analyzed as described (7). In Table I selected (unlabeled) Strecker degradation products from cysteine and methionine are summarized. Pyruvat (1), 2- and 3-mercaptopropionic acids (2, 3) from cysteine as well as 2-oxo-5-thiahexanoic acid (4) and 5-thiahexanoic acid (5) from methionine,

Table I. Strecker Degradation Products from [1-^{13}C]-D-Glucose with Cysteine or Methionine

No.	compound	formation [ppm] with Cys	Met
1	CH$_3$-CO-CO$_2$H	50	-
2	CH$_3$-CH(SH)-CO$_2$H	600	-
3	HS-CH$_2$-CH$_2$-CO$_2$H	180	-
4	CH$_3$-S-CH$_2$-CO-CO$_2$H	-	10
5	CH$_3$-S-CH$_2$-CH$_2$-CO$_2$H	-	1050
6	CH$_3$-S-CH$_2$-CO-CH$_3$	-	410
7	2,4-dimethyl-1,3-dithiolane	20	-
8	2,4-dimethyl-1,3-dithiane	15	-
9	CH$_3$-S-CH$_2$-CHO	-	1300
10	CH$_3$-S-CH$_2$-CH$_2$-S-CH$_3$	-	10
11	methylpyrazine	8	80
12	2-(3-methylthiopropyl)pyrazine	-	90
13	5-methyl-2-formylpyrrole	70	20
14	2-acetylpyrrole	110	100

contain the intact carbon skeletons of the amino acids. They are formed as unlabeled compounds in [1-^{13}C]- and [6-^{13}C]-D-glucose Maillard systems. Therefore, compounds **3**, **4**, and **5** correspond to a transamination of cysteine and methionine and subsequent reduction of the α-keto acids (Scheme 2).

The identification of unlabled pyruvate and 2-mercaptopropionic acid in [^{13}C]-D-glucose/L-cysteine Maillard systems clearly indicates a degradation pathway via ß-elimination. In the corresponding [1-^{13}C]-D-glucose/L-methionine model experiment, 2-methylthiopropionic acid was generated as a 20:80 mixture of unlabeled and singly labeled isotopomers. The formation of compound **6** from methionine is obscure. The oxidative decarboxylation of cysteine and methionine leads to mercaptoacetaldehyde and 3-mercaptothiopropanal, respectively. The Strecker aldehydes undergo reductive cleavages into acetaldehyde/H2S and propanal/CH3SH which are further transformed into sulfur compounds. In addition, 3-methylthiopropanal is degraded into propenal and CH3SH via retro Michael type reaction. During the reductive cleavage of mercaptoacetaldehyde and the reduction of α-keto acids, the enaminols are oxidized. Therefore, the formation of pyrazines is a minor pathway in cysteine Maillard systems. The labeling experiments indicate four degradation pathways:

(a) formation of pyrroles (**13,14**) and pyridinoles,
(b) ß-elimination of H2S and formation of pyruvate and 2-mercaptopropionic acid,
(c) transamination to 2-oxo-3-mercaptopropionic acid and subsequent reduction to 3-mercaptopropionic acid,
(d) oxidative decarboxylation to mercaptoacetaldehyde and reductive cleavage into acetaldehyde and H2S.

The pathways (a), (c) and (d) are also operative in the methionine/D-glucose Maillard systems. The degradation pathways indicate analogy to the enzymatic transformations and are reported for the first time.

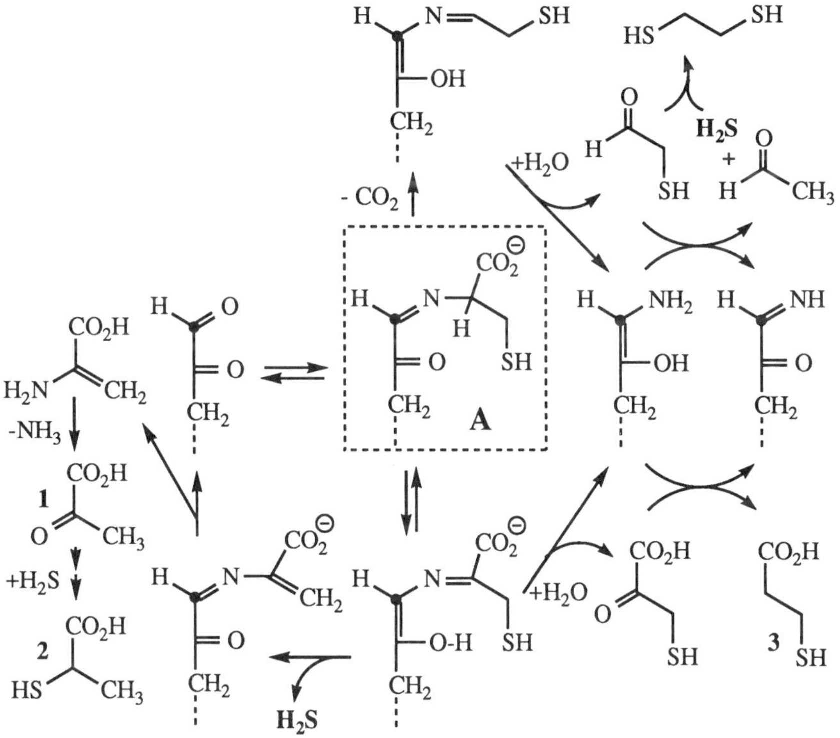

Scheme 2. Strecker degradation, ß-elimination and transamination in glucose/cysteine model systems

Formation of Compounds by Degradation via 3-Deoxyaldoketose

During heating of [1-^{13}C]-D-glucose with mixtures of primary α-amino acids, the main part of the amino acids is incorporated into pyrrole lactones (*10*). In Table II [^{13}C]-labeled flavor compounds are summarized which are formed via 3-deoxyaldoketose (**15-18**) and 1-deoxydiketose (**28, 32-34**) as intermediates. All C6-compounds are 100% singly labeled. Furfurylalcohol (**18**) is generated from the C2-C6 skeleton of [1-^{13}C]-D-glucose. The addition of cysteine strongly influences the generation of 3-deoxyosone products by a nucleophilic addition of H2S to 3,4-dideoxyaldoketose (*11*). Flavor compounds which are formed via 1-deoxydiketose are not correlated to this

Table II. Formation of Pyrrole Lactones, Furans, Furanones, and Pyranones from [1-^{13}C]-D-Glucose

No.	compound	formation [ppm] with AS / Pro	AS / Cys
15	AS = Val, Ile, Leu, Phe	880	650
16		40	1100
17		230	3420
18		170	330
28		120	125
32		310	300
33		400	230
34		35	90

inhibition. Cysteine specific compounds, which are formed from D-glucose, were identified by MS-and 1H-NMR-spectroscopy (5) (Table III). The C6-compounds from [1-^{13}C]-and [6-^{13}C]-D-glucose experiments were characterized as 100% singly labeled isotopomers (Table IV). The unexpected generation of *unlabeled* C5-compounds from [1-^{13}C]-D-glucose and 100% *singly labeled* isotopomers from the corresponding [6-^{13}C]-D-glucose and [1-^{13}C]-D-arabinose model experiments indicates different pathways from glucose and arabinose to furfurylalcohol (18), 2-hydroxymethyl-thiophene (22), 5-hydroxymethyltetrahydrothiophene-3-one (23) and 3-thiolanone (24). In [^{13}C]-D-glucose/cysteine Maillard experiments, which were carried out comparable to the proline/hydroxyproline model experiments (7), 19, 20, and 21 were identified as 100% singly labeled isotopomers. Compounds [^{13}CH2SH]-19 and [^{13}CH2SH]-20 indicate a formation pathway via Strecker degradation of 3,4-dideoxy-aldoketose and the subsequent dehydration product, respectively. In L-methionine/[^{13}C]-D-glucose experiments the corresponding methylsulfides were identified (5).

The formation pathways to furfurylmercaptans and sulfides (Scheme 3) include the incorporation of sulfur during Strecker degradation as an important step. The mass spectrometric fragmentation of [2-^{13}C]-21, formed in [1-^{13}C]-D-glucose/L-cysteine model experiments, clearly indicates the intact carbon chain by M(0%) : M+1(100%) : M+2(0%) and an unlabeled acetyl group by m/z=43 and 112 (COCH3 and M-COCH3).

Table III. MS (m/z, Relative Intensity) and ^{1}H NMR Data of Selected Furans, Thiophenes, and Thiophenones, Characterized in Cysteine Model Experiments with [1-^{13}C]- and [6-^{13}C]-D-Glucose

furfurylalcohol (18)
  unlabeled: 98(100), 97(50), 81(48), 70(28), 69(28), 53(30), 42(30) 41(35), 39(25)
  [^{13}CH2]-18 from [6-^{13}C]-D-glucose: 99(100), 98(57), 82(55), 71(20), 70(35), 69(20), 54(45), 41(60), 39(50)

5-hydroxymethylfurfurylmercaptan (19)
  unlabeled: 144(25), 111(100), 94(10), 83(20)
  [^{13}CH2SH]-19 from [1-^{13}C]-D-glucose: 145(20), 112(100), 95(10), 84(17), 66(15), 56(25)

5-methylfurfurylmercaptan (20)
  unlabeled: 128(15), 95(100), 67(6), 55(8), 45(6), 43(43), 41(18), 40(11)
  [^{13}CH2SH]-20 from [1-^{13}C]-D-glucose: 129(20), 96(100), 68(5)

5-acetyltetrahydrothiophene-3-one (21)
  unlabeled: 144(60), 111(46), 101(61), 100(20), 98(28), 73(100), 71(22), 55(25), 46(16), 43(67);
  1H-NMR (CDCl3): δ = 2.38 (s, 3H; CH3); 2.52 (dd, J=18Hz, 7.5Hz, 1H; 4-H); 2.88 (dd,J=18Hz, 1.7Hz, 1H; 4-H'); 3.21, 3.24 (dd, J=18Hz, 2H; 2-H, 2-H'); 4.00 (dd, J=7.5Hz, 1.7Hz, 1H; 5-H)
  [2-^{13}C]-21 from [1-^{13}C]-D-glucose: 145(40), 112(25), 102(55), 101(20), 98(25), 74(100), 71(20), 55(20), 47(25), 43(80)

2-hydroxymethylthiophene (22)
  unlabeled: 144(100), 113(35), 97(60), 85(77), 81(22), 69(5), 58(6), 53(8)
  [^{13}CH2]-22 from [6-^{13}C]-D-glucose: 115(100), 114(40), 98(62), 85(75), 82(30), 69(5), 59(15), 58(14), 54(13), 53(10)

5-hydroxymethyltetrahydrothiophene-3-one (23)
  unlabeled: 132(100), 101(76), 73(87), 57(59), 47(23), 46(67), 45(75), 41(27)
  1H-NMR (CDCl3): δ = 2.00 (br S, 1H; OH); 2.57 (dd, J=18.0Hz, 3.8Hz, 1H; 4-H); 2.77 (dd, J=18Hz, 7.6Hz, 1H; 4-H'); 3.28, 3.38 (AB, J=17.8Hz, 2H; 2-H, 2-H'); 3.63 (mc, 1H; 5-H); 3.72 (mc, 2H; CH2O)
  [2-^{13}C]-32 from [1-^{13}C]-D-arabinose: 133(100), 102(85), 74(70), 57(55), 47(60), 45(80), 41(32).

2,5-dimethyl-4-hydroxy-3(2H)-furanone (32)
  unlabeled: 128(35), 85(17), 57(50), 43(100)
  [5-^{13}CH3]-32/[2-^{13}CH3]-32 from [1-^{13}C]-D-glucose: 129(70), 86(20), 85(19), 73(5), 58(50), 57(63), 44(95), 43(100)
  [5-^{13}C]-32/[2-^{13}C]-32 from [6-^{13}C]-D-glucose: 129(65), 86(12), 85(22), 73(5), 58(52), 57(58), 44(100), 43(93)

Continued on next page

Table III. Continued

2,4-dihydroxy-2,5-dimethyl-3(2H)-thiophenone (or thioacetylformoin) (**39**)
unlabeled (from D-glucose/cysteine): 160(25), 132(15), 117(35), 101(5), 89(20), 59(35), 55(25), 43(100)

[5-^{13}CH$_3$]-**39**/[2-^{13}CH$_3$]-**39** from [1-^{13}C]-D-glucose: 161(40), 133(22), 118(35), 117(27), 102(5), 101(7), 90(20), 89(18), 60(38), 59(37), 56(25), 44(88), 43(100)

[5-^{13}CH$_3$]-**39**/[2-^{13}CH$_3$]-**39** from [6-^{13}C]-D-glucose: 161(43), 133(20), 118(27), 117(35), 90(15), 89(22), 60(32), 59(40), 44(100), 43(85)

2,5-dimethyl-4-hydroxy-3(2H)-thiophenone (**40**)
unlabeled (from D-glucose/cysteine): 144(53), 114(6), 113(8), 85(45), 61(13), 60(27), 59(100), 45(20)

^{1}H-NMR (CDCl$_3$): δ = 1.55 (d, J=7Hz, 3H; 2-CH$_3$); 2.25 (d, J=1.2Hz, 3H; 5-CH$_3$); 3.70 (dq, J=7.5Hz, 1.2Hz, 1H; 2-H)

[5-^{13}CH$_3$]-**40**/[2-^{13}CH$_3$]-**40** from [1-^{13}C]-D-glucose: 145(70), 114(12), 113(12), 87(40), 86(38), 62(15), 61(40), 60(100), 59(85), 45(40)

(5-^{13}CH$_3$)-**40**/(2-^{13}CH$_3$)-**40** from (6-^{13}C)-D-glucose: 145(55), 144(10), 113(8), 86(24), 85(30), 61(26), 60(91), 59(100), 45(50)

5-hydroxymethyl-2-methyltetrahydrothiophene-3-one (**42**)
unlabled mixture of two diastereomers (from D-glucose/cysteine): 146(52), 115(20), 87(44), 61(33), 60(100), 59(35), 45(42)

[^{13}CH$_3$]-**42** from [1-^{13}C]-D-glucose: 147(53), 116(22), 88(45), 62(28), 61(100), 60(32), 45(38)

[^{13}CH$_2$OH]-**42** from [6-^{13}C]-D-glucose: 147(45), 115(20), 87(40), 61(48), 60(100), 59(40), 45(50)

Table IV. Formation of 100% Singly Labeled and Unlabeled Compounds in [1-^{13}C]-, [6-^{13}C]-D-Glucose, and [1-^{13}C]-D-Arabinose/Cys Model Experiments

No.	isotopomers from Cys and		No.	isotopomers from Cys and	
	[1-^{13}C]-glucose	[6-^{13}C]-glucose		[1-^{13}C]-glucose	[6-^{13}C]-glucose or [1-^{13}C]-arabinose
19			18		
20			22		
21			23		
14			24		

The formation of compound **21** is competitive to **20** (Scheme 3). The mercaptan which is generated during Strecker degradation undergoes cyclization to **20** or Michael type addition to **21**. In the corresponding [1-^{13}C]-D-arabinose/L-cysteine experiments, 5-hydroxymethyltetrahydrothiophene-3-one (**23**), furfurylmercaptan, and 2-hydroxymethylthiophene (**22**) were identified with intact carbon skeletons. They are generated by Strecker degradation via 3,4-dideoxy-pentosone, comparable to the [1-^{13}C]-D-glucose/L-cysteine Maillard system (Scheme 3). All C$_5$-compounds were identified as 100% singly labeled isotopomers in [6-^{13}C]-D-glucose/L-cysteine and as unlabeled compounds in [1-^{13}C]-D-glucose/L-cysteine model experiments.

In the Maillard reaction of [1-^{13}C]-D-glucose with hydroxypoline (**7**) and in the corresponding 4-aminobutyric acid Maillard system (blocked Strecker) (**9**) we observed two fragmentations into C$_1$+C$_5$ compounds. The α-dicarbonyl-fragmentation of 3-deoxyaldoketose into 2-deoxypentose and [^{13}C]-labeled formic acid generates unlabeled furfurylalcohol. In the [6-^{13}C]-D-glucose/L-cysteine Maillard system this fragmentation leads to 100% singly labeled C$_5$-products as outlined in Scheme 4. The pathway to **22** and **23** from D-glucose via 2-deoxypentose is different from that of the D-arabinose Maillard system. In addition, the incorporation of sulfur in the D-glucose system (nucleophilic addition, keto-enol tautomerism, cyclization) is different from that of the pentose Maillard system (Strecker degradation, cyclization).

Formation of Compounds by Degradation via 1-Deoxydiketoses

The important flavor compounds from proline (**7**) and cysteine (**10**) Maillard systems are generated via 1-deoxydiketoses as reactive intermediates (Scheme 5). This pathway is operative in the pH range of 5 - 7. All compounds from this pathway possess ^{13}CH$_3$-[1-^{13}C]-D-glucose Maillard systems and are generated via Amadori rearrangement and allylic deamination of the 2,3-endiol into 1-deoxy-2,3-diketose. The conversion of 1-deoxyosone into furanones, pyranones and cyclopentenones (Scheme 5) was investigated in proline (**7**) and 4-aminobutyric acid (**9**) Maillard systems. As 100% singly labeled compounds [^{13}CH$_3$]-**28**, [^{13}CH$_3$]-**32** (mixture of two singly labeled isotopomers) and [4-^{13}C]-cyclotene (**33**) from [1-^{13}C]-D-glucose/proline were identified.

In the corresponding [1-^{13}C]-and [6-^{13}C]-D-glucose/L-cysteine Maillard experiments (Table V), the thio-analogous compounds **39** and **40** were identified with intact carbon chains. Comparable to the proline/glucose experiments, thioactylformoin (**39**) and thiofuraneol (**40**) were generated as mixtures of two 100% singly labeled isotopomers. The mass spectrometric analysis clearly indicates ratios of [5-^{13}CH$_3$]-**39** to [2-^{13}CH$_3$]-**39** of 55% to 45%. The opposite distribution was obtained in the corresponding [6-^{13}C]-D-glucose/cysteine model experiments. This ratio was also determined for the isotopomers of **40**. In L-cysteine/D-glucose model systems thiofuraneol (**40**) is generated as a major product. The labeling experiments indicate that **39** and **40** might be generated by a cysteine specific pathway. The mass spectra of 2,4-dihydroxy-2,5-dimethyl-3(2H)-thiophenones from [1-^{13}C]-D-glucose and [6-^{13}C]-D-glucose are seen to be 100% singly labeled, which unmistakably demonstrates an exclusive formation of the intact hexose carbon chain. The ratio of [5-^{13}CH$_3$]-**39** to [2-^{13}CH$_3$]-**39** was calculated by comparing the fragment ion intensities (M-COCH$_3$)/(M-CO^{13}CH$_3$) and m/z = 43/44 (COCH$_3$/(CO^{13}CH$_3$) (corrected to their natural ^{13}C content). Thiofuraneol (**40**) was identified by MS- and 1H-NMR-spectroscopy from L-cysteine/D-glucose experiments (Table III). The mass spectrum of the unlabeled compound showed a base peak at m/z = 59 (CSCH$_3$) and a fragmentation of (M-59) at m/z = 85. This sequence is comparable to that of furaneol with m/z = 43 and (M-43) at m/z = 85. By comparison of these fragments a ratio of 54% [5-^{13}CH$_3$]-**40** and 46% [2-^{13}CH$_3$]-**40** was determined. In contrast to that, the distribution of the isotopomers of

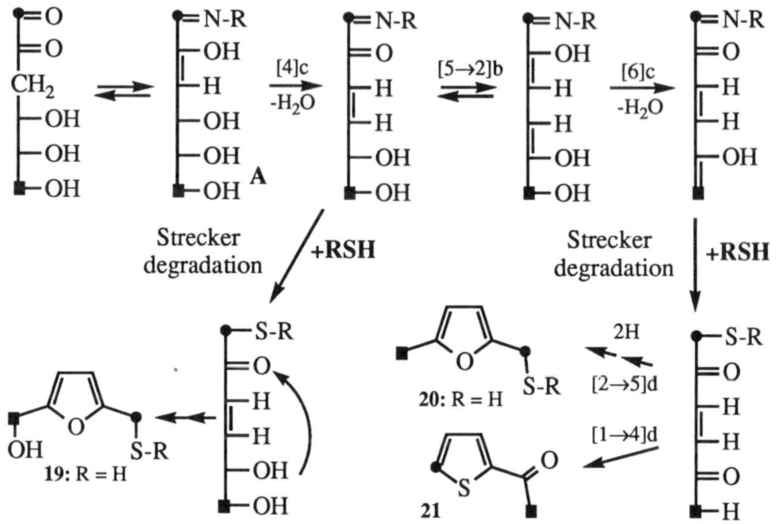

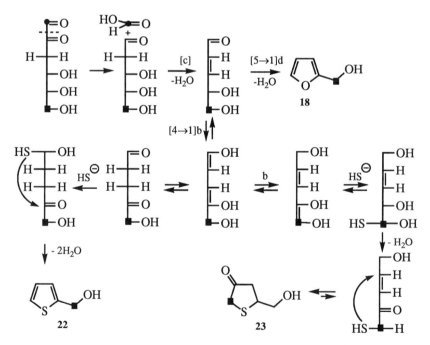

Scheme 3. Formation of furfurylmercaptans and sulfides in [1(or 6)-^{13}C]-D-glucose/ Cys (or Met) model experiments (d = cyclization)

Scheme 4. α-Dicarbonyl fragmentation of 3-deoxyaldoketose in [6-^{13}C]-D-glucose/ Cys Maillard experiments

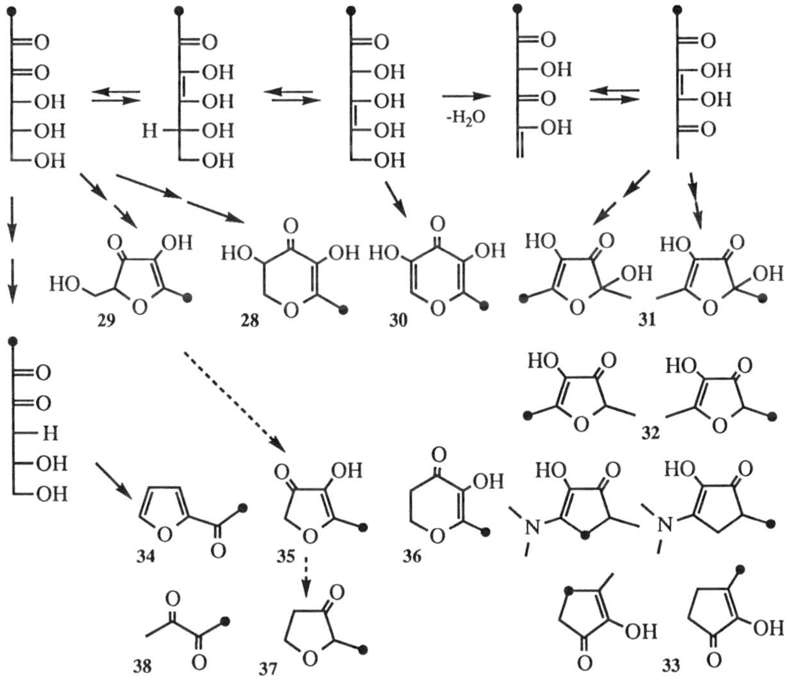

Scheme 5. General pathway of the conversion of 1-deoxyosone into furanones, pyranones, and cyclopentenones

Table V. 100% Singly Labeled Compounds Generated in D-Glucose/L-Cysteine Maillard Experiments

No.	isotopomers from Cys and [1-^{13}C]-D-glucose		[6-^{13}C]-D-glucose	
28	(structure)		(structure)	
39	55%	45%	45%	55%
40	54%	46%	46%	54%
32	48%	52%	52%	48%
41	54%	46%		
42/43	(structure)	(structure)	(structure)	(structure)

[5-^{13}CH3]-32 and [2-^{13}CH3]-32 was found to be 48:52 (opposite distribution). The correlated 31 was not detectable in the labeling experiments.

Comparable to the unexpected results of proline and hydroxyproline Maillard systems (leading to [4-^{13}C]-33 and [^{13}CH$_3$]-33, respectively) the 100% singly labeled isotopomers of 39, 40 and 32 were not generated in a ratio of 50:50 as would be expected, if these compounds were formed simply along the route 1-deoxyosone → diacetylformoin → 39, 40, 32. Therefore, we propose a pathway to 39 and 40 (Scheme 6) starting with two subsequent keto-enol tautomerizations into intermediate D, which can undergo an allylic dehydration to two isomeric bis-enol forms of diacetylformoin. These can be transformed into isomeric enol-diones, and nucleophilic addition of SH⁻, intramolecular cyclization, and dehydration generate the isotopomers of 39. The labeling experiments support a pathway to thiofuraneol (40) via thioacetylformoin and 39 by reduction. Kinetic effects or hindered interchangeability of the isomers by participation of the amino acid (Scheme 6) might be responsible for the observed ratios of isotopomers.

Compound 42 was analyzed as a mixture of two diastereomers from [1-^{13}C]- and [6-^{13}C]-D-glucose, respectively. The mass spectrometric analysis of [^{13}CH$_3$]-42 from [1-^{13}C]-D-glucose indicates two 100% singly labeled compounds with identical mass spectra and different retention indices. [^{13}CH$_3$]-42 from [1-^{13}C]-D-glucose/L-cysteine was identified as 5-hydroxymethyl-2-[^{13}CH$_3$]methyltetrahydrothiophene-3-one (Table III). The fragmentations (M-CH2OH) at m/z = 116 and a base peak at m/z = 61 clearly indicate an unlabeled hydroxymethyl group. The formation pathway to 42 is correlated to that of 2-acetylthiophene (43) via Strecker degradation (or reduction) of 1-deoxyosone to 1,4-dideoxy-2,3-diketose and subsequent allylic dehydration (*10*).

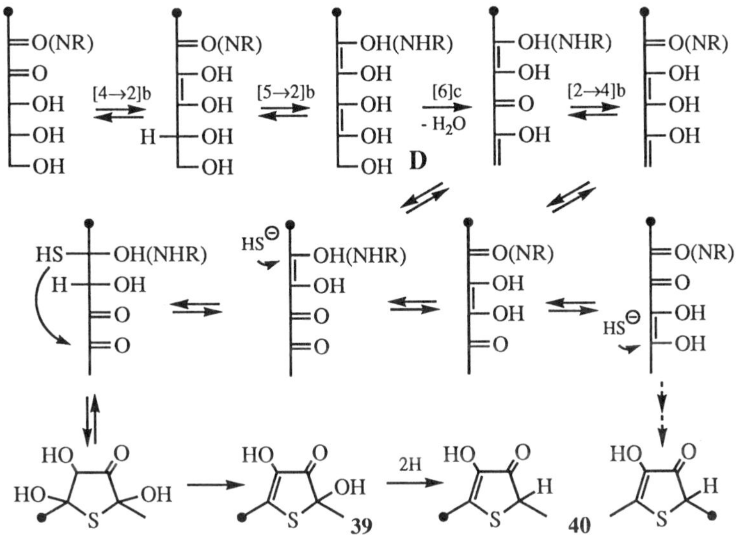

Scheme 6. Formation of thiofuraneol in [1-^{13}C]-D-glucose/Cys Maillard experiments

Acknowledgements

This work was supported by the AIF (Köln).

Literature cited

1. Tressl, R.; Helak, B.; Martin, N.; Kersten, E. In *Thermal Generation of Aromas*; Parliment, T.H., McGorrin, R.J., Ho, C.-T., Eds.; ACS Symp.Series 409; American Chemical Society, Washington, DC, 1989, 156-171.
2. Kobayasi, N.; Fujimaki, M. *Agric. Biol. Chem.* **1963**, *29*, 698.
3. Shu, C.-K.; Hagedorn, M.L.; Mookherjee, B.D.; Ho, C.-T. *J. Agric. Food Chem.* **1985**, *33*, 438.
4. de Rijke, D.; van Dort, J.M.; Boelens, H. In *Flavour '81*; Schreier, P., Ed.; Proceedings of the 3rd Weurman Symposium, Munich 1981; Walter de Gruyter, Berlin-New York, 1981, 417-431.
5. Martin, N. Ph.D. Thesis, Technische Univ. Berlin, 1989.
6. Nyhammar, T.; Olsson, K.; Pernemalm, P.-A. *Acta Chem. Scand.* **1983**, *B 37*, 879.
7. Tressl, R.; Helak, B.; Kersten, E.; Rewicki, D. *J. Agric. Food Chem.* **1993**, *41*, 547.
8. Werkhoff, P.; Brüning, J.; Emberger, R.; Güntert, M.; Köpsel, M.; Kuhn, W.; Surburg, H . *J. Agric. Food Chem.* **1990**, *38*, 777.
9. Tressl, R.; Kersten, E.; Rewicki, D. *J. Agric. Food Chem.*, in press.
10. Tressl, R.; Helak, B.; Kersten, E.; Nittka, C. In *Recent Developments in Flavor and Fragrance Chemistry*; Hopp, R., Mori, K., Eds.; Proceedings of the 3rd International Haarmann & Reimer Symposium, Kyoto 1992; Verlag Chemie, Weinheim, New York, Basel, Cambridge, 1993; 165-181.
11. Anet, E.F.L. *J. Adv. Carbohydr. Chem.* **1964**, *19*, 181.

RECEIVED March 23, 1994

Functional Properties

Chapter 19

Sulfur Compounds in Wood Garlic (*Scorodocarpus borneensis* Becc.) as Versatile Food Components

Kikue Kubota and Akio Kobayashi

Department of Nutrition and Food Science, Ochanomizu University, Tokyo 112, Japan

The volatile flavor components of the fruit of *Scorodocarpus borneensis* Becc. which is named "wood garlic" due to its garlic-like smell, were investigated. Although the volatiles contained a large amount of ethanal, most of the components were sulfur-containing. Methyl methylthiomethyl disulfide (I) and bis(methylthiomethyl) disulfide (II), two polysulfides not previously identified in the *Allium* genus, were determined to be potent odor compounds of *S. borneensis* by a sensory evaluation. At a dilute concentration, I produced the smell of freshly cut fruit, while II exhibited a slightly unpleasant odor which develops over time after cutting the fruit. In addition, the antimicrobial activity of the fruit was examined. Relatively strong activity was observed in the ethanol extract of the fruit; II and methylthiomethyl (methylsulfonyl)methyl disulfide (III) were isolated as the active components. II exhibited relatively strong antifungal activity, while III, a novel compound, exhibited broader activities than II against bacteria and fungi. These results show that the fruit of *S. borneensis* possesses useful properties for use as a natural preservative.

Scorodocarpus borneensis Becc. is a tall tree belonging to the family Olacaceae. It grows naturally in Sumatra, the Malay peninsula, and Borneo, and has been named "wood garlic" by the natives due to its characteristic smell. The garlic smell is present in the leaves, flowers and fruit (*1*). The fruit, after ripening and falling away from the tree, consists of a hard outer nutshell and pulp similar to the walnut (see Figure 1), but having much less fat, *ca.* 3% of dry weight, and a unique smell. The natives of Sabah in Malaysia sometimes use the pulp of the fruit as a seasoning when they cook fish. The fruit can be consumed fresh, shortly after picking, or dried.

Garlic of the *Allium* genus, one of the most widely known spices in the world, is noted for the enzymatic formation of a large amount of organic sulfur compounds when it is crushed. The powerful flavor and possible medical application of the

Figure 1. Fruit pulp and fruit with the nutshell of wood garlic.

decomposed sulfur products have attracted the attention of chemists for more than a century, and many reviews have been published (2,3). In contrast, chemical studies on "wood garlic" have not been done because the trees are grown only in very limited areas. Since the smell and utility for cooking are very similar to those of garlic, the volatile flavor components and antimicrobial activity were investigated. A few unusual polysulfides, different from those of the *Allium* species, were identified as flavor components or antimicrobial substances. In this paper, we summarize the determination of the potent odor compounds and the isolation of antimicrobial substances from the fruit of wood garlic (see also Kubota *et al.*, *Biosci. Biotech. Biochem.*, in press).

Materials

The fruits of wood garlic (*S. borneensis*) were collected in Sabah, Malaysia. Unless otherwise mentioned, after removing the outer shell of each fresh fruit, the pulp was used for the experiments. The volatiles were also collected from frozen leaves.

Potent Odor Compound in Wood Garlic

The fruit covered with the outer shell does not emit a strong odor, but when the nutshell has been removed, or the pulp is cut, a strong odor results. This suggests that the odor of wood garlic might be formed enzymatically in the same way as in the *Allium* species. The flavor constituents were compared to those of *Allium*.

Isolation and Identification of the Volatile Flavor Compounds. The volatiles were isolated via a dynamic headspace sampling technique using a commercial purge and trap injector (Chrompack, The Netherlands) mounted on a Hitachi G-3000 gas chromatograph. Identification was done by GC-MS on a JEOL DX-300/JA-5000 instrument. The volatiles from the fruit were led into a cryogenic trapping tube prechilled to -130°C, and then injected directly into the GC column in the splitless mode by heating the trapping tube. This automatic purge and trap injection method (PTI) isolated the low-boiling-point components very well with good reproducibility. However, PTI dose not provide good recovery of the higher-boiling components because the condenser used to remove excess water at -15°C traps some of the higher-boiling volatiles. Consequently, we additionally used steam distillation under reduced pressure. Since the leaves exhibited the garlic smell with a green odor when they were injured, the volatile components of the leaves were also investigated.

The identified components and their peak areas by GC are listed in Tables I and II. For both the fruits and leaves, the composition of the volatiles was relatively simple. In the headspace volatiles, ethanal was the main component (96.4%). This compound is probably produced in the fruit during storage from pyruvic acid by decarboxylation under anaerobic respiration (4). Among the other compounds were dimethyl sulfide, dimethyl disulfide, and methane- and propane-thiols, together with two unusual polysulfides which have not been found in the *Allium* species. The sulfur compounds in the leaves were similar to those found in the fruit, but the concentrations of (*E*)-2-hexenal and (*Z*)-3-hexenol, which are well known to be responsible for a green note (5), were much higher than found in the fruit.

Table I. Composition of the Headspace Concentrate Obtained by the PTI Method from Wood Garlic Fruit

Compound	Peak area (%) by GC
(Sulfur-containing compounds)	
methanethiol	0.04
propanethiol	0.03
dimethyl sulfide	0.3
dimethyl disulfide	0.09
1,3-dithiethane*	0.2
methyl methylthiomethyl disulfide	0.5
(Others)	
ethanal	96.4
acetone	0.3
methanol	0.4
ethanol	0.05
(E)-2-hexenal	0.05

* Tentative identification from MS data.
Reproduced with permission from reference 19. Copyright 1994 American Chemical Society.

Table II. Main Components of the Volatile Concentrate Obtained by Steam Distillation from the Fruit and Leaves of Wood Garlic

Compound	Peak area (%) by GC	
	Fruit	Leaf
(Sulfur-containing compounds)		
1,3-dithiethane*	0.2	-
methyl methylthiomethyl disulfide	2.4	21.3
tris(methylthio)methane*	0.1	3.8
bis(methylthiomethyl) disulfide	94.2	27.4
(Others)		
1-penten-3-ol	-	3.0
(Z)-3-hexenol	-	0.7
(E)-2-hexenal	0.1	25.1
benzaldehyde	-	0.6
3-hydroxy-2-butanone	0.1	-
α-ionone	-	0.4
ß-ionone	-	0.1

* Tentative identification from MS data.
Reproduced with permission from reference 19. Copyright 1994 American Chemical Society.

Since sulfur compounds usually exhibit very powerful and characteristic odors, it seemed that they might contribute to the smell of wood garlic. We focused our attention on two components which constituted approximately 97% and 49%, respectively, of the volatiles from the fruits and leaves obtained by steam distillation. They were identified as methyl methylthiomethyl disulfide, $CH_3SCH_2SSCH_3$ (I), and bis(methylthiomethyl) disulfide, $CH_3SCH_2SSCH_2SCH_3$ (II). I has been previously identified in cooked cabbage, broccoli and cauliflower (6), *shiitake* mushroom (7), Beaufort cheese (8), and hop oil (9), and both I and II have been identified in the essential oil from *Gallesia integrifolia* (Sprengel) Harms (10). There was, however, no available data on their organoleptic properties. Therefore, we synthesized both compounds from chloromethyl methyl sulfide, thioacetic acid and dimethyl disulfide by the published procedure (11). After distillation and preparative TLC their odor characteristics were examined by sensory evaluation.

Odor Significance of Compounds I and II in Wood Garlic. At high concentrations, I and II presented a considerably different aroma from that of the original wood garlic. On the other hand, the odor of each compound in the eluate from a capillary GC was reminiscent of the original wood garlic. We next determined their odor threshold values and re-evaluated them at dilute concentrations.

The odor threshold values were determined in aqueous solutions by a published method (12) using polyethylene bottles. The concentration was reduced by 2-fold dilutions, and the samples and a blank (water) were presented to the subjects. The threshold concentration was the lowest concentration at which more than 50% of the subjects could discriminate the odor between the sample being tested and the blank. The odor thresholds determined were (I) 40 ppb and (II) 3.5 ppb, as shown in Table III. The volatiles from the fruit obtained by steam distillation included I and II in concentrations of approximately 21 ppm and 840 ppm, respectively. Since each amount exceeded the threshold concentration by a large extent, it is clear that both of them significantly contributed to the smell of the fruit.

Furthermore, at concentrations ranging from 40 ppb to 20 ppm in water, compound I reproduced the freshly cut fruit smell of wood garlic. The smell was somewhat different from the fresh garlic smell of *Allium* and similar to the smell of long-standing grated garlic or onion. Compound II possessed the smell of wood garlic at a very low concentration (under 200 ppb), although it also had a very

Table III. Odor Threshold Values of the Main Volatile Constituents

Compound	Odor Threshold in Water (ppb)	Odor Characteristic at Dilute Concentration
$CH_3SCH_2SSCH_3$	40	long-standing garlic onion
$CH_3SCH_2SSCH_2SCH_3$	3.5	onion with irritating effect

strongly irritating odor. It must be emphasized that this irritating odor also existed in the fruit and increased with time after the fruit had been crushed. One crushed fruit could fill a room with the onion-like smell. From these results, it is suggested that compound **I** was the principal fresh flavor component of wood garlic, while compound **II** enhanced the strength of the garlic-like smell with an irritating odor.

Formation of Polysulfides in Wood Garlic. The foregoing observation strongly suggests that **II** was enzymatically formed. Gmelin *et al.* (*13*) have isolated a dipeptide precursor, $CH_3SCH_2SOCH_2CH(COOH)NHCOCH_2CH_2CH(NH_2)COOH$, with a garlic odor in some mushrooms. They also suggested a pathway leading to the odorous substances, including two-step enzymatic cleavage in the same way as observed in the *shiitake* mushroom. They further postulated that the corresponding sulfinic ester was very unstable and decomposed rapidly into compound **II** and $CH_3SCH_2SO_2SCH_2SCH_3$, although these structures have not yet been defined. It is possible that wood garlic contains the same or similar precursor. On the other hand, it has been suggested for hop oil that compound **I** may be produced by the oxidation of methylthiomethanethiol in the presence of methanethiol (*9*), although there is no data to confirm this. The formation of the odor of wood garlic, including the formation of **I**, is under investigation.

Antimicrobial Activity

The first systematic research concerning the medical properties of garlic probably started with studies on antibiotic activity about fifty years ago. The investigators isolated allicin as the active compound in garlic and reported that allicin was effective in the range from 1:85,000 to 1:125,000 against a number of Gram-positive and Gram-negative organisms (*14-16*). It is still very important to develop new natural food ingredients which have effective antibiotic activity against various microorganisms. Therefore, we began studying the effect of wood garlic on microorganisms. Assays by the paper disk method indicated that an ethanol or water extract of the fruit had relatively strong antibacterial activity.

Isolation of the Antimicrobial Compounds. Grated fruit pulp was soaked overnight in ethanol and then treated by sonicating for 30 min to extract the active components. The ethyl acetate-soluble fraction of the concentrate of the ethanol extract, which exhibited strong activity, was submitted to silica gel chromatography. Two active compounds were isolated with the yields shown in Table IV. One was $CH_3SCH_2SSCH_2SCH_3$, which was also identified as one of the potent odor components of wood garlic as already described. The other compound was obtained as crystals and then recrystallized with ethyl ether to afford colorless needles. This

Table IV. Isolated Antimicrobial Compounds in the Fruit of Wood Garlic

Compound	Yield (mg/100 g)	Appearance
$CH_3SCH_2SSCH_2SCH_3$	140.0	colorless oil
$CH_3SO_2CH_2SSCH_2SCH_3$	35.0	colorless crystal (mp 55.0°C)

was an unknown compound, and its chemical structure was next examined by spectroscopic methods.

Identification of $CH_3SO_2CH_2SSCH_2SCH_3$ (III) as a New Antimicrobial Compound. The empirical molecular formula of the crystal was deduced as $C_4H_{10}S_4O_2$ by an elemental analysis. The IR spectrum (KBr pellet) showed strong absorption bands at 1140 and 1300 cm^{-1} that can be assigned to the sulfonyl group. The mass spectrum of **III** is shown with those of $CH_3SCH_2SSCH_3$ (I) and $CH_3SCH_2SSCH_2SCH_3$ (II) in Figure 2. These three mass spectra show very similar fragmentation profiles and suggest that the structures of the three compounds are closely correlated. The base peak of m/z 61 suggests that the crystal also contained a CH_3SCH_2 moiety. Four singlet signals at δ_H 2.24 (CH_3S), 3.04 (CH_3SO_2), 4.07 (SCH_2SS) and 4.19 (SO_2CH_2SS) in the ^{1}H-NMR spectrum, and, in the ^{13}C-NMR spectrum, four signals at δ_C 15.45 (CH_3S), 39.27 (CH_3SO_2), 45.29 (SCH_2SS) and 62.15 (SO_2CH_2SS) could be reasonably assigned by referring to the literature (*17,18*). From these data, the chemical structure of the unknown compound was determined to be methylthiomethyl (methylsulfonyl)methyl disulfide (**III**). **III** is a novel antimicrobial compound (see Kubota *et al., Biosci. Biotech. Biochem.*, in press). It is interesting when considering the formation pathway of this compound that the structure closely resembles one of the garlic odor components of mushroom as already described (*13*).

Assay Method Employed. The antimicrobial activity of each of the three polysulfides (**I**, **II** and **III**) against 6 bacteria and 4 fungi was examined by determining the minimum inhibitory concentration (MIC). A sample was dissolved in dimethyl sulfoxide, and a bioassay was performed by the 2-fold agar dilution method. The bacteria were cultured on nutrient agar (pH 7.2) at 37°C for 24 hr, while the fungi were cultured on potato dextrose agar (pH 5.6) at 27°C for 48 hr. The MIC is defined as the lowest concentration of the compound at which no growth of the organism was observed in the culture.

Antimicrobial Activity. The MICs are shown in Table V. Compound **I** exhibited no activity against any of the organisms tested. Compound **II** showed strong activity with a MIC range of 12.5-25 μg/ml against the fungi, but action against the bacteria was weak. Compound **III** showed the broadest and strongest activity among the three polysulfides against the microorganisms tested here. It is notable that only compound **III** exhibited strong antibacterial activity, although it was several tens of times weaker than allicin, which was effective in the range of 1:125,000 against various bacteria (*16*).

Conclusion

In this study, a potent odor component of freshly cut wood garlic was identified, methyl methylthiomethyl disulfide. This compound exhibited a garlic or onion-like flavor at dilute concentrations. Two more polysulfides, bis(methylthiomethyl) disulfide and methylthiomethyl (methylsulfonyl)methyl disulfide, were also isolated in high concentrations (35 mg to 140 mg/100 g yield) from the grated fruit. The existence of these compounds strongly suggests that wood garlic possesses a polysulfide formation pathway similar to that of the *Allium*, including enzymatic

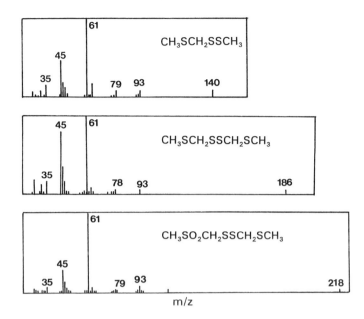

Figure 2. Mass spectra of the polysulfides isolated from wood garlic.

Table V. Antimicrobial Activity of the Polysulfides* Isolated from Wood Garlic

Microorganism Tested	MIC (μg/ml)		
	I	II	III
(Gram positive bacteria)			
Staphylococcus aureus FDA 209P	>100	50	12.5
Micrococcus luteus PCI 1002	>100	>100	25
Bacillus subtilis PCI 219	>100	50	12.5
Microbacterium smegmatis ATCC 607	>100	50	12.5
(Gram negative bacteria)			
Escherichia coli NIHJ	>100	50	12.5
Pseudomonus aeruginosa IFO 3080	>100	>100	>100
(Yeasts)			
Candida albicans KF 1	>100	12.5	25
Saccharomyces cerevisiae ATCC 9763	>100	12.5	25
(Molds)			
Mucor racemosus IFO 4581	>100	25	12.5
Aspergillus niger KF 105	>100	12.5	25

* I: $CH_3SCH_2SSCH_3$, II: $CH_3SCH_2SSCH_2SCH_3$, III: $CH_3SO_2CH_2SSCH_2SCH_3$
Reproduced with permission from references 19 and 20. Copyright 1994 American Chemical Society.

cleavage. In addition, the latter two compounds showed relatively strong antimicrobial activity. From these results, it was concluded that wood garlic may serve not only as a flavor additive, but also as a natural preservative when used as a seasoning.

Acknowledgments

This work was conducted in cooperation with Mika Matsumoto (Shimojima), Ms. S. Ohhira, and M. Ueda. We are grateful to Dr. M. Murata of Ochanomizu University for kindly providing the microorganisms, and to Dr. N. Tominaga of Ochanomizu University for her helpful discussions on the antimicrobial assay.

Literature Cited

1. *A Dictionary of the Economical Products of the Malay Peninsula;* Burkill, I. H., Ed.; London, 1935, pp 1985-1986.
2. Carson, J. F. *Food Rev. Int.* **1987**, *3*, 71-103.
3. Block, E. *Angew. Chem. Int. Ed. Engl.* **1992**, *31*, 1135-1178.
4. Weichmann, J. In *Postharvest Physiology of Vegetables;* Weichmann, J., Ed.; Food Science and Technology 24; Marcel Dekker: New York, NY, 1987, pp 231-237.
5. Hatanaka, A; Kajiwara, T; Matsui, K. In *Progress in Flavour Precursor Studies;* Schreier, P.; Winterhalter, P, Eds.; Proceedings of the International Conference; Allured Publishing: Carol Stream, IL, 1993, pp 151-170.
6. Buttery, R. G.; Guadagni, D. G.; Ling, L. C.; Seifert, R. M.; Lipton, W. *J. Agric. Food Chem.* **1976**, *24*, 829-832.
7. Kameoka, H.; Higuchi, M. *Nippon Nogei Kagaku Kaishi* **1976**, *50*, 185-186.
8. Dumont, J. P.; Adda, J. *J. Agric. Food Chem.* **1978**, *26*, 364-367.
9. Moir, M.; Seaton, J. C.; Suggett, A. *Phytochemistry* **1980**, *19*, 2201.
10. Akisue, M. K.; Wasicky, R.; Akisue, G.; de Oliveira, F. *Rev. Farm. Bioquim. Univ. S. Paulo* **1984**, *20*, 145-157.
11. Dubs, P.; Stüsi, R. *Helv. Chim. Acta.* **1978**, *61*, 2351-2359.
12. Buttery, R. G.; Guadagni, D. G.; Ling, L. C. *J. Agric. Food Chem.* **1973**, *21*, 198-201.
13. Gmelin, R.; Luxa, H-H.; Roth, K.; Höfle, G. *Phytochemistry* **1976**, *15*, 1717-1721.
14. Cavallito, C. J.; Bailey, J. H. *J. Am. Chem. Soc.* **1944**, *66*, 1950-1951.
15. Cavallito, C. J.; Buck. J. S. *J. Am. Chem. Soc.* **1944**, *66*, 1952-1954.
16. Small, L. V.; Bailey, J. H.; Cavallito, C. J. *J. Am. Chem. Soc.* **1947**, *69*, 1710-1713.
17. Höfle, G. *Tetrahedron Letters* **1976**, *36*, 3129-3132.
18. Takazawa, H.; Tajima, F; Miyashita, C. *Yakugaku Zasshi* **1982**, *102*, 489-491.
19. Kubota, K.; Ohhira, S.; Kobayashi, A. *Biosci. Biotech. Biochem.* 1994, 58, 644-646.
20. Kubota, K.; Matsumoto, M.; Ueda, M.; Kobayashi, A. *Biosci. Biotech. Biochem.* 1994, 58, 430-431.

RECEIVED April 26, 1994

Chapter 20

Sulfur-Containing Heterocyclic Compounds with Antioxidative Activity Formed in Maillard Reaction Model Systems

Jason P. Eiserich and Takayuki Shibamoto

Department of Environmental Toxicology, University of California, Davis, CA 95616-8588

Extracts of microwave heated glucose/cysteine Maillard reaction systems and headspace samples of heated peanut oil/cysteine and peanut oil/methionine mixtures were shown to possess strong antioxidative activity. Several common classes of sulfur-containing volatile heterocyclic compounds known to be formed in Maillard-type reactions were evaluated for antioxidative activity. Antioxidative activity was measured by a recently developed method involving the oxidation of heptanal to heptanoic acid in a dichloromethane solution. Alkyl thiophenes, thiazoles, thiazolidine, and 1,3-dithiolane inhibited heptanal oxidation for various periods of time. The degree of unsaturation in the heterocyclic ring, as well as the substituent type had variable effects on the antioxidative capacity of these compounds. Thiophenes substituted with electron donating alkyl groups showed stronger activity than the unsubstituted thiophene. However, thiophene substituted with an electron withdrawing substituent such as the acetyl group showed little or no antioxidative activity. Saturated cyclic sulfides showed higher activity than either the thiazoles or thiophene derivatives. Cyclic sulfides reacted with *t*-butylhydroperoxide (*t*-BuOOH) and *m*-chloroperoxybenzoic acid (*m*-CPBA) to form S-oxides. Mechanisms of antioxidative activity are proposed. These results present the potential role of sulfur-containing heterocyclic compounds for inhibiting the oxidative degradation of lipid-rich foods.

Lipid peroxidation is the primary mechanism by which food deteriorates upon storage in the presence of oxygen. This process of oxidation can be initiated enzymatically, by metal ion catalysis, or by photochemical processes, to name a few. Free radicals including peroxyl, alkoxyl, and hydroxyl have been implicated in the mechanism of lipid peroxidation. The changes in the quality of processed foods are manifested by deterioration of flavor, aroma, color, texture, nutritive value, and the formation of toxic components (*1, 2*) including aldehydes and epoxides. These changes in food quality are therefore of significance to both the food industry and the consumer. Traditionally, the effects of lipid peroxidation are eliminated or suppressed with the use of synthetic antioxidants such as BHA and BHT. Recently, however, the safety of these compounds with respect to human health has been questioned, hence the search for "natural" alternatives.

Maillard reaction model systems, usually involving the heat treatment of a sugar or lipid, and an amino acid, has given significant insight into the thermal interactions of food constituents, and the subsequent formation of flavor and aroma. Aside from the complex flavor and aroma that is generated in model and real food systems, Maillard reaction products (MRPs) have been shown to possess some very interesting chemical and biological properties, including both antimutagenic and antioxidative activity (*3*).

The antioxidative activity of the MRPs was first observed by Franzke and Iwainsky (*4*), when they reported on the oxidative stability of margarine following the addition of products from the reaction of glycine and glucose. It has further been shown that antioxidative materials are formed as the result of heating foods. Yamaguchi et al. (*5*) found that the addition of glucose and amino acids to cookie dough prior to baking significantly increases oxidative stability. Shin-Lee (6) found that diffusate from beef browned extensively when heated at 180 °C for 2 hours, and inhibited lipid oxidation (TBA values) by 95% in cooked beef during storage.

Identification of Maillard reaction antioxidants has focused primarily on the higher molecular weight melanoidins. The Maillard reaction, however, also produces hundreds of volatile compounds that are responsible for the aroma of cooked food. Recently, volatile MRPs prepared by heating a glucose-glycine solution were found to slow the oxidative degradation of soybean oil by increasing the oxidation induction period and by decreasing the oxidation rate constant (*7, 8*). Macku and Shibamoto (*9*) identified 1-methylpyrrole as an antioxidant generated in the headspace of a heated corn oil/glycine model reaction system. Similarly, Eiserich et al. (*10*) showed that thiazoles, oxazoles, and furanones formed in an L-cysteine/D-glucose Maillard model system possessed antioxidative activity. These previous studies clearly show the potential of volatile MRPs to inhibit the oxidative degradation of lipid-rich foods.

Natural sulfur-containing compounds such as cysteine, glutathione, and lipoic acid, as well as synthetic compounds including N-acetylcysteine and a-mercaptopropionylglycine have historically been known to protect against oxidative stress in biological systems through the scavenging and reduction of various oxidants (*11*). The particular chemical and physcial characteristics including hypervalency, multiplicity of oxidation state, and the potential role of 3d orbitals in the stabilization of free radicals (*12*) implicates sulfur-containing compounds in foods as potential antioxidants. With these facts in mind, the objective of this research was to identify volatile heterocyclic sulfur-containing MRPs from various model systems that possessed antioxidative activity; compounds which have previously been overlooked as potential antioxidants.

Experimental Procedures

Maillard Reaction Sample Preparation.

Peanut Oil/Cysteine and Peanut Oil/Methionine Systems. Peanut oil (100 g) and 10.0 g of cysteine or methionine were mixed and placed in a 500-mL two-neck round-bottom flask, which was interfaced to a simultaneous purging and solvent extraction (SPE) apparatus developed by Umano and Shibamoto (13). The mixture was heated at 200°C for 5 hr while stirring. The headspace volatiles were purged into 250 mL of deionized water by a purified nitrogen stream at a flow rate of 10 mL/min. The volatiles trapped by the water were continuously extracted with dichloromethane (50 mL) for 6 hr. The water temperature was kept at 10°C by a Brinkman RM6 constant-temperature water circulator. The dichloromethane extract was dried over anhydrous sodium sulfate, and the extract was then concentrated to 2.0 mL by fractional distillation with a Vigreux column at atmospheric pressure. The concentrated extract was placed in a vial and stored under argon at -4°C until tested for antioxidative activity.

Glucose/Cysteine Systems. The method of sample preparation was adapted from Yeo and Shibamoto (14). L-Cysteine (0.05 mol) and D-glucose (0.05 mol) were dissolved in 30 mL of deionized water, and the pH of each solution was adjusted to 2, 5, 7, and 9 with either 6N HCl or 6N NaOH. The solutions were then brought to a final volume of 50 mL with deionized water and covered with Saran Brand plastic wrap. The four solutions were irradiated at the high setting of a 700-W microwave oven for 15 min. At 4-min intervals, the irradiation was interrupted and the samples rotated 90° to ensure uniform irradiation. The irradiation time coincided with the onset of browning, whereas further irradiation led to charring and sample combustion.

After microwave irradiation, each brown mass was dissolved in approximately 100 mL of deionized water and allowed to cool to room temperature. The resulting solutions were adjusted to pH 8 with 6N NaOH to enhance the extraction efficiency of nitrogen-containing heterocyclic compounds. The aqueous solution was extracted with 50 mL of dichloromethane using a liquid-liquid continuous extractor for 6 h and then dried over anhydrous sodium sulfate for 12 h. After removal of sodium sulfate, the dichloromethane extract was concentrated to 1 mL by fractional distillation with a Vigreux column at atmospheric pressure.

Measurement of Antioxidative Activity.
The antioxidative activity of the Maillard reaction products was evaluated by a method similar to that developed by Macku and Shibamoto (9) and later modified by Eiserich et al. (10). The antioxidative activities of 25 μL-aliquots of peanut oil/cysteine and peanut oil/methionine dichloromethane extracts and 5 μL-aliquots of the glucose/cysteine extracts were measured. 2-Alkyl-thiophenes, 2-thiophenethiol, 2-methyl-3-furanthiol, furfuryl mercaptan, thiazolidine, and 1,3-dithiolane were tested for antioxidative activity at a concentration of 1 mM. The above extracts and standards were added to dichloromethane solutions containing 25 mg of heptanal. Nonadecane (400 mg) was added as a gas chromatographic internal standard, and the resulting solutions were brought to a 5-mL final volume with

dichloromethane. The solutions were transferred to small vials and stored at room temperature. The headspace of each vial was purged with air every 2 days. Controls containing only heptanal, the internal standard, and dichloromethane were prepared for each experiment. The experimental vials and controls were periodically analyzed by GLC/FID.

A Hewlett-Packard (HP) Model 5890 gas chromatograph equipped with a flame ionization detector (FID) and a 30 m x 0.25 mm i.d. DB-Wax bonded-phase fused silica capillary column (J&W Scientific, Folsom, CA) was used to quantitate heptanoic acid formed from heptanal oxidation. The GC peak areas of heptanoic acid obtained at various time intervals were divided by the GC peak area of the internal standard (nonadecane, 400 mg) to calculate a relative peak area (RPA).

Reactions of Sulfur-Containing Compounds with *m*-CPBA and *t*-BuOOH. *m*-CPBA (3 mM) and *t*-BuOOH (3 mM) were reacted separately with equal molar concentrations of 2-butylthiophene and 1,3-dithiolane. The reactions were conducted in sealed glass vials at 25 and 50°C for *m*-CPBA and *t*-BuOOH, respectively. The reactions were allowed to proceed for 12 hr (*m*-CPBA) and 24 hr (*t*-BuOOH). The reaction products were subsequently analyzed by GC.

Identification of *m*-CPBA- and *t*-BuOOH-Induced Oxidation Products. The oxidation reaction products of the above listed sulfur-containing compounds were analyzed using a HP Model 5890 gas chromatograph equipped with a flame photometric detector (FPD) set in the sulfur mode. The injector and detector temperatures were both set at 250°C. A 30 m x 0.25 mm i.d. DB-1 bonded-phase fused silica capillary column (J&W Scientific, Folsom, CA) was used to separate the reaction products. The GC oven temperature was programmed from 80 to 250°C at a rate of 5°C/min. The linear velocity of the helium carrier gas flow was 26.5 cm/s, with a split ratio of 1:25.

Mass spectral (MS) identification of the oxidation products was carried out on a HP Model 5971 series mass selective detector (MSD) interfaced to a HP Model 5890 gas chromatograph. Mass spectra were obtained by electron impact ionization at 70 eV and a source temperature of 250°C. The capillary column and GC conditions were the same as described above.

Results and Discussion

Antioxidative Activity of Maillard Reaction Extracts. In the antioxidative assay system utilized in this study heptanal was readily oxidized to heptanoic acid in the dichloromethane solutions. However, the presence of a-tocopherol (Figure 1) inhibited this transformation in a concentration dependent manner. This system was then used to evaluate the antioxidative activity of dichloromethane extracts of several Maillard reaction model systems. Figure 2 shows the antioxidative activity of 5-μL aliquots of each pH extract from a microwave heated glucose/cysteine model system. The order of antioxidative effect of the extracts from the samples was as follows: pH 9 > pH 5 > pH 2 > pH 7. The Maillard reaction is catalyzed under both slightly basic and acidic conditions and may explain this trend. Volatiles from sugar/cysteine Maillard reaction

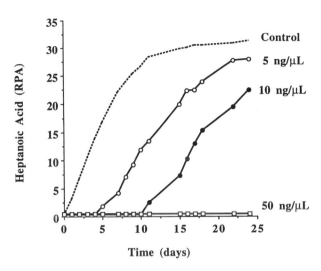

Figure 1. Antioxidative activity of various concentrations of α-tocopherol. The relative peak area (RPA) is equal to the GC peak area of heptanoic acid divided by the GC peak area of the internal standard, nonadecane.

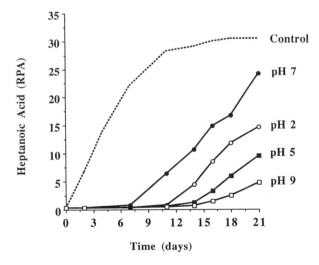

Figure 2. Antioxidative activity of L-cysteine/D-glucose extracts (5 μL) formed at different pHs. The relative peak area (RPA) is as defined in Figure 1.

model systems have been characterized and have been shown to give rise to an abundance of sulfur-containing heterocyclic compounds. Volatiles produced in lipidic model Maillard reaction systems give rise to similar compounds, including a variety of long chain alkyl-substituted heterocyclic compounds such as thiophenes. Aliquots (25 μL) of dichloromethane extracts of headspace volatiles collected from heated peanut oil/cysteine and peanut oil/methionine mixtures were similarly shown to possess antioxidative activity as illustrated in Figure 3. The volatiles derived from cysteine appeared to possess stronger activity than those formed from methionine.

Antioxidative Activity of Thiophene Derivatives. One of the most common sulfur-containing volatiles identified in Maillard reaction model systems and in real food systems such as cooked meat are thiophene derivatives. The high prevalence of thiophene derivatives in the Maillard reaction systems shown to possess antioxidative activity thus far prompted an investigation into their possible role as the active species. Figure 4 shows the antioxidative activities of various thiophene derivatives at 1 mM concentrations. Compared to the control vial, all of the thiophene derivatives inhibited the formation of heptanoic acid from heptanal in a dichloromethane solution, however, to varying degrees. Unsubstituted thiophene was slightly effective in inhibiting heptanoic acid formation, however, substitution with a methyl, ethyl, or butyl group at the 2-position greatly improved its antioxidative activity. These particular alkyl groups are effective electron donating substituents and increase the p-electron excessive character of carbons in the heteroaromatic ring. The length of the alkyl substituent, however, had no observable effects on the activity of these compounds. Conversely, substituting the thiophene with an acetyl group, an electron withdrawing substituent, decreased the antioxidative activity as compared to the unsubstituted thiophene. This type of trend was evident in a previous study with thiazole derivatives (*10*). These types of five-membered aromatic heterocyclic compounds are considered to be p-electron excessive. Figure 5 shows the electron densities of various heterocyclic compounds. The higher p-electron densities of the five-membered heterocyclic compounds render them open to electrophilic addition by radicals. It is interesting to note that in pyrazine, a six-membered heterocyclic ring, the two nitrogen atoms withdraw electrons from the carbon atoms rendering them p-electron deficient and not susceptible to addition by radicals; consistent with the fact that pyrazine derivatives do not show antioxidative activity (*10*).

Antioxidative Activity of Thiazolidine and 1,3-Dithiolane. Saturated cyclic sulfides represent another common class of sulfur-containing heterocyclic compounds formed in model systems and cooked foods. We have selected thiazolidine and 1,3-dithiolane as model compounds for which to evaluate the antioxidative activity of this class of compounds and the results are shown in Figure 6. 1,3-Dithiolane appears to exhibit slightly higher activity than that of thiazolidine. Both, however, inhibit heptanal oxidation for a longer period than do the thiophene derivatives; presumably due to their different structural characteristics and hence potentially different mechanisms of antioxidative action.

Reactions with *m*-CPBA and *t*-BuOOH. In an attempt to characterize and identify the mechanism(s) by which these various sulfur-containing heterocylic compounds behave as antioxidants, 2-butylthiophene and 1,3-dithiolane were reacted with the electrophilic

20. EISERICH & SHIBAMOTO *Sulfur-Containing Heterocyclic Compounds* 253

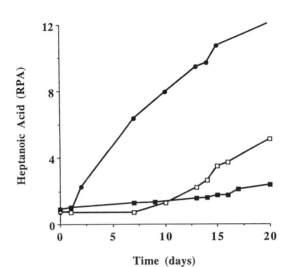

Figure 3. Antioxidative activity of extracts (25 µL) of headspace volatiles formed from heated peanut oil/cysteine (shaded squares) and peanut oil/methionine (open squares) systems. The relative peak area (RPA) is as defined in Figure 1. The control is indicated by the shaded circles.

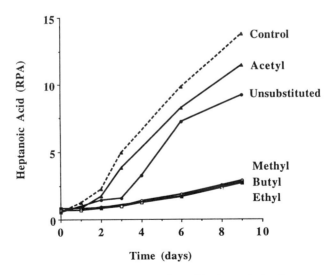

Figure 4. Antioxidative activity of various thiophene derivatives at concentrations of 1 mM. The relative peak area (RPA) is as defined in Figure 1.

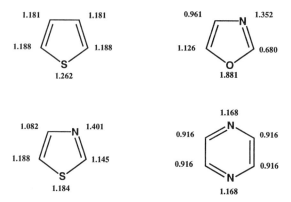

Figure 5. p-Electron densities of some common five- and six-membered heterocyclic compounds formed in Maillard reaction model systems. Values were calculated by Mahanti (20) using the Hückel molecular orbital (HMO) method.

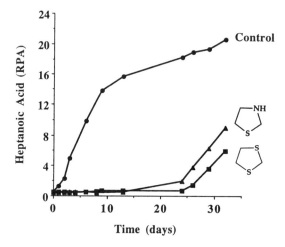

Figure 6. Antioxidative activity of thiazolidine and 1,3-dithiolane at concentrations of 1 mM. The relative peak area (RPA) is as defined in Figure 1.

oxidants m-CPBA and t-BuOOH in a dichloromethane solution. 1,3-Dithiolane was oxidized to two products, specifically the 1,3-dithiolane 1-oxide (sulfoxide) and the 1,3-dithiolane 1,3-oxide (disulfoxide) by both the peracid m-CPBA and the lipid soluble peroxide t-BuOOH. Mass spectra of the oxidized products are shown in Figure 7. 2-Butylthiophene, on the other hand, was unreactive with either m-CPBA or

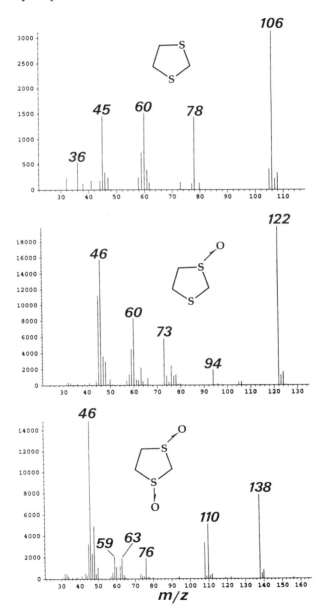

Figure 7. Mass spectra of the products obtained by the oxidation of 1,3-dithiolane with m-CPBA and t-BuOOH.

t-BuOOH; the concentration of the parent species did not decrease over the reaction time and no products were observed. 1,3-Dithiolane reacts with organic hydroperoxides and peracids by nucleophilic attack. Lipid hydroperoxides are known to be involved in Fenton-type reactions producing oxygen-centered peroxyl and alkoxyl radicals which are capable of inducing lipid peroxidation. Consumption of these reactive oxygen precursors, then, may be their mechanism of antioxidant protection. Another possible mechanism by which these cyclic sulfides may behave as antioxidants is through the quenching of the high energy singlet oxygen ($^1O_2$). Photooxidation of vegetable oils is a major concern of the food industry, in that they contain natural photosensitizers and are commercially sold under light. 1,3-Dithiolane 1-oxide has been efficiently synthesized via singlet oxidation of 1,3-dithiolane (18). Quenching of $^1O_2$, then, may be an interesting mechanism of cyclic sulfide antioxidant protection deserving further investigation.

Conclusions

Identification of volatile sulfur-containing heterocyclic compounds as antioxidants derived from Maillard reaction systems is a novel finding. Synthetic meat flavor mixtures, consisting of a multitude of sulfur-containing aroma compounds have previously been shown to protect desirable flavor and oxidative stability of cooked meat during storage (19). Our findings suggest that volatile sulfur-containing heterocyclic compounds including thiophenes, thiazoles, and saturated cyclic sulfides may be in part responsible for the antioxidative phenomena observed with Maillard reaction products.

Acknowledgments

The authors thank Dr. Tae Yung Chung for his assistance in preparation of the Maillard reaction samples.

Literature Cited

1. Kanner, J. In *Lipid Oxidation in Food*; St. Angelo, A. J., Ed.; ACS Symposium Series 500; American Chemical Society: Washington, DC, 1992; pp. 55-73.
2. Kanner, J.; German, J. B.; Kinsella, J. E. *CRC Crit. Rev. Food Sci. Nutr.* **1987**, *25*, 317.
3. Namiki, M. *Adv. Food Res.* **1988**, *32*, 115-184.
4. Franzke, C.; Iwainsky, H. *Dtsch. Lebensm. Rundsch.* **1954**, *50*, 251-254.
5. Yamaguchi, N.; Yokoo, Y.; Koyama, Y. *Nippon Shokuhin Kogyo Gakkaishi* **1964**, *11*, 431-435.
6. Shin-Lee, S. Y. Warmed-Over Flavor and its Prevention by Maillard Reaction Products; PhD Thesis; University of Missouri, Columbia, 1988.

7. Elizalde, B. E.; Dalla Rosa, M.; Lerici, C. R. *J. Am. Oil Chem. Soc.* **1991**, *68*, 758-762.
8. Elizalde, B. E.; Bressa, F.; Dalla Rosa, M. *J. Am. Oil Chem. Soc.* **1992**, *69*, 331-334.
9. Macku, C.; Shibamoto, T. *J. Agric. Food Chem.* **1991**, *39*, 1990-1993.
10. Eiserich, J. P.; Macku, C.; Shibamoto, T. *J. Agric. Food Chem.* **1992**, *40*, 1982-1988.
11. Murphy, M. E.; Scholich, H.; Sies, H. *Eur. J. Biochem.* **1992**, *210*, 139-146.
12. Oae, S. In *Organic Sulfur Chemistry: Structure and Mechanism*; Oae, S., Doi, J. D., Eds.; CRC Press: Boca Raton, FL, 1991.
13. Umano, K.; Shibamoto, T. *J. Agric. Food Chem.* **1987**, *35*, 14-18.
14. Yeo, H. C. H.; Shibamoto, T. *J. Agric. Food Chem.* **1991**, *39*, 370-373.
15. Mac Leod, G.; Ames, J. M. *Chem. Ind.* **1986**, 175-177.
16. Ames, J. M.; Mac Leod, G. *J. Food Sci.* **1985**, *50*, 125-135.
17. Werkoff, P.; Brüning, J.; Emberger, R.; Güntert, M.; Köpsel, M.; Kuhn, W.; Surburg, H. *J. Agric. Food Chem.* **1990**, *38*, 777-791.
18. Pandey, B.; Bal, S. Y.; Khire, U. R.; Rao, A. T. *J. Chem. Soc. Perkin Trans. I* **1990**, 3217-3218.
19. Bailey, M. E.; Um, K. W. In *Lipid Oxidation in Food*; St. Angelo, A. J., Ed.; ACS Symposium Series 500; American Chemical Society: Washington, DC, 1992; pp. 122-139.
20. Mahanti, M. K. *Indian J. Chem.* **1977**, *15B*, 168-174.

RECEIVED March 23, 1994

Chapter 21

Mechanisms of Beneficial Effects of Sulfur Amino Acids

Mendel Friedman

Food Safety and Health Research Unit, Western Regional Research Center, Agricultural Research Service, U.S. Department of Agriculture, 800 Buchanan Street, Albany, CA 94710

Sulfhydryl (thiol) compounds such as cysteine, N-acetylcysteine, reduced glutathione, and mercaptopropionylglycine interact with disulfide bonds of plant protease inhibitors and lectins via sulfhydryl-disulfide interchange and oxidation-reduction reactions. Such interactions with inhibitors from soybeans and with lectins from lima beans facilitate heat inactivation of the potentially toxic compounds, resulting in beneficial nutritional effects. Related transformations of protease inhibitors in soy flour are also beneficial. Since thiols are potent nucleophiles, they have a strong affinity for unsaturated electrophilic centers of several dietary toxicants, including aflatoxins, sesquiterpene lactones such as elephantropin and parthenin, urethane, carbonyl compounds, quinones, and halogen compounds. Such interactions may be used *in vitro* to lower the toxic potential of the diet, and *in vivo* for prophylactic and therapeutic effects against oxidative damage. A number of examples are cited to illustrate the concepts and mechanisms of using sulfur amino acids to reduce the antinutritional and toxic manifestations of food ingredients.

The sulfhydryl (SH) group of the naturally occurring sulfur amino acid cysteine (CySH) and of the naturally occurring tripeptide reduced glutathione (GSH), as well as the disulfide group (SS) of the corresponding oxidized forms, cystine and oxidized glutathione, respectively, participate in many chemical and biochemical processes in plants, animals, and foods. Free SH groups appear in plasma as so-called non-protein SH groups, mainly in the form of reduced glutathione. Protein SH groups are found in many structural proteins, enzymes, and nucleoproteins. They have physiological and biochemical functions in cell division (*1-3*). Disulfide bonds associated with cystine residues are structural units essential in most proteins and enzymes. Disulfide bridges in proteins, together with hydrogen-bonding and hydrophobic interactions, are largely responsible for the conformations and configurations that proteins assume in solution or in the solid state.

This chapter not subject to U.S. copyright
Published 1994 American Chemical Society

Since SH groups are among the most reactive functional groups in chemical and biochemical systems, it is not surprising that many reactive groups in food ingredients and in biological molecules interact with and modify SH groups *in vivo* and *in vitro*. Such sites include disulfide bonds, which can participate with SH groups in sulfhydryl-disulfide interchange and related oxidation-reduction reactions, and carbon-carbon and carbon-oxygen double bonds, which can participate with SH groups in nucleophilic addition reactions.

This brief review describes the chemical and mechanistic basis for reducing the adverse effects of several classes of potentially toxic food ingredients based on the interaction of the SH group of cysteine and related thiols with the toxicants reactive sites. The cited references (*1-96*) offer an entry into the widely scattered literature on the subject.

Protease Inhibitors

Feeding rats raw soybeans or pure soybean inhibitor of the digestive enzymes chymotrypsin and trypsin, the so-called Bowman-Birk and Kunitz protease inhibitors, induce hypertrophy and hypersecretion of chymotrypsin and trypsin by the pancreas. Raw soy flour diets also potentiate the carcinogenicity of azaserine and nitrosamines in rats. A so-called biofeedback mechanism has been postulated to explain these consequences of soybean nutrition (*4-17*).

These harmful effects suggest a need to find ways to inactivate the inhibitors before soybeans are consumed. Although heat is often used to inactivate protease inhibitors, such inactivation is often incomplete. High temperature may also destroy lysine and sulfur amino acids and thus damage protein nutritional quality.

Most inhibitors of chymotrypsin and trypsin contain disulfide bonds which are needed to maintain active conformations and configurations. Since thiols are expected to react with inhibitor disulfide bonds via sulfhydryl-disulfide interchange and oxidation-reduction reactions, illustrated in Figure 1, we carried out extensive studies on the ability of cysteine, N-acetylcysteine, and reduced glutathione to synergize heat inactivation of soybean inhibitors. Figures 2 and 3 depict some of our findings.

With reference to Figure 1, if the inactivation is a thiol-disulfide interchange, then the excess thiolate anion is expected to force the equilibrium to the right. Although the inhibitor may tend to regenerate its original conformation, this process may be suppressed by excess thiol or the mixed disulfides. Because of altered structure, the modified inhibitors do not readily combine with the active sites of chymotrypsin or trypsin to form tight-fitting complexes.

Our results show that thiols and sulfites, which also cleave disulfide bonds, make it possible to lower protease inhibitor content of soybean flour to near zero levels. The accompanying improvement in nutritional quality as measured by rat feeding studies may result from one or more of the following events: (a) inactivation of inhibitors of digestive enzymes; (b) improvement in digestibility of the soy and

R-SH + In-S-S-In ⇌ R-S-S-In + HS-In

R-SH + Pr-S-S-Pr ⇌ R-S-S-Pr + HS-Pr

In-SH + Pr-S-S-Pr ⇌ In-S-S-Pr + HS-Pr

In-SH + HS-Pr + ½O₂ → In-S-S-Pr + H₂O

In-SH + HS-In + ½O₂ → In-S-S-In + H₂O

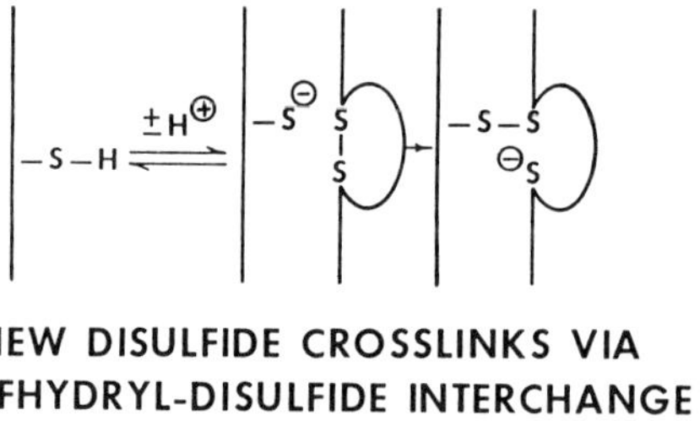

$$-S-H \underset{}{\overset{\pm H^{\oplus}}{\rightleftharpoons}} -S^{\ominus} \quad \overset{S}{\underset{S}{\big|}} \quad -S-S \atop \ominus S$$

**NEW DISULFIDE CROSSLINKS VIA
SULFHYDRYL-DISULFIDE INTERCHANGE**

Figure 1. Sulfhydryl-disulfide-interchange and oxidation pathways between thiols (RSH), protease inhibitors (In-S-S-In), and structural proteins (P-S-S-P). The net result is a new network of disulfide bonds and changed structure of the inhibitors.

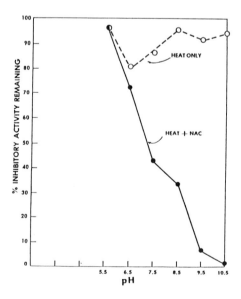

Figure 2. Effect of pH and N-acetylcysteine (NAC) on trypsin inhibitory activity of lima bean inhibitor at 45°C. Upper plot, no NAC; lower plot, with NAC.

protease inhibitor proteins; and (c) introduction of new cystine residues in sulfur-poor legumes.

This approach should generally be useful for inactivating related disulfide-containing toxic plant and animal proteins.

Lectins

Legumes are a rich source of lectins or phytohemagglutinins. Inadequately cooked legumes may cause lectin-induced disturbances and adverse nutritional effect in humans. Lima bean lectin contains exposed cysteinyl groups necessary for its activity. We assessed the ability of N-acetylcysteine to act synergistically with heat in destabilizing and inactivating this lectin in lima bean flour.

Our study (*18*) revealed that the cysteine derivative facilitated inactivation of lima bean lectin in the temperature range 25 to 85°C and in the pH range 4.4 to 10.0. As with the protease inhibitors described above, the beneficial action of N-acetylcysteine is postulated to involve formation of mixed disulfide bonds between the cysteine derivative, lectins, and structural proteins in the flour.

In contrast to the findings with lima bean lectins soybean lectins which contain no cysteinyl residues, were not affected by the thiol treatments.

Aflatoxin B_1

Aflatoxin B_1 (AFB_1) is a precarcinogen that is transformed *in vivo* to an active epoxide which then alkylates DNA, inducing liver and other tumors (Figure 4). Because thiols are strong nucleophiles, they may be expected to react with electrophilic sites of AFB_1 and thus competitively inhibit the interaction of these sites with DNA or other cell components, as illustrated in Figures 4 and 5. The resulting products are therefore expected to be inactive when evaluated in the Ames *Salmonella typhirmurium in vitro* mutagenicity assay. To test this hypothesis, we evaluated the effectiveness of twelve thiols in inhibiting the mutagenicity of AFB_1 (*19*). Our results revealed that N-acetylcysteine (but not cysteine), reduced glutathione, and mercaptopropionylglycine were highly effective in suppressing the mutagenic activity observed with untreated AFB_1.

It is instructive to analyze possible mechanisms of reaction between a thiol and four electrophilic sites in the aflatoxin molecule shown in Figure 5. In pathway 1, the thiolate anion adds to the double bond of the protonated furan ring of AFB_1. In pathway 2, a thiol cleaves the lactone ring to produce a thiol ester. In pathway 3, a thiol adds to the double bond of the lactone ring, which is activated by protonation of the methoxy group. Pathway 4 represents addition of the thiol to the other end of the same double bond, where it is activated by the protonated carbonyl group. The last two processes are less likely to occur than the first two because addition to the double bond of the lactone ring would disrupt conjugation and would therefore not be favored energetically. The most likely event is reaction of the thiol with the double bond of the furan ring (pathway 1). Such an interaction would prevent the double bond of the furan ring from forming a mutagenic epoxide. This would explain the effectiveness of thiols in suppressing the mutagenic activity of

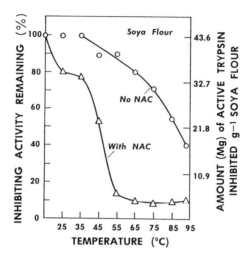

Figure 3. Effect of NAC on trypsin inhibitory activity of soybean inhibitors in soybean flour. Upper plot, no NAC; lower plot, with NAC.

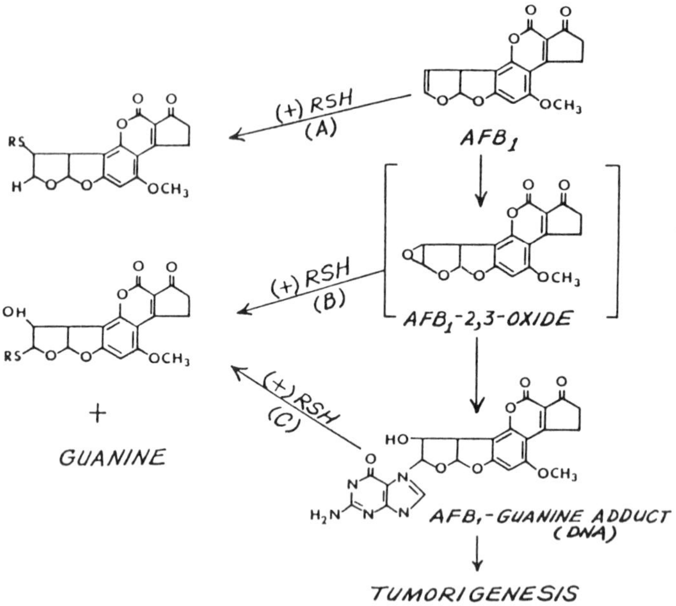

Figure 4. Some possible aflatoxin-thiol interactions: A. addition to the double bond of AFB_1 to form an inactive thiol adduce; B. interaction with AFB_1 epoxide which may prevent the epoxide from interacting with DNA; C. displacement of an AFB_1-DNA adduct blocking tumor formation.

AFB$_1$. Figure 6 shows experimental evidence for the reaction of N-acetylcysteine and AFB$_1$.

Related studies on beneficial effects of thiols in reducing mycotoxin toxicity are described by Mandel et al. (*20*) and Shetty et al. (*21*).

Sesquiterpene Lactones

Kinetic and synthetic studies showed that naturally occurring potentially toxic sesquiterpene lactones such as elephantropin and parthenin participate in alkylation reactions with cysteine forming inactive mono- and bis- cysteine adducts, as shown in Figures 7 and 8 (*22-24*).

Urethane

Urethane (ethyl carbamate) is found in low levels in some foods (*25-29*). The compound is reported to induce lung cancer. DeFlora *et al.* (*25*) showed that the incidence of lung tumors in mice fed urethanes as part of a standard diet reached 94.1%. Feeding N-acetylcysteine 15 days before and 4 months after injection of the carcinogen drastically and significantly reduced the tumor incidence to 58.7%. The spontaneous tumor rate in untreated animals was 27.8%.

A possible mechanism for the prevention of tumor formation is illustrated below. The thiol or thiolate anion can participate with ethyl carbamate in a nucleophilic displacement reaction (equation 1). The resulting thioamide can be cleaved hydrolytically to regenerate the thiol and produce NH_3 and CO_2 (equation 2). The net effect is a destruction of urethane, preventing it from modifying DNA or other cellular sites.

$$RSH + NH_2CO\text{-}OC_2H_5 \rightarrow NH_2CO\text{-}SR + C_2H_5OH \quad (1)$$
$$\text{thiol} \quad\quad \text{urethane} \quad\quad\quad \text{thioamide} \quad\quad \text{ethanol}$$

$$NH_2CO\text{-}SR + H_2O \rightarrow RSH + NH_3 + CO_2 \quad (2)$$

Lysinoalanine

Lysinoalanine (HOOCCH(NH$_2$)CH$_2$NH(CH$_2$)$_4$CH(NH$_2$)COOH) is an unnatural crosslinked amino acid formed from cystine and serine residues and the ϵ-NH$_2$ group of lysine. Protein-bound and free lysinoalanine, when fed to rats, induce cytotoxic changes in kidney cells. These changes are characterized by enlargement of the nucleus and cytoplasm and disturbances in DNA metabolism. These changes suggest disruption of normal regulatory function of kidney cells (*30-56*).

A postulated mechanism for lysinoalanine formation is a two-step process. First, hydroxide ion-catalyzed elimination reactions of serine, threonine, and cystine give rise to a dehydroalanine intermediate, illustrated in Figures 9 and 10 for cystine. The dehydroalanine residue, which contains a conjugated carbon-carbon double bond, then reacts with the ϵ-NH$_2$ group of lysine to form a lysinoalanine crosslink.

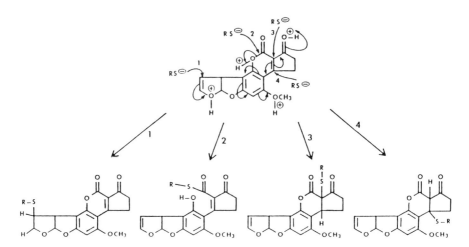

Figure 5. Some possible reactions between RS⁻ and four electrophilic sites of AFB_1.

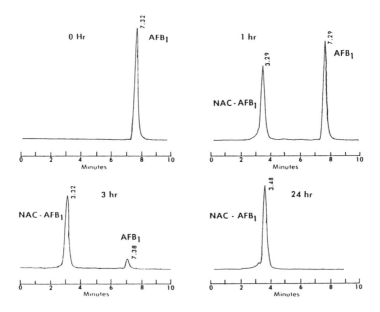

Figure 6. HPLC of AFB_1 and its N-acetylcysteine derivative (NAC-AFB_1).

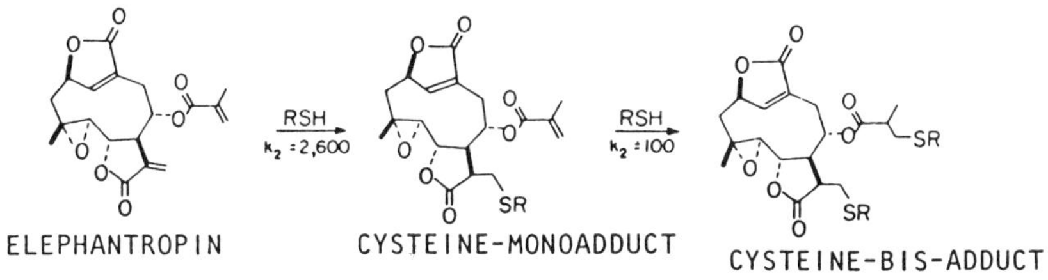

Figure 7. Cysteine adducts of the sesquiterpene lactone, elephantropin.

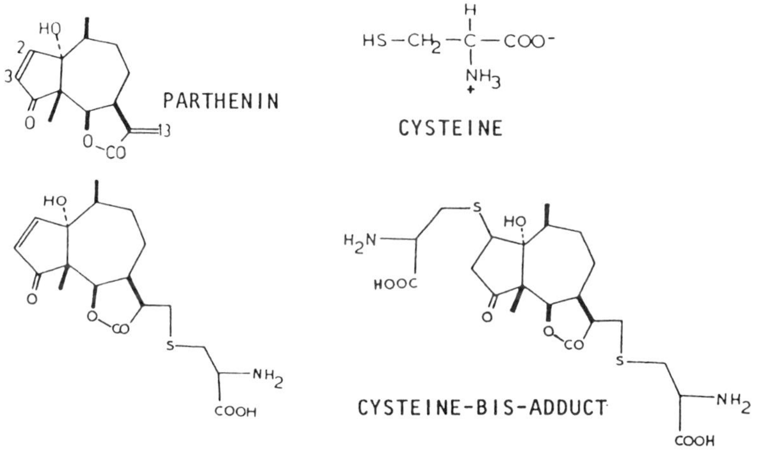

Figure 8. Cysteine adducts of the sesquiterpene lactone, parthenin.

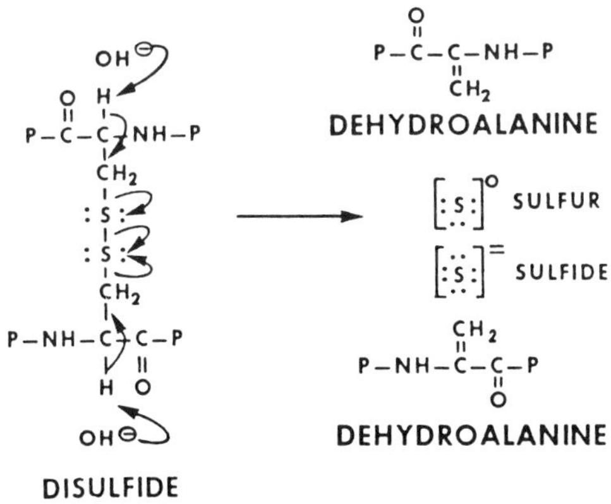

Figure 9. Mechanism for base-catalyzed transformation of a protein disulfide bond to two dehydroalanine side chains, a sulfide ion, and sulfur.

Since SH groups react more readily with double bonds than do NH_2 groups (*43,44,49,54*), addition of thiols should trap the residue of dehydroalanine as described below. These competitive reactions should minimize lysinoalanine formation. These expectations were realized.

Added thiols or other nucleophiles such as sulfite ions can inhibit lysinoalanine formation by at least three mechanisms (Figure 11). First, direct competition from the added thiol can trap dehydroalanine residues from protein amino acid side chains (Figure 11, equations 1 and 2). Second, the added nucleophile can cleave protein disulfide bonds (Figure 11, equations 3 and 4) and generate free SH groups, which may, in turn, combine with dehydroalanine residues blocking lysinoalanine formation (Figure 11, equation 5). Third, the added nucleophile can eliminate a potential source of dehydroalanine by cleaving disulfide bonds, inasmuch as cystine residues (P-S-S-P) of a protein would be expected to undergo ß-elimination reactions more easily than negatively charged cysteine (P-S⁻) residues.

The cited evidence for the ß-elimination mechanism leading to dehydroalanine formation merits further comment. Nashef et al. (*41*) report that alkali-treatment of lysozyme ribonuclease and several other proteins resulted in loss of cystine and lysine residues and the appearance of new amino acids lysinoalanine, lanthionine, and ß-aminoalanine. Alkali-treatment of the proteins induced an increase in absorbance at 241 nm, presumably from the formation of dehydroalanine residues. The dehydroalanine side chain can participate in nucleophilic addition reactions with the ε-NH_2 group of lysine to form lysinoalanine, with the SH groups of cysteine to form lanthionine, and with ammonia to form ß-aminoalanine.

Although the results are consistent with a ß-elimination reaction leading to formation of dehydroalanine, the conclusions are based on the assumption that the absorbance at 241 nm is associated with dehydroalanine side chains derived from cystine residues. This assumption may not always be justified for the following reasons. First, alkali treatment of casein which has very few or no disulfide bonds also yields significant amounts of dehydroalanine residues (*52*). These presumably arise from serine side chains. Second, Nashef et al. (*41*) cite evidence that other functionalities may contribute to the 241 nm absorption. These considerations suggest that there is a need to directly measure dehydroalanine in proteins. This is now possible with our method (*52*), whereby the dehydroalanine residues are first transformed to S-pyridylethyl side chains by reaction with 2-mercaptopyridine (Figure 12). Amino acid analysis of the acid-hydrolyzed protein permits estimation of the dehydroalanine content as S-ß-(2-pyridylethyl)-L-cysteine along with the other amino acids.

Browning Prevention

Reaction of amino groups of amino acids, peptides, and proteins with carbonyl groups of carbohydrates or with tyrosine-derived quinone groups encompass changes respectively called nonenzymatic and enzymatic browning reactions (*57-75*). Such changes may adversely affect the appearance, taste, nutritional quality, and safety of many foods. Until recently, sodium sulfite has been used to inhibit browning

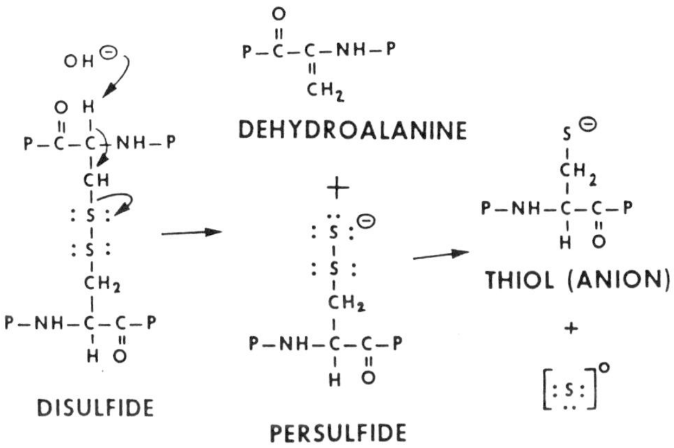

Figure 10. Mechanism for base-catalyzed formation of one dehydroalanine residue, and one persulfide ion from a protein disulfide bond. The persulfide can decompose to a thiolate anion and elemental sulfur.

Direct Competition

$R\text{-}S^- + CH_2=C(NHCOP)\text{-}P + H^+ \rightleftarrows R\text{-}S\text{-}CH_2\text{-}CH(NHCOP)\text{-}P$ (1)

$SO_3^{-2} + CH_2=C(NHCOP)\text{-}P + H^+ \rightleftarrows {}^-O_3S\text{-}CH_2\text{-}CH(NHCOP)\text{-}P$ (2)

Suppression and First Step of Indirect Competition

$2(R\text{-}S^-) + P\text{-}S\text{-}S\text{-}P \rightleftarrows 2(P\text{-}S^-) + R\text{-}S\text{-}S\text{-}R$ (3)

$SO_3^{-2} + P\text{-}S\text{-}S\text{-}P \rightleftarrows P\text{-}S^- + P\text{-}S\text{-}SO_3^-$ (4)

Indirect Competition

$P\text{-}S^- + CH_2=C(NHCOP)\text{-}P + H^+ \rightleftarrows P\text{-}S\text{-}CH_2\text{-}CH(NHCOP)\text{-}P$ (5)

R = thiol side chain
P = protein side chain

Figure 11. Mechanistic pathways for trapping dehydroalanine residues by RS$^-$ and SO$_3^{-2}$, thus preventing lysinoalanine formation.

a. ENZYMATIC BROWNING:

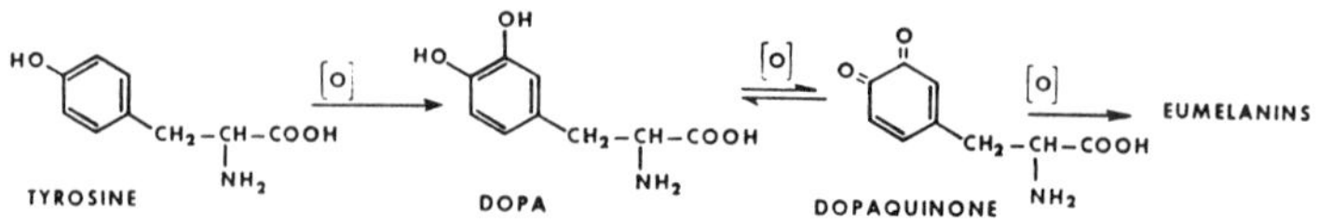

TYROSINE → DOPA ⇌ DOPAQUINONE → EUMELANINS

b. BROWNING PREVENTION:

1. MODIFICATION OF ACTIVE SITE:

   a. Enzyme-Cu-Histidine + HS-R → Enzyme-Cu-S-R + Histidine

   b. Enzyme-Cu-Histidine + 2 HS-R → Enzyme + R-S-Cu-S-R + Histidine

   c. Oxygen-mediated reduction of Cu(II) to Cu(I):

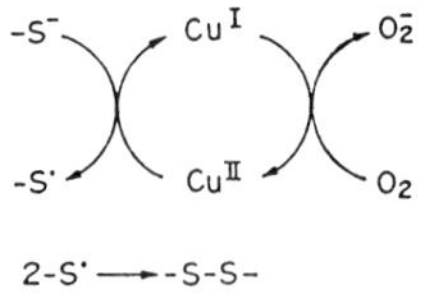

   $2\text{-S}^{\cdot} \longrightarrow \text{-S-S-}$

2. ADDITION TO AND/OR REDUCTION OF QUINONE INTERMEDIATE:

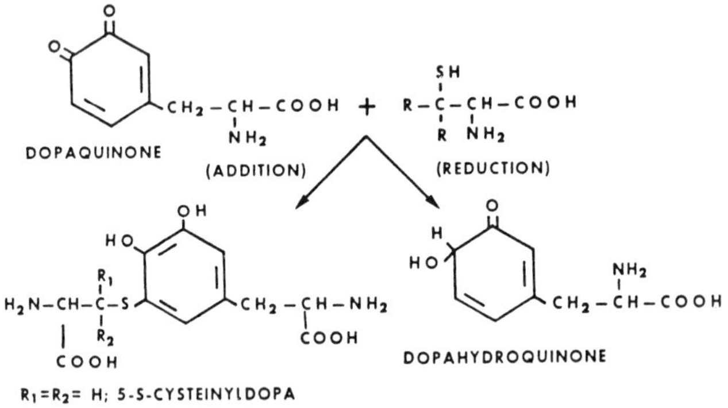

   DOPAQUINONE + R-C(SH)(R)-CH(NH$_2$)-COOH

   (ADDITION) → 5-S-CYSTEINYLDOPA ($R_1 = R_2 = H$)

   (REDUCTION) → DOPAHYDROQUINONE

Figure 12. Transformation of dehydroalanyl to lysinoalanine (LAL), S-ß-(2-pyridylethyl)-L-cysteine (2-PEC), and lanthionine (LAN) residues in a protein. Hydroxide ions induce elimination reactions in cysteine and serine to form dehydroalanine. The double bond of dehydroalanine then interacts with the ε-NH$_2$ group of lysine to form LAL, with the SH group of cysteine to form LAN, and with the SH group of added 2-mercaptoethylpyridine to form 2-PEC. The latter is identical to the compound obtained from cysteine and 2-vinylpyridine.

reactions. Since some individuals are sensitive to sulfites, their use is being discontinued.

Our studies show that N-acetylcysteine and reduced glutathione are nearly as effective as sodium sulfite in preventing browning of heated amino acid and protein-carbohydrate mixtures, apples, potatoes, and fruit juices (64, 68, 69, 73-75).

The mechanism of thiols' prevention of nonenzymatic browning may involve preferential reaction of the thiol with the carbonyl of carbohydrates, preventing them from interacting with amino groups under the influence of heat (65).

Golan-Goldhirsh and colleagues (71, 72) report that sodium bisulfite, reduced glutathione, and ascorbic acid caused a 50% inactivation of mushroom polyphenol oxidase after 28, 106, and 130 min, respectively, at 5 mM concentration. Dithiothreitol was a more effective inhibitor, causing 50% inactivation at 0.1 mM after only 70 min.

These authors suggest that (a) the observed effect of reductants on mushroom polyphenol oxidase is dependent on whether a polarographic or spectroscopic method is used to assay the enzyme; (b) the spectrophotometric method is useful to determine the optimum reductant concentration required to inhibit browning; and (c) the polarographic method is useful to establish whether the inactivation of the enzyme is irreversible.

Matheis and Whitaker (59) reviewed the chemistry of the modification of proteins by polyphenol oxidase and peroxidase. Monophenols such as tyrosine are first hydroxylated to dihydroxyphenylalanine (DOPA). The DOPA is then oxidized to an ortho-quinone. Ortho-quinones can undergo at least two types of reactions with thiols. Two molecules of a thiol can participate in an oxidation-reduction reaction with a quinone to form the corresponding hydroquinone and a disulfide, or the thiol anions may add to the conjugated system of a quinone to form a substituted hydroquinone. An excess of quinone in the reaction mixture usually oxidizes the monosubstituted quinone, which may then participate in another nucleophilic addition, and so on (Figure 13). The cysteine adduct(s) cannot further react with amino groups to form brown products.

Structurally different thiols may have different effects on the two inactivation pathways of enzymatic browning (71, 75; Table I).

Elsewhere, I have proposed the name "Oxidation-Addition-Substitution Reaction" to describe the oxidation of hydroquinones to quinones followed by nucleophilic addition of SH or other groups to the quinone (1). Analogous oxidation-addition-substitution reactions can also take place when porphyrins are first oxidized to dehydroporphyrin intermediates, which can then participate with nucleophiles in addition reactions to form substituted porphyrins (76).

Recent efforts to characterize thiol adducts formed during enzymatic and non-enzymatic browning of foods are described in the references (77-80).

Photochemistry of Proteins

In a study on the influence of dimethyl sulfoxide (DMSO) on the photochemistry of cereal proteins such as wheat gluten, we found that methionine and cystine residues were more sensitive to ultraviolet radiation in DMSO than in aqueous media (81).

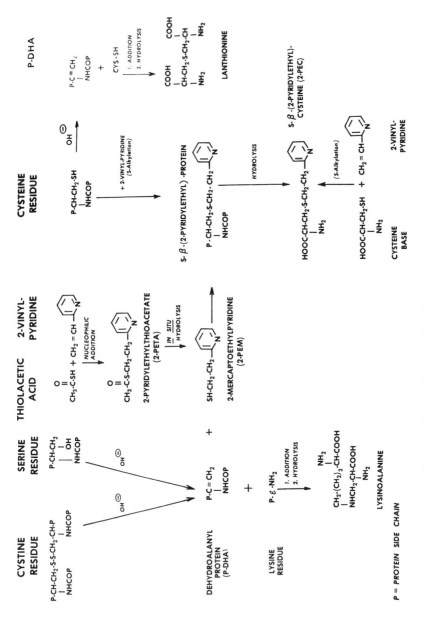

Figure 13. Possible mechanisms for the action of PPO and its inhibition by thiols.

Table I. Millimolar Concentration of Inhibitors Required to Reduce PPO Activity at 25°C by 50% (I_{50}), and Potencies Relative to Cysteine. The Lower the I_{50} Value, the More Potent the Inhibitor.

	I_{50}	potency relative to L-cysteine
L-cysteine	0.35	1.00
L-cysteinyl-glycine	0.43	0.81
N-acetyl-L-cysteine	0.27	1.30
homocysteine	0.23	1.52
sodium bisulfite	0.21	1.67
2-mercaptoethane sulfonic acid	0.14	2.50
L-cysteine ethyl ester	0.12	2.92
reduced glutathione	0.12	2.92
2-mercaptopyridine	0.10	3.50
L-cysteine methyl ester	0.099	3.53
N-(2-mercaptopropionyl)-glycine	0.058	6.03

If the primary step in the photochemical transformation of sulfur amino acids involves photo-ejection of an electron to form a radical cation, as illustrated in equation 3, any photochemical transformation of disulfide bonds that proceeds through such a radical cation intermediate should be sensitized in DMSO, possibly as illustrated below. The stabilization of the radical cation by DMSO would favor its formation, thus explaining the enhancement of photochemical destruction of sulfur amino acids by dipolar aprotic solvents.

$$\text{P-S-S-P} + h\nu \rightarrow [\text{P-S-}\overset{\bullet}{\text{S}}\text{-P}]^+ + e^- \quad (3)$$

protein disulfide   disulfide radical cation

$$[\text{P-S-}\overset{\bullet}{\text{S}}\text{-P}]^+ \cdot \text{O-S}^+\Big\langle \begin{matrix} \text{CH}_3 \\ \text{CH}_3 \end{matrix}$$

stabilized radical cation

In related studies (82-85), we discovered that styrene in a protein-DMSO solution prevents destruction of sensitive amino acids such as tyrosine, methionine, and cystine by ultraviolet radiation. The basis of this effect is not established. On irradiation, styrene probably forms a low-energy triplet state. Consequently, no transfer of energy occurs by a triplet state of styrene to the singlet state of amino acid residues. Another possibility is that radiation may be absorbed by styrene before it can attack sensitive amino acid residues. This suggests that styrene may be a useful radioprotectant of proteins.

Detoxification Reactions

Sulfhydryl compounds such as reduced glutathione react *in vivo* with aliphatic and aromatic halides, conjugated compounds, epoxides, and numerous other xenobiotics to form soluble derivatives which can be more readily eliminated from the body than the unreacted compounds (*86-88*). These so-called "detoxification" reactions are catalyzed by the enzyme glutathione S-transferase. They seem to have a common mechanism. The organic substance alkylates the SH group of glutathione. Then, successively, the γ-glutamyl and glycine residues are removed and the amino group of the cysteine derivative is acetylated. The resulting mercapturic acid is excreted (Figure 14).

Recent studies on glutathione-glutathione S-transferase interrelationships important in such detoxifications of xenobiotics are described by Beutler (*89*), Henning et al. (*90*), Jones (*91*), Rushmore and Pickett (*92*), and Smith (*93*).

Antioxidative Effects

According to Lii and Hendrich (*94*), two pathways exist for the *in vivo* S-glutathiolation of protein SH groups: (a) glutathione disulfide (GSSG)-protein mixed disulfide exchange (equation 4); and (b) direct combination of thiol free radicals to mixed disulfides, possibly as shown in equations 5-7.

$$P\text{-}SH + GS\text{-}SG \rightarrow PS\text{-}SG + GSH \quad (4)$$

$$GSH + t\text{-}BuOOH \rightarrow GS\cdot \quad (5)$$

$$PSH + t\text{-}BuOOH \rightarrow PS\cdot \quad (6)$$

$$PS\cdot + GS\cdot \rightarrow PS\text{-}SG \quad (7)$$

The formation of protein-glutathione mixed disulfide prevents oxidation of protein SH groups to, for example, a sulfonic acid derivative (PSO_3). The GS· and PS· radicals derived from the action of *t*-butyl peroxide (*t*-BuOOH) on GSH and PSH (equations 5,6) can dimerize to form PS-SG (equation 7). The net effect is protection by glutathione of sensitive protein SH groups against oxidative damage.

In an interesting application of these concepts, Lii and Hendrich (*94*) showed that both glutathione and selenium participate in protecting SH groups of the enzyme carbonic anhydrase III against oxidative damage by t-butyl peroxide.

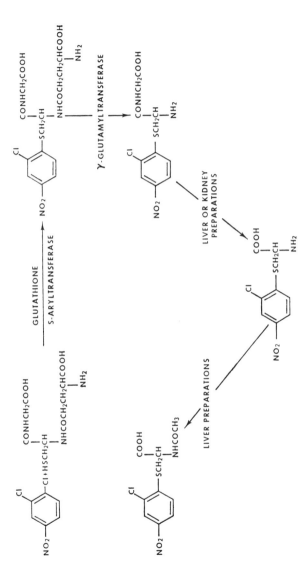

Figure 14. Pathway to mercapturic acid from dichloronitrobenzene and reduced glutathione.

In a related study (95), we showed, with the aid of a nuclear magnetic resonance technique, that the SH groups of cysteine can reduce methionine sulfoxide, but not methionine sulfone, back to methionine, possibly as shown in equation 8.

$$\underset{\substack{\text{methionine} \\ \text{sulfoxide}}}{\text{MetSO}} + \underset{\text{cysteine}}{\text{2CySH}} \rightarrow \underset{\text{methionine}}{\text{Met}} + \underset{\text{cystine}}{\text{CyS-SCy}} \quad (8)$$

Finally, recent reports (96,97) that reduced glutathione inhibits oral carcinogenesis in Syrian hamsters and that glutathione potentiates the action of other antioxidants such as ascorbic acid, ß-carotene, and α-tocopherol reinforce the cited findings of beneficial action of sulfur amino acids.

In summary, beneficial effects of sulfur amino acids are due to their ability to act as strong nucleophiles that trap electrophilic compounds and free radicals derived from biologically active compounds and nutrients. Future research should investigate the practical application of these transformations to improve the quality and safety of the diet and the health of the consumer.

Acknowledgment

I thank Gary McDonald for constructive contributions and helpful comments.

Literature Cited

1. Friedman, M. *The Chemistry and Biochemistry of the Sulfhydryl Group in Amino Acids, Peptides and Proteins.* Pergamon Press: Oxford, England, 1973; 485 pages.
2. Schwimmer, S.; Friedman, M. *Flavour Industry* **1972**, 137-145.
3. Smolin, L.A.; Benevenga, N.J. In *Absorption and Utilization of Amino Acids*; Friedman, M., Ed.; CRC: Boca Raton, FL, 1989; Vol. 1, pp 157-187.
4. Schneeman, B.O.; Gallaher, D. In *Nutritional and Toxicological Significance of Enzyme Inhibitors in Foods; Adv. Exp. Med. Biol.*; Friedman, M., Ed.; Plenum: New York, 1986; Vol. 199, 167-184.
5. Liener, I.E.; Kakade, M. In *Toxic Constituents of Plant Foodstuffs*; Liener, I.E., Ed.; Academic Press: New York, 2d ed.,1980; pp. 7-71
6. Liener, I.E.; Goodale, R.L.; Deshmukh, A.; Satterberg, T.L.; DiPietro, C.M.; Bankey, P.E.; Borner, J.W. *Gastroenterology* **1988**, *94*, 419-527.
7. Toskes, P.P. In *Nutritional and Toxicological Significance of Enzyme Inhibitors in Foods; Adv. Exp. Med. Biol.*; Friedman, M., Ed.; Plenum: New York, 1986; Vol. 199, 143-152.
8. Brandon, D.L.; Bates, A.H; Friedman, M. In *Protease Inhibitors as Potential Cancer Chemopreventive Agents*, Troll, W., Kennedy, Eds.; Plenum Press: New York, 1993; pp 107-129.
9. Friedman, M. *ACS Symp. Ser.* **1992**, *284*, 429-462.
10. Friedman, M. (Editor). *Nutritional and Toxicological Significance of Enzyme Inhibitors in Foods. Adv. Exp. Med. Biol.* Plenum: New York; 1986; Vol. 199, 570 pages.

11. Friedman, M. (Editor). *Nutritional and Toxicological Aspects of Food Safety. Adv. Exp. Med. Biol.* Plenum: New York; 1984; Vol. 177, 596 pages.
12. Friedman, M.; Gumbmann, M.R. In *Nutritional and Toxicological Significance of Enzyme Inhibitors in Foods; Adv. Exp. Med. Biol.*; Friedman, M., Ed.; Plenum: New York, 1986; Vol. 199, 357-390.
13. Friedman, M.; Grosjean, O.K.; Gumbmann, M.R. *J. Nutr.* **1984**, *114*, 2241-2246.
14. Friedman, M.; Grosjean, O.K.; Zahnley, J.C. *J. Sci. Food Agric.* **1982**, *33*, 165-172.
15. Gumbmann, M.R.; Friedman, M. *J. Nutr.* **1987**, *117*, 1018-1023.
16. Friedman, M.; Gumbmann, M.R. *J. Food Sci.* **1986**, *51*, 1239-1241.
17. Liener, I.E. *Crit. Rev. Food Sci. Nutr.* **1994**, *34*, 31-67.
18. Wallace, J.M.; Friedman, M. *Nutr. Repts. Intern.* **1984**, *32*, 743-748.
19. Friedman, M.; Wehr, C.M.; Schade, J.E.; MacGregor, J.T. *Food Chem. Toxicol.* **1982**, *20*, 887-892.
20. Mandel, H.G.; Judah, D.J.; McNeal, G.E. *Carcinogenesis* **1993**, *13*, 1853-1857.
21. Shetty, T.R.; Francis, A.R.; Bhattacharya, R.R. *Mutat. Res.* **1989**, *222*, 403-407.
22. Pickman, A.K.; Rodriguez, E.; Towers, G.H.N. *Chem.-Biol. Interactions* **1979**, *32*, 1567-1573.
23. Hill, D.W.; Camp, B.J. *J. Agric. Food Chem.* **1979**, *27*, 682-685.
24. Kupchan, S.M. *Intra-Science Chem. Reports* **1974**, *8*(4), 56-57.
25. De Flora, S.; Benicelli, C.; Serra, D.; Izzotti, A.; Cesarone, C.F. In *Absorption and Utilization of Amino Acids*; Friedman, M., Ed.; CRC: Boca Raton, FL, 1989; Vol. 3, pp 19-53.
26. Malkinson, A.M.; Beer, D.S. *J. Natl. Cancer Inst.* **1984**. *73*, 925-933.
27. Monteiro, F.F.; Trousdale, E.K.; Bisson, L.F. *Am. J. Enol.* **1989**, *40*, 1-8.
28. Fauhl, C.; Catsburg, R.; Wittkowski, R. *Food Chem.* **1993**, *48*, 313-316.
29. Gold, L.S.; Slone, T.H.; Stern, B.R.; Manley, N.B.; Ames, B.N. *Science* **1992**, *258*, 261-265.
30. Bohak, Z. *J. Biol. Chem.* **1964**, *239*, 2878-2887.
31. Sokolovsky, M.; Sadah, T.; Patchornik, A. *J. Am. Chem. Soc.* **1964**, *86*, 1212-1217.
32. Ziegler, K. *J. Biol. Chem.* **1964**, *239*, PC2713-2714.
33. Ziegler, K.; Melchert, I.; Lurken, C. *Nature* **1967**, *214*, 404-405.
34. Sternberg, M.; Kim, C.Y.; Schwende, F. J. *Science* **1975**, *190*, 992-994.
35. Sternberg, M.; Kim, C.Y. In *Protein Crosslinking: Nutritional and Medical Consequences; Adv. Exp. Med. Biol.*; Friedman M., Ed.; Plenum: New York; 1977, Vol. 86B, 73-84.
36. Asquith, R.S.; Otterburn, M.S. In *Protein Crosslinking: Nutritional and Medical Consequences; Adv. Exp. Med. Biol.*; Friedman M., Ed.; Plenum: New York; 1977, Vol. 86B, 93-122.
37. Gross, E. In *Protein Crosslinking: Nutritional and Medical Consequences; Adv. Exp. Med. Biol.*; Friedman M., Ed.; Plenum: New York; 1977, Vol. 86B, 131-153.

38. Whitaker, J.R.; Feeney, R.E. In *Protein Crosslinking: Nutritional and Medical Consequences; Adv. Exp. Med. Biol.*; Friedman M., Ed.; Plenum: New York; 1977, Vol. 86B, 155-175.
39. Feairheller, S.H.; Taylor, M.; Bailey, D.G. In *Protein Crosslinking: Nutritional and Medical Consequences; Adv. Exp. Med. Biol.*; Friedman M., Ed.; Plenum: New York; 1977, Vol. 86B, 177-186.
40. Spande, T.F.; Witkop, B.; Degani, Y.; Patchornik, A. *Adv. Protein Chem.* **1980**, *24*, 97-126.
41. Nashef, A.S.; Osuga, D.T.; Lee, H.S.; Ahmed, A.I.; Whitaker, J.R.; Feeney, R.E. *J. Agric. Food Chem.* **1977**, *25*, 245-251.
42. Anonymous. *Nutr. Rev.* **1989**, *47*, 362-364.
43. Cavins, J.F.; Friedman, M. *Biochemistry* **1967** *6*, 3766-3770.
44. Cavins, J.F.; Friedman, M. *J. Biol. Chem.* **1968**, *243*, 3357-3360.
45. Finley, J.W.; Snow, J.T.; Johnston, P.; Friedman, M. *J. Food Sci.* **1978**, *43*, 619-621.
46. Friedman, M. (Editor). *Nutritional Improvement of Food and Feed Proteins. Adv. Exp. Med. Biol.* Plenum: New York, 1978; Vol. 105, 882 pages.
47. Friedman, M. *ACS Symp. Ser.* **1982**, *206*, 231-273.
48. Friedman, M.; Pearce, K.N. *J. Agric. Food Chem.* **1989**, *37*, 123-127.
49. Friedman, M.; Cavins, J.F.; Wall, J.S. *J. Amer. Chem. Soc.* **1965**, *87*, 3672-3680.
50. Friedman, M. Gumbmann, M.R.; Masters, P.M. In *Nutritional and Toxicological Aspects of Food Safety. Adv. Exp. Med. Biol.*; Plenum: New York; 1984; Vol. 177, 367-412.
51. Friedman, M.; Levin, C.E.; Noma, A.T. *J. Food Sci.* **1984**, *49*, 1282-1288.
52. Masri, M.S.; Friedman, M. *Biochem. Biophys. Res. Commun.* **1982**, *104*, 321-325.
53. Pearce, K.N.; Friedman, M. *J. Agric. Food Chem.* **1988**, *36*, 707-717.
54. Snow, J.T.; Finley, J.W.; Friedman, M. *Int. J. Peptide Protein Res.* **1976**, *8*, 57-64.
55. Woodard, J.C.; Short, D.D.; Alvarez, M.R.; Reyniers, J. In *Protein Nutritional Quality of Foods and Feeds*; Friedman, M., Ed.; Marcel Dekker: New York, 1975; Part 2, pp 595-616.
56. Friedman, M. *J. Agric. Food Chem.* **1994**, *42*, 3-20.
57. Sapers, G.M. *Food Technol.* **1993**, No 10, 75-84.
58. Matheis, G. Chem. *Mikrobiol. Technol. Lebensm.* **1984**, *11*, 33-41.
59. Matheis, G.; Whitaker, J.R. *J. Food Biochem.* **1984**, *8*, 137-162.
60. Mayer, A.M. *Phytochemistry* **1987**, *26*, 11-20.
61. Wong, D.W.S. *Mechanism and Theory in Food Chemistry*; AVI-Van Nostrand Rheinhold: New York, 1989.
62. Dudley, E.D.; Hotchkiss, J.H. *J. Food Biochem.* **1989**, *13*, 65-75.
63. Takagi, T.; Isemura, T. *J. Biochem. (Tokyo)* **1964**, *56*, 344-349.
64. Friedman, M. *Diabetes* **1982**, *31*, 5-14.
65. Friedman, M. *Adv. Exp. Med. Biol.* **1991**, *289*, 171-215.
66. Friedman, M. (Editor). *Nutritional and Toxicological Consequences of Food Processing. Adv. Exp. Med. Biol.* Plenum: New York, 1991; Vol. 289, 542 pages.
67. Friedman, M. *Annu. Rev. Nutr.* **1992**, *12*, 119-127.

68. Friedman, M.; Molnar-Perl, I. *J. Agric. Food Chem.* **1990**, *38*, 1642-1647.
69. Friedman, M.; Molnar-Perl, I.; Knighton, D.K. *Food Add. Contam.* **1992**, *9*, 499-503.
70. Friedman, M.; Grosjean, O.K.; Zahnley, J.C. *Food Chem. Toxicol.* **1986**, *24*, 897-902.
71. Golan-Goldhirsh, A.; Whitaker, J.R. *J. Agric. Food Chem.* **1984**, *32*, 1003-1009.
72. Golan-Goldhirsh, A.; Whitaker, J.R.; Kahn, V. In *Nutritional and Toxicological Aspects of Food Safety; Adv. Exp. Med. Biol.*; Friedman, M., Ed.; Plenum: New York, 1984; Vol. 177, 457-495.
73. Molnar-Perl, I.; Friedman, M. *J. Agric. Food Chem.* **1990**, *38*, 1648-1651.
74. Molnar-Perl, I.; Friedman, M. *J. Agric. Food Chem.* **1990**, *38*, 1652-1656.
75. Friedman, M.; Bautista, F.F. 1994, manuscript in preparation.
76. Friedman, M. *Oxidation and Reduction of Porphyrins and Metalloporphyrins.* Ph. D. Thesis, University of Chicago, 1962.
77. Edwards, A. S.; Wedzicha, B.L. *Food Add. Contam.* **1992**, *9*, 461-469.
78. Naim, M.; Wainish, Z.; Zehavi, U.; Peleg, H.;Rousseff, R.L.; Nagy, S. *J. Agric. Food Chem.* **1993**, *41*, 1355-1358.
79. Naim, M.; Zuker, E.; Zehavi, U.; Rouseff, R.L. *J. Agric. Food Chem.* **1993**, *41*, 1359-1361.
80. Richard, F.C.; Goupy, P.M.; Nicolas, J.J.; Lacombe, J.M.; Pavia, A.A. *J. Agric. Food Chem.* **1991**, *39*, 841-847.
81. Eskins, K.; Friedman, M. *Photochem. Photobiol.* **1970**, *12*, 245-247.
82. Eskins, K.; Friedman, M. *J. Macromol. Sci.* **1971**, *A5*, 543-548.
83. Krull, L.H.; Friedman, M. *J. Polym. Sci.* **1967**, A-1, *5*, 2535-2546.
84. Krull, L.H.; Friedman, M. *Biochem. Biophys. Res. Commun.* **1967**, *29*, 373-377.
85. Wall, J.S. *J. Agric. Food Chem.* **1971**, *19*, 619-625.
86. Meister, A. *J. Biol. Chem.* **1988**, *263*, 17205-17208.
87. Boyland, E.; Chasseud, L.F. *Adv. Enzymol.* **1969**, *32*, 173-219.
88. White, R.D.; Wilson, D.M.; Glossen, J.A.; Madsen, D.C.; Rowe, D.W.; Goldberg, D.I. *Toxicology Letters* **1993**, *69*, 15-24.
89. Beutler, E. *Annu. Rev. Nutr.* **1989**, *9*, 287-302.
90. Henning, S.M.; Zhan, J.Z.; McKee, R.W.; Swenseid, M.A.; Jacob, R.A. *J. Nutr.* **1991**, *121*, 1669-1675.
91. Jones, D.P. *Nutrition and Cancer* **1992**, *17*, 55-75.
92. Rushmore, T.H.; Pickett, C.B. *J. Biol. Chem. - Minireview Compendium* **1993**, 11475-11478.
93. Smith, T.K.; Bray, T.M. *J. Anim. Sci.* **1992**, *70*, 2510-2515.
94. Lii, C.K.; Hendrich, S. *J. Nutr.* **1993**, *123*, 1480-1486.
95. Snow, J.T.; Finley, J.W.; Friedman, M. *Biochem. Biophys. Res. Commun.* **1975**, *64*, 441-447.
96. Trickler, D.; Shklar, G.; Schwartz, J. *Nutr. Cancer* **1993**, *20*, 139-144.
97. Shklar, G.; Schwartz, J.; Trickler, D.; Cheverie, S.R. *Nutr. Cancer* **1993**, *20*, 145-151.

RECEIVED March 23, 1994

Chapter 22

Inhibition of Chemically Induced Carcinogenesis by 2-n-Heptylfuran and 2-n-Butylthiophene from Roast Beef Aroma

Luke K. T. Lam, Jilun Zhang, Fuguo Zhang, and Boling Zhang

LKT Laboratories, Inc., 2233 University Avenue West, St. Paul, MN 55114–1629

2-n-Butylthiophene (BT) and 2-n-heptylfuran (HF) are two of the numerous furan and thiophene containing compounds present in roast beef aroma. They have been found to induce increased activity of the detoxifying enzyme, glutathione S-transferase (GST). Since furan containing natural products that induce increased activity of GST are known to inhibit tumorigenesis in laboratory animals, the inhibitory effects of BT and HF were investigated in various target tissues. Benzo(a)pyrene-induced forestomach tumors and 4-(methylnitrosamino)-1-(3-pyridyl)-1-butanone-induced lung tumors in A/J mice were reduced by BT and HF. Their relative inhibitory effects on 1,2-dimethylhydrazine-induced aberrant crypts formation and tumorigenesis in the colon of mice indicated that BT reduced both preneoplastic and neoplastic lesions while HF did not. These results suggest furan and thiophene containing compounds may be effective inhibitors of chemical carcinogenesis in selective tissues.

Cooked meat contains a large number of furan and thiophene compounds that constitute part of the flavor and aroma. (1-3) The composition of these compounds varies with the cooking method and water contents in the meat. Two representative compounds, 2-n-butylthiophene (BT) and 2-n-heptylfuran (HF) (Fig. 1), have been found to induce increased activity of a major detoxification enzyme system called glutathione S-transferase (GST). Chemicals that induce increased activity of GST are often inhibitors of chemical-induced tumorigenesis in laboratory animals. In this study, the inhibitory effects of BT and HF on chemically induced preneoplastic and neoplastic changes were examined.

0097–6156/94/0564–0278$08.00/0
© 1994 American Chemical Society

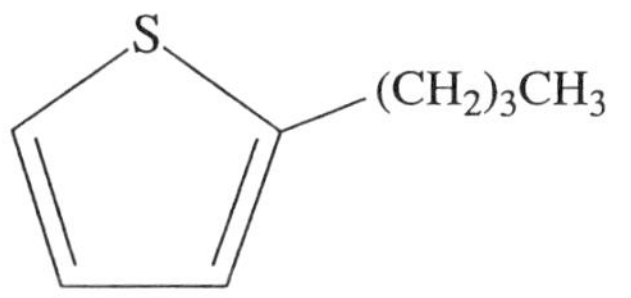

2-n-Butylthiophene

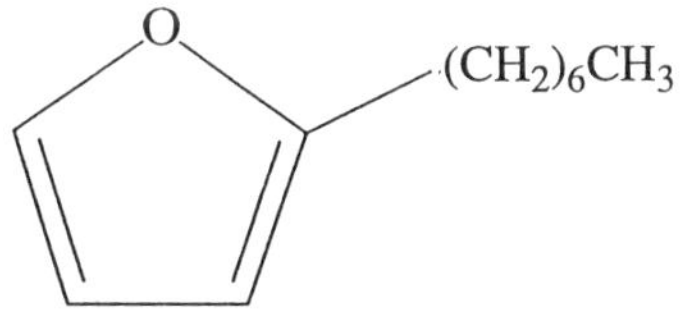

2-n-Heptylfuran

Figure 1. Structures of 2-n-butylthiophene and 2-n-heptylfuran.

Glutathione S-transferase Induction

Many furan-containing natural products have been shown to induce the detoxifying enzyme system, GST (4-7). GST is a major detoxifying enzyme system that catalyzes the conjugation of glutathione (GSH) with electron deficient centers of xenobiotics that include activated carcinogens (8). The GSH conjugates are generally less reactive and more water soluble thus facilitating excretion. An increase of GST activity indicates an enhancement of detoxification potential to eliminate carcinogens. It has been found that many chemicals that are GST inducers are capable of inhibiting chemically induced carcinogenesis.

The furanoids, kahweol and cafestol, isolated from green coffee beans (4) have been found to induce increased GST activity and inhibit 7,12-dimethylbenz(a)anthracene (DMBA) induced mammary tumor formation (9) and hamster cheek pouch carcinogenesis (10,11). The presence of the furan moiety was determined to be essential for enzyme induction. When the furan ring in cafestol was saturated by hydrogenation, its activity as a GST enzyme inducer was lost (5).

Citrus limonoids are furanoid natural products found in citrus fruits. The most abundant limonoids, limonin and nomilin, have been determined to induce increased activity of GST and inhibit chemically induced carcinogenesis (6,7). These findings with furan containing natural products led to the investigation of mono and disubstituted furan and the corresponding sulfur containing thiophene compounds that are found in complex mixtures of beef aroma.

The determination of GST activity involved control and experimental groups that contained 3 to 5 animals per group (12). The furan, HF, or thiophene, BT, were dissolved in cottonseed oil and administered by gavage once every other day for a total of three doses. Twenty-four hours after the last administration, the mice were killed by cervical dislocation. The liver, lung, forestomach, and the mucosal layer of the proximal 1/3 of the small bowel were removed for cytosol preparation. The activity of cytosolic GST was determined according to the method of Habig et. al.(13) using chloradinitrobenzene (CDNB) as the substrate.

The GST inducing activity of BT and HF in the liver, small bowel mucosa, forestomach and lung of A/J mice at 4 dose levels is shown in Table 1 (12). Significant increase of GST activity was found in the liver and small bowel mucosa of mice treated with 45 and 90 μmol per dose of BT. At the highest dose (90 μmol), an 18% increase in enzyme activity was observed in the lung. No appreciable increase in GST activity was found in the forestomach at all 4 dose levels.

With HF, greater than 28 % induction of GST activity was observed at all 4 dose levels in three of the four tissues examined. The highest activity was obtained in the small bowel mucosa, which was 6.1 times higher than that of the control. The GST activity in the lung was apparently unaffected by this compound.

The acid soluble sulfhydryl level, a good measure of GSH contents in tissues, was examined in tissue homogenates from mice treated with BT and HF. The results are shown in Table 2. BT induced significant increase of GSH in the liver and lung at the higher doses. No change was observed in either the small bowel mucosa or the forestomach.

Table 1. Effects of BT and HF on Glutathione S-Transferase Activity in Tissue Cytosols of Mice[a]

Compound	Dose,[b] μmol	Glutathione S-Transferase Activity[c], μmol/min/mg protein			
		Liver	Small bowel mucosa	Forestomach	Lung
Control		2.32 ± 0.28	1.12 ± 0.28	1.30 ± 0.21	0.91 ± 0.08
BT	11	2.42 ± 0.43	1.13 ± 0.15	1.48 ± 0.03	0.90 ± 0.08
	22	2.74 ± 0.34	1.20 ± 0.25	1.49 ± 0.20	0.98 ± 0.12
	45	3.69 ± 0.40*	1.96 ± 0.41√	1.38 ± 0.04	0.91 ± 0.19
	90	4.19 ± 0.60*	2.49 ± 0.39*	1.51 ± 0.18	1.07 ± 0.03√
HF	12	3.07 ± 0.55◊	1.60 ± 0.05√	1.67 ± 0.11√	1.01 ± 0.09
	25	4.37 ± 0.84*	1.74 ± 0.62	1.94 ± 0.06*	0.91 ± 0.12
	50	4.72 ± 1.07*	3.01 ± 0.32*	1.90 ± 0.10*	0.96 ± 0.07
	80	3.28 ± 0.62◊	6.87 ± 0.85*	2.10 ± 0.24**	0.73 ± 0.09

a. Abbreviations are as follows: BT, 2-n-butylthiophene; HF, 2-n-heptylfuran.
b. Dose was administered by gavage every other day for a total of 3 doses.
c. Statistical significance is as follows: significantly different from control; *,$p < 0.005$;√, $p < 0.025$; ◊, $p < 0.05$;**, $p < 0.01$

SOURCE: Reprinted with permission from reference 12. Copyright 1992 American Chemical Society.

Table 2. Effects of BT and HF on Acid-Soluble Sulfhydryl Level in Tissues of Mice[a]

		Acid-Soluble Sulfhydryl Level,[b] μmol/g tissue			
Compound	Dose,[b]	Liver	Small bowel mucosa	Forestomach	Lung
Control		5.6 ± 1.0	7.2 ± 1.3	2.5 ± 0.3	2.1 ± 0.2
BT	11	6.5 ± 1.3	7.0 ± 0.6	2.6 ± 0.4	2.2 ± 0.2
	22	7.5 ± 1.9	7.8 ± 1.0	2.4 ± 0.3	2.6 ± 0.3*
	45	9.6 ± 0.8√	8.1 ± 1.5	2.6 ± 0.4	2.9 ± 0.7◊
	90	10.5 ± 1.1√	8.7 ± 1.3	2.9 ± 0.6	3.0 ± 0.7◊
HF	12	6.1 ± 0.9	7.8 ± 1.4	3.0 ± 0.2◊	2.8 ± 0.3*
	25	9.2 ± 1.5*	8.1 ± 1.4	3.0 ± 0.5	3.8 ± 0.5√
	50	10.9 ± 0.8√	9.7 ± 0.8◊	3.0 ± 0.2◊	4.5 ± 0.7√
	80	7.4 ± 2.0	9.3 ± 0.7	2.6 ± 0.6	3.4 ± 1.1

a. Abbreviations are as follows: BT, 2-n-butylthiophene; HF, 2-n-heptylfuran.
b. Dose was administered by gavage every other day for a total of 3 doses.
c. Statistical significance is as follows: significantly different from control: *,$p < 0.01$; √,$p < 0.001$; ◊, $p < 0.05$.

SOURCE: Reprinted with permission from reference 12. Copyright 1992 American Chemical Society.

At 50 µmol dose HF was found to increase GSH level in all four tissues studied. At lower doses the significance of the increase was not consistent. At the 80 µmol dose the increase of GSH contents was lower than those of the 50 µmol dose in all four tissues. At this dose level toxicity may become a factor that influenced the GSH level.

The GST enzyme system consists of three main classes of isoenzymes. They are the α, μ, and π class. Using a chromatofocusing fast protein liquid chromatographic (FPLC) technique some of the isoenzymes in the α and μ classes can be separated as distinct peaks (14). Figure 2 shows the effects of BT and HF on the FPLC profile of isoenzymes isolated from the liver of rats.

The control shows intense α (peaks # 1-1, 1-2, 2-2) class isoenzymes with the μ (peaks # 3-3, 3-4, 4-4)class at lower intensity. BT and HF treatments altered the profile in such a way that the relative intensity of the α and μ isoenzymes has changed. Both compounds induced the μ class of isoenzyme, in particular 3-3 and 3-4 at high concentration. It is important to note the relative intensity of these peaks. It appears that the induction of the μ class of isoenzyme may be more closely associated with the inhibition of carcinogenesis than that of the α class.

Inhibition of Benzo[a]pyrene-induced forestomach and lung tumors

The inhibition of benzo(a)pyrene-induced lung and forestomach tumorigenesis was carried out with A/J mice (Table 3) (12). At the start of the experiment, the control and each of the experimental groups had 20 animals per group. The carcinogen, BP, was administered by gavage at 1 mg(4µmol)/0.2 mL of corn oil, twice a week for 4 weeks. BT and HF, at various doses, were given by gavage three times a week on days other than the carcinogen treatment. Three additional doses of inhibitors, 2 before the first dose and 1 after the last dose of BP, were administered. Eighteen weeks after the first dose of BP, the experiment was terminated.

At high doses, the toxicity of BT (90 µmol) and HF (50 µmol) reduced the number of surviving animals to less than 13. For the remaining groups, the casualty was mainly the result of gavage technique. The carcinogen treated group had 14 out of 17 animals (82.4 %) develop forestomach tumors and 100 % of the animals developed lung adenomas. The average number of tumors per mouse were 2.94 and 21.8, respectively, for the forestomach and lung. BT treatment at 45 and 90 µmol per dose reduced the tumor incidence to 50.0 and 69.2 %, respectively, in the forestomach. No reduction of tumor incidence in the lung was observed. The number of tumors per mouse, however, was reduced significantly in the forestomach and lung at both dosages employed.

The potency of HF as inhibitor of carcinogenesis was similar to that of BT in the forestomach of mice. The tumor incidence at the two dosages (50 and 25 µmol) employed were 75.0 and 66.7 % respectively. The reduction of the tumor multiplicity was essentially the same as that by BT. In the lung, HF was found to be ineffective as an inhibitor of pulmonary adenoma formation.

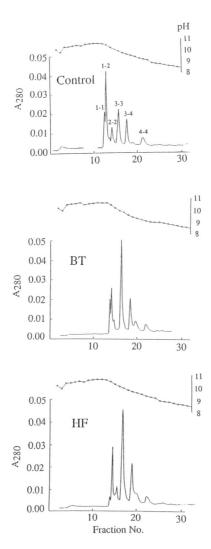

Figure 2. The fast protein liquid chromatographic (FPLC) analysis of isoenzymes isolated from the liver of rats treated with cottonseed oil (Control), 2-n-butylthiophene (BT), and 2-n-heptylfuran (HF).

Table 3. Inhibition of BP-Induced Forestomach and Lung Tumors in A/J Mice by BT and HF

Chemicals[a]	Dose μmol	No. of Mice	Weight Gain, g	Forestomach Tumors[b]			Lung Tumors		
				No. of mice with tumors	% of mice with tumors	No. of tumors/mouse	No. of mice with tumors	% of mice with tumors	No. of tumors/mouse
None		13	3.8	0	0	0	1	7.7	0.1 ± 0.0
BT	90	11	3.3	0	0	0	3	27.3	0.3 ± 0.0
HF	50	12	3.3	0	0	0	3	25.0	0.2 ± 0.0
BP		17	4.1	14	82.4	2.94 ± 0.63	17	100.0	21.8 ± 2.8
BP-BT	90	13	3.5	9	69.2	0.92 ± 0.24√	13	100.0	9.8 ± 1.5◊
BP-BT	45	18	3.4	9	50.0*	0.83 ± 0.20**	18	100.0	12.1 ± 2.6
BP-HF	50	12	4.2	9	75.0	1.08 ± 0.30#	12	100.0	17.9 ± 2.3
BP-HF	25	15	4.1	10	66.7	1.13 ± 0.38#	15	100.0	20.7 ± 2.3

a. Benzo[a]pyrene (BP) was given by gavage at 1 mg (4μmol)/0.2 ml cottonseed oil twice a week for 4 wks; BT and HF were dissolved in 0.2 ml cottonseed oil and given 3 times a week starting 1 day before the 1st dose of BP and terminating 1 day after the last dose of BP.
b. Statistical significance is as follows: significantly different from control: *,p < 0.05 (chi-square analysis); √, p < 0.02; ◊, p < 0.005; **, p < 0.01; #, p < 0.05 (Student's t test).

SOURCE: Reprinted with permission from reference 12. Copyright 1992 American Chemical Society.

Inhibition of 4-(methylnitrosamino)-1-(3-pyridyl)-1-butanone-induced lung tumors

The tobacco specific nitrosamine, 4-(methylnitrosamino)-1-(3-pyridyl)-1-butanone (NNK) is the most potent carcinogenic nitrosamine found in tobacco and tobacco smoke.(15). It has been found to induce lung tumors in all animal species tested regardless of the route of administration (16-19). In this experiment, a single 2 mg (9.7 µmol) dose of NNK was administered to the animals 2 hr. after the 3rd dose of BT or HF. The effect of BT and HF treatment on lung tumor formation was determined 18 weeks after the NNK administration (20).

A single 9.7 µmol dose of NNK produced an average of 7.33 tumors in 100% of the animals (Table 4). The inhibitor, BT, given at 50 and 25 µmol per dose for a total of 3 doses reduced the number of tumors per mouse to 2.6 and 5.0, respectively. The tumor incidence was also lowered with the high dose group, but was close to control level with the low dose group. The size of the largest tumor in both groups was significantly reduced indicating a reduction of tumor burden for these animals.

At the same dose levels, HF, reduced both the tumor incidence (60, 50%) and the number of tumors/animal to lower values (1.55, 1.35) than those of the BT group. The tumor burden was also lowered with respect to the NNK and the BT treated groups.

Inhibition of 1,2-dimethylhydrazine-induced aberrant crypts and tumors in the colon

The precancerous lesions termed aberrant crypts (AC) have been developed as an early marker for colon carcinogenesis (21). Upon treatment with colon carcinogens mice and rats develop early preneoplastic changes in the crypts of the colon that are easily observable under light microscope after methylene blue staining. Histological examination of AC foci confirmed their dysplasia nature, and are considered precursors to colon cancer (22-25).

In a previous study, we observed a reduction of AC formation in the colon of CF1 mice by a number of chemopreventive agents (26). Colon AC were induced by 1,2-dimethylhydrazine (DMH) in CF1 mice. The known inhibitor of carcinogenesis, 3-butyl-4-hydroxyanisole (BHA), and phenyl propyl isothiocyanate, were found to reduce DMH-induced AC formation by 46 and 40%, respectively. Difluoromethylornithine, a known inhibitor of colon carcinogenesis, was determined to reduce AC formation when given after the carcinogen. Thus the inhibition of AC formation is positively correlated with chemopreventive agents that are effective in the colon.

Using DMH as the carcinogen, the effects of BT and HF on AC formation in the colon of CF_1 mice were investigated (27). In this experiment, the control group had 5 animals and the experimental groups had 10 mice per group. The inhibitors, BT and HF, 20 µmol each dissolved in cottonseed oil, were given by gavage once a day for 8 days. One hour after the inhibitors administration on the

Table 4. Effects of 2-n-butylthiophene (BT) and 2-n-heptylfuran (HF) on 4-(methylnitrosamino)-1-(3-pyridyl)-1-butanone (NNK)-induced lung tumors

Group	Chemicals (μmol)[a]	No. of Animals at risk[b]	No. of Animals with Tumors[c] (%)	No. of Tumors/ Animal[d]	Ave. Size of Largest tumor[d]
1	None	14	3 (21.4)	0.21±0.11	1.03±0.13
2	BT (50)	14	0	0	0
3	HF (50)	10	0	0	0
4	NNK (9.7)	24	24 (100)	7.33±0.66	2.06±0.10
5	BT (50) NNK (9.7)	20	16 (80.0)*	2.60±0.57***	1.63±0.12**
6	BT (25) NNK (9.7)	19	18 (94.7)	5.00±0.73*	1.28±0.08***
7	HF (50) NNK (9.7)	20	12 (60.0)**	1.55±0.45***	1.12±0.10***
8	HF (25) NNK (9.7)	20	10 (50.0)**	1.35±0.44***	1.16±0.06***

a. The inhibitors, BT and HF in two concentrations (50μmol or 25μmol/ 0.3 mL cottonseed oil), were administered by gavage every 2 days for a total of 3 doses. Single dose of the carcinogen, NNK (9.7 μmol/0.1 mL saline solution) was given by i.p. injection 2 hr after the last inhibitor administration.
b. Number of animals at risk at the end of the experiment.
c. Values significantly different from group 4 (NNK) by chi-square analysis, *P<0.01, **P<0.005
d. Values are mean ± S.E, significantly different from group 4 (NNK) by Student's t-test, *P<0.025, **P<0.01, ***P<0.001

4th and 8th day 0.4 mg of DMH was given by gavage. Twenty one days after the first dose of DMH the animals were sacrificed, and the colon aberrant crypts determined.

The results are given in Table 5. The number of aberrant crypt foci (ACF)/animal and AC/animal in the DMH treated group were 16.60 and 19.00, respectively. With BT as the inhibitor, the number of ACF/animal and AC/animal were reduced to 10.40 and 12.60, respectively. In contrast to the inhibitory nature of BT, HF treatment increased the two values to 25.75 and 31.20, respectively. Both the reduction by BT and the enhancement by HF were found to be significantly different from the control. These results suggested that BT may be an effective anticarcinogen in the colon and that HF is likely not active or may even be cocarcinogenic in the same system.

To confirm the findings described above an experiment aimed to investigate the anticarcinogenicity of BT and HF in the colon was carried out with CF_1 mice. DMH was used as the carcinogen and the duration of the experiment was 40 weeks instead of the 3 weeks required for the formation of AC. The vehicle and inhibitor only control groups had 12 animals per group. The carcinogen only group had 25 animals and the inhibitor plus carcinogen groups had 20 animals per group. The inhibitors, BT and HF, were given by gavage at 40 and 20 µmol/0.2 mL cottonseed oil per animal three times a week for 20 weeks. The carcinogen used to induce colon tumors were multiple doses of 0.4 mg DMH by subcutaneous injection once a week for 20 weeks. Twenty weeks after the last dose of DMH the animals were sacrificed, and the number of tumors in the colon were examined by histopathological readings.

The results are shown in Table 6. As in the case of inhibition of AC formation, BT reduced tumor formation in the colon in a dose dependent manner. At 40 µmol of BT per dose, 11.1 % of the animals developed colon tumors compared to 50 % for the carcinogen control. The tumor multiplicity was also reduced from 1.4 for the control to 0.3 for this group, a 78.6 % inhibition. At a lower dose of 20 µmol no significant reduction of tumor incidence was observed. The tumor multiplicity, however, was lowered to 0.9, a 35 % reduction.

The furan compound, HF, on the contrary, was not effective as an inhibitor of DMH-induced tumorigenesis. At 40 µmol per dose both the incidence (40%) and the number of tumors/animal (1.7) was indistinguishable from the values of the control. At 20 µmol, HF was shown to increase the number of tumors/animal to 2.4, over 70 % higher than that of the control. This increase was similar to the enhancement observed in the aberrant crypts experiment.

Summary

In summary, two of the representative furan and thiophene compounds from cooked meat have been determined to induce the enzyme activity of GST. They have been investigated as potential inhibitors of chemically induced carcinogenesis in three different tumor systems. Their effectiveness to reduce tumors depends on the kind of carcinogens used and the organ tissues where the tumors developed. Whereas HF was determined to reduce tumors in the forestomach and lung, BT was found to be effective in the forestomach, lung, and colon models.

Table 5. Effects of 2-n-butylthiophene (BT) and 2-n-heptylfuran (HF) on 1,2-dimethyl-hydrazine-induced aberrant crypt formation in the colon of CF1 mice

Group	Chemicals[a]	No. of Foci/animal[b,c]	No. of AC/Animal[b,c]	No. of AC/Focus
1	None	0.00	0.00	0.00
2	BT	0.00	0.00	0.00
3	HF	0.00	0.00	0.00
4	DMH	16.60 ± 0.97	19.0 ± 1.00	1.14
5	BT-DMH	10.40 ± 1.70*	12.6 ± 2.30*	1.21
6	HF-DMH	25.75 ± 3.10***	31.2 ± 5.90**	1.21

a) BT and HF (20 µmol/ 0.2 mL cottonseed oil) were given by gavage once daily for 8 consecutive days. DMH (0.4 mg, 0.2 mL 0.001M EDTA, pH 6.5) was given one hour after the inhibitor on the fourth and last day.
b) The animals were killed 21 days after the first dose of DMH. The colons were excised, fixed in formalin and stained with methylene blue. The AC were read under light microscopy.
c) Values are mean ± S.E. *$p < 0.025$, **$p < 0.01$, ***$p < 0.005$

Table 6. Effects of 2-n-butylthiophene (BT) and 2-n-heptylfuran (HF) on 1,2-dimethylhydrazine (DMH)-induced colon carcinogenesis

Group	Chemicals (µmol)[a]	No. of Animals at risk[b]	No. of Animals with Tumors(%)	No. of Tumors/ Animal[c]
1	None	13	0	0
2	BT (40)	13	0	0
3	HF (40)	13	0	0
4	DMH (3)	18	9 (50.0)	1.4±0.40
5	BT (40) DMH (3)	18	2 (11.1)[d]	0.3±0.23[e]
6	BT (20) DMH (3)	17	8 (47.1)	0.9±0.29
7	HF (40) DMH (3)	15	6 (40.0)	1.7±0.72
8	HF (20) DMH (3)	20	10 (50.0)	2.4±0.63

a. The inhibitors, BT and HF in two concentrations (40µmol or 20µmol/ 0.2 mL cottonseed oil), were administered by gavage 3 times a week for 20 weeks. The carcinogen, DMH (3 µmol/0.2 mL 0.001M EDTA, pH 6.5) was given by s.c. injection once a week for 20 weeks.
b. Number of animals at risk at the end of the experiment.
c. Values are mean ± S.E.
d. Chi-square analysis was used to compare 2BT(40)DMH(3) vs DMH(3) control, $p<0.01$.
e. Student t-test was used to compare 2BT(40)DMH(3) vs DMH control, $p<0.05$.

Acknowledgment: This work was supported by grants from the Minnesota Beef Council, the National Live Stock and Meat Board, and the National Cancer Institute (USPHS CA54037).

Literature Cited

1. Min, D.B.S.; Ina, K.; Peterson, R.J.; Chang, S.S. *J. Food Sci.* **1979**, *44*, 639-642.
2. MaCleod, G.; Ames, J.M. *J. Food Sci.* **1986**, *51*, 1427-1434.
3. Umano, K.; Shibamoto, T. *J. Agric. Food Chem.* **1987**, *35*, 14-18.
4. Lam, L. K. T.; Sparnins, V. L.; Wattenberg, L. W. *Cancer Res.* **1982**, *42*, 1193-1198.
5. Lam, L. K. T.; Sparnins, V. L.; Wattenberg, L. W. *J. Med. Chem.* **1987**, *30*, 1399-1403.
6. Lam, L. K. T.; Li, Y.; Hasegawa, S. *J. Agric. Food Chem.* **1989**, *37*, 878-880.
7. Lam, L. K. T.; Hasegawa, S. *Nutr. Cancer* **1989**, *12*, 43-47.
8. Chasseaud, L. F. *Adv. Cancer Res.* **1979**, *29*, 175-274.
9. Wattenberg, L.W.; Lam, L. K. T. *Banbury Rep.* **1984**, *17*, 137-145.
10. Miller, E.G.; Formby, W.A.; Rivera-Hidalgo, F.; Wright, J.M. *Oral Surg.* **1988**, *65*, 745-749.
11. Miller, E. G.; McWhorter, K.; Rivera-Hidalgo, F.; Wright, J.M.; Hirsbrunner, P.; Sunahara, G.I. *Nutr. Cancer*, **1991**, *15*, 41-46.
12. Lam, L. K. T.; Zheng, B. L. *Nutr. Cancer* **1992**, *17*, 19-26.
13. Habig, W. H., Pabst, M. J.; Jakoby, W. B. *J. Biol. Chem.* **1974**, *249*, 7130-7139.
14. Vos, R.M.E.; Snoek, M.C.; van Berkel, W.J.H.; Muller, F.; van Bladeren, P.J.; Biochem. Pharm. **1988**, *37*, 1077-1082.
15. Hecht, S. S.; Trushin, N.; Castonguy, A.; Rivenson, A. *Cancer Res.* **1986**, *46*, 498-502.
16. Morse, M. A.; LaGreca, S. D.; Amin, S. G.; Chung, F.-L. *Cancer Res.* **1990**, *50*, 2613-2617.
17. Hecht, S. S.; Hoffmann, D. *Carcinogenesis,* **1988**, *9*, 875-884.
18. Hoffmann, D.; Hecht, S. S. *Cancer Res.* **1985**, *45*, 935-944.
19. Rivenson, A.; Hoffmann, D.; Prokopczyk, B.; Amin, S.; Hecht, S. S. *Cancer Res.* **1988**, *48*, 6912-6917.
20. Lam, L. K. T.; Zhang, J.; Zhang, F. *Nutr. Cancer*, submitted.
21. Bird, R.P. *Cancer Letters* **1987**, *37*, 147-151.
22. Chang, W.W.L. Scand. *J Gastroenterology*, **1984**, *19 (s104)*, 27-44.
23. Day, D.W.; Morson, B.C., In:BC Morson (ed.), *The pathogenesis of Colorectal Cancer*, , WB Saunders, Philadelphia, **1978**, pp58-71.
24. Bird, R.P.; McLellan, E.A.; Bruce, W. R. *Cancer Surveys*, **1989**, *8*, 189-200.
25. McLellan, E.A.; Bird, R.P, *Cancer Res.*, **1988**, *48*, 6183-6186.
26. Lam, L.K.T.; Zhang, J. *Carcinogenesis*, **1991**, *12*, 2311-2315.
27. Lam, L.K.T.; Zhang, J. *Carcinogenesis*, submitted.

RECEIVED April 22, 1994

Author Index

Ames, Jennifer M., 147
Bertram, H.-J., 199
Block, Eric, 63
Calvey, Elizabeth M., 63
Chan, F., 127
Chin, Hsi-Wen, 90
Eiserich, Jason P., 247
Emberger, R., 199
Friedman, Mendel, 258
Grimm, Casey C., 49
Güntert, Matthias, 199
Ho, Chi-Tang, 138,188
Hopp, R., 199
Jasper, B. L., 8
Keelan, Mary E., 1
Kersten, E., 224
Kobayashi, Akio, 238
Kubota, Kikue, 238
Lam, Luke K. T., 278
Leach, David N., 36
Lindsay, Robert C., 90
Madruga, Marta S., 180
Miller, James A., 49
Mistry, B. S., 8
Mottram, Donald S., 180
Mussinan, Cynthia J., 1
Nagy, S., 80
Naim, M., 80
Nittka, C., 224
Onodenalore, Akhile C., 171
Parliment, Thomas H., 160
Reineccius, G. A., 8,127
Rewicki, D., 224
Rouseff, R. L., 80
Shahidi, Fereidoon, 106,171
Shewfelt, Robert L., 36
Shibamoto, Takayuki, 247
Sommer, H., 199
Spanier, Arthur M., 49
Stahl, Howard D., 160
Steely, Jeffrey S., 22
Synowiecki, Jozef, 171
Tressl, R., 224
Wang, Youming, 36
Werkhoff, P., 199
Wu, Chung-May, 188
Wyllie, S. Grant, 36
Yu, Tung-Hsi, 188
Zehavi, U., 80
Zhang, Boling, 278
Zhang, Fuguo, 278
Zhang, Jilun, 278
Zheng, Yan, 138
Zuker, I., 80

Affiliation Index

Agricultural Research Service, 49,258
Coca-Cola Foods, 8
Florida Department of Citrus, 80
Food Industry Research and Development Institute, 188
Freie Universität Berlin, 224
Haarmann & Reimer GmbH, 199
Hebrew University of Jerusalem, 80
Hershey Foods Corporation, 22
International Flavors and Fragrances, 1
Kraft General Foods, 160
LKT Laboratories, Inc., 278
Memorial University of Newfoundland, 106,171
Ochanomizu University, 238
Rutgers University, 138,188
State University of New York at Albany, 63
Technische Universität Berlin, 224
University of California, 247

University of Florida, 80
University of Georgia, 36
University of Minnesota, 8,127
University of Reading, 147,180
University of Western Sydney, 36
University of Wisconsin, 90
U.S. Department of Agriculture, 49,258
U.S. Food and Drug Administration, 63

Subject Index

A

N-Acetyl-L-cysteine
  browning, 81,83f
  2,5-dimethyl-4-hydroxy-3(2H)-furanone formation, 81,83f,84,85f
  stored orange juice aroma, 86,87f,88
2-Acetylthiopene, kinetics of formation via Maillard reaction, 127–136
Aflatoxin B_1, mechanism of beneficial effects, 261–263,264f
Alkanol–ammonia–hexane extraction process, description, 116,117–119f
Alk(en)ylcysteine sulfoxides, role in garlic flavor, 188–189
Allicin
  decomposition, 67–68
  identification, 65–66
  role in antibacterial principle of garlic, 64–66
Alliin–inosine 5′-monophosphate thermal interactions, 188–197
Allium chemistry
  allicin decomposition, 67–68
  previous studies, 64–66
  natural flavor identification, 63–64
  product formation from chopped garlic, 65
  volatile characterization methods, comparison, 67,69–78
Antibacterial principle of garlic, 64–66
Antioxidative activity, sulfur-containing heterocyclic compounds formed in Maillard reaction model systems, 247–256
Antioxidative effects, sulfur amino acids, 272,274

Antioxidative properties, sulfur compounds in foods, 4
Aqueous washing, role in heat-induced changes of sulfhydryl groups of muscle foods, 174,175t
Aroma of muskmelon, sulfur volatile sensory evaluation by GC–olfactometry, 36–47
Aroma threshold, determination, 81
Aroma volatiles in meat, sulfur-containing, *See* Sulfur-containing aroma volatiles in meat
Artifacts, sulfur compounds in foods, 3
Atomic emission detector
  comparison to flame photometric and sulfur chemiluminescence detectors, 17,21
  dynamic range, 17,18f
  minimum detectable level, 17
  principle, 15,17
  schematic representation, 15,16f
  selectivity, 17,19–20f

B

Beef, effect of analysis on sulfur-containing flavor compounds, 49–60
Benzo[a]pyrene-induced forestomach and lung tumors, inhibition by 2-N-butylthiophene and 2-N-heptylfuran, 283,285t
Brassica, source of goitrogens, 106
Brassica oilseeds
  alternative methods of processing, 113,116
  detoxification, 113
  glucosinolates, 107,110f,111

Brassica oilseeds—*Continued*
  processing, 111,112f
  thioglucosides, 106–122
Broccoli, modulation of objectionable
  flavors, 98,100–102t
Browning prevention, sulfur amino acids,
  266,269,270f,271t
2-*N*-Butylthiophene
  inhibition of chemically induced
    carcinogenesis, 278–290
  structure, 278,279f

C

Canola, description, 111,113
Caraway seed, suppression of undesirable
  sauerkraut flavors, 95–99
Character-impact sulfur compounds, 1–2
Chemically induced carcinogenesis
  inhibition by 2-*N*-heptylfuran and
    2-*N*-butylthiophene from roast beef
    aroma
  benzo[*a*]pyrene-induced forestomach
    and lung tumor inhibition, 283,285t
  1,2-dimethylhydrazine-induced aberrant
    colon crypts and tumors, 286,288,
    289–290t
  glutathione *S*-transferase induction,
    280–283,284f
  4-(methylnitrosamino)-1-(3-pyridyl)-1-
    butanone-induced lung tumors,
    286,287t
Chemiluminescence detection of sulfur
  compounds in cooked milk
  chromatograms, 28–30
  compounds identified, 33
  description, 24–25
  detector, 25–27
  experimental procedure, 27–28
  flavor fade, 33
  flavor fade determination, 30
  stripping temperature, 30,32f,33
  sulfur peak identifications, 30,31f
  temperature, 33

Citrus products, L-cysteine and
  N-acetyl-L-cysteine effect on
  off-flavor formation, 80–88
^{13}C-labeled sugars, sulfur-containing
  flavor compound formation, 224–235
Coffee, roasted, furfuryl mercaptan
  generation, 160–169
Colon crypts and tumors,
  1,2-dimethylhydrazine-induced,
  inhibition by 2-*N*-butylthiophene and
  2-*N*-heptylfuran, 286,288–290t
Cooked foods, sulfur compound
  content, 1
Cooked milk, chemiluminescence
  detection of sulfur compounds, 22–33
Cruciferous vegetables
  anticarcinogenic properties, 90,92
  compounds affecting flavor, 90
Cucumis melo cv. Makdimon aroma,
  sulfur volatile sensory evaluation by
  GC–olfactometry, 36–47
Cysteine
  kinetics of HS release during thermal
    treatment, 138–145
  sulfur-containing flavor compound
    formation, 224–235
L-Cysteine
  2,5-dimethyl-4-hydroxy-3(2*H*)-furanone
    formation, 81,83f,84,85f
  stored orange juice aroma, 86,87f,88
  use in browning prevention, 80,81,83f
Cysteine–pentose model systems,
  furfuryl mercaptan generation, 160–169

D

3-Deoxyaldoketose, degradation of
  sulfur-containing flavor compound
  formation from ^{13}C-labeled sugars,
  cysteine, and methionine, 227–232
Deoxyalliin–inosine 5′-monophosphate
  thermal interactions, volatile
  compounds, 188–197

INDEX

1-Deoxydiketoses, degradation of sulfur-containing flavor compound formation from ^{13}C-labeled sugars, cysteine, and methionine, 231,233–235

Detoxification reactions, sulfur amino acids, 272,273f

Dimethyl disulfide
  kinetics of formation via Maillard reaction, 127–136
  role in broccoli odor, 92

Dimethyl trisulfide, role in broccoli odor, 92

1,2-Dimethylhydrazine-induced aberrant colon crypts and tumors, inhibition by 2-N-butylthiophene and 2-N-heptylfuran, 286,288–290t

2,5-Dimethyl-4-hydroxy-3(2H)-furanone
  L-cysteine and N-acetyl-L-cysteine effect on formation, 81,83f,84,85f
  generation, 81,82f

Disulfide bonds
  alkali effect on degradation, 172
  role in functional and structural properties of heat-processed meat, 171–172

1,3-Dithiolane antioxidative activity, sulfur-containing heterocyclic compounds with antioxidative activity formed in Maillard reaction model systems, 252,254f

F

Flame photometric detector comparison to sulfur chemiluminescence and atomic emission detectors, 17,21
  diagram, 10,11f
  drawbacks, 10
  dynamic range, 10,12f
  minimum detectable level, 10,13t
  principle, 10
  selectivity, 10,13
  sulfur compounds in cooked milk, 25

Flavor(s) associated with cooked meats, compounds identification, 180–181

Flavor compounds in beef, sulfur-containing, analysis, 51–60

Flavor precursors of garlic, formation of Maillard-type flavor compounds, 189

Flavor quality of meat, 49–50

Flavor threshold, description, 81

Forestomach tumors, benzo[a]pyrene-induced, inhibition by 2-N-butylthiophene and 2-N-heptylfuran, 283,285t

Formation, sulfur compounds in foods, 3–4

Functional properties, sulfur compounds in foods, 4–5

Furan compounds in cooked meat, 278,279f

Furfuryl mercaptan, identification in coffee, 160

Furfuryl mercaptan generation in cysteine–pentose model systems in relation to roasted coffee
  activation energy determination, 165–169
  coffee precursors, 161
  experimental procedure, 162–163
  generation in roasted coffee, 163,164f
  model systems, 160–161
  pH effect on formation, 165,166f
  rate constant determination in model systems, 163,164f,165t,166f
  reaction kinetics, 161–162
  roasting process, 162

2-Furfurylthiol, See Furfuryl mercaptan

G

Garlic
  antibacterial principle, 64–66
  interaction of flavor precursors with nucleotides in food, 189

Garlic flavor, role of alk(en)ylcysteine dipeptides and sulfoxides, 188–189

Garlic of *Allium* genus, sulfur compound formation, 238
Gas chromatography, detector evaluated for analysis of volatile sulfur compounds in foods
  atomic emission detector, 15–21
  comparison of flame photometric, sulfur chemiluminescence, and atomic emission detectors, 17,21
  experimental procedure, 9
  flame photometric detector, 10–13
  sulfur chemiluminescence detector, 13–15
Gas chromatography–mass spectrometry, *Allium* chemistry determination, 63–78
Gas chromatography–olfactometry, sulfur volatile sensory evaluation by muskmelon, 36
D-Glucose–L-cysteine Maillard system, nonspecific initial phase, 224–225
[^{13}C]Glucose–cysteine–methionine reactions, sulfur-containing flavor compound formation from ^{13}C-labeled sugars, cysteine, and methionine, 225–227
Glucosinolates
  Brassica oilseeds
    antinutritional effects, 107,111
    beneficial effects, 111
    chemical transformation, 116,120–124
    content of canola varieties, 111,113,114–115
    toxicity, 111
  chemical structure, 106–107,108*f*
  enzymatic hydrolysis products, 107,109*f*
  examples, 107,108*t*
  flavor, 107
γ-Glutamyl alk(en)yl cysteine dipeptides, role in garlic flavor, 188–189
Glutathione, kinetics of HS release during thermal treatment, 138–145
Glutathione *S*-transferase induction, 2-*N*-butylthiophene and 2-*N*-heptylfuran, 280–283,284*f*

Goitrin, production, 107,110*f*
Goitrogens, *Brassica* as source, 106
Guanosine 5′-monophosphate, use as flavor enhancer, 189

H

Heat-induced changes of sulfhydryl groups of muscle foods
  aqueous washing effect, 174,175*t*
  experimental procedure, 172–174
  heating at 20–99 °C, 174,176*f*,177
  thermal denaturation, 177–179
Heating of meat, nutritional effects, 171
2-*N*-Heptylfuran, inhibition of chemically induced carcinogenesis, 278–290
Heterocyclic compounds, sulfur-containing, *See* Sulfur-containing heterocyclic compounds with antioxidative activity formed in Maillard reaction model systems
High-performance liquid chromatography, *Allium* chemistry determination, 63–78
Hydrogen sulfide
  kinetics of release from cysteine and glutathione during thermal release, 139–145
  precursors, 138–139
  role in meat flavor, 138

I

Inosine 5′-monophosphate, use as flavor enhancer, 189
Inosine 5′-monophosphate–alliin or deoxyalliin thermal interactions, volatile compounds, 188–197
Isothiocyanates, formation, 90,91*f*

K

Kinetics
  release of HS from cysteine and glutathione during thermal treatment
    activation energies, 144,145t,
    experimental procedure, 138–140
    first-order kinetics
      cysteine, 140,141–142f,143t
      glutathione, 140,144t
    pH–rate constant relationship
      cysteine, 140,143f
      glutathione, 144,145t
    pH effect
      cysteine, 140,141–142f,143t
      glutathione, 140,144t
  Strecker aldehyde formation via Maillard reaction
    experimental procedure, 128–129
    heating temperature, 129
    kinetic analysis method, 131,134
    model system complexity, 131
    pH, 129,130–133
    reaction rates vs. pH and temperature, 134f
    time of heating, 129–132
    zero-order kinetics, 134,135f,136t

L

Lectins, mechanism of beneficial effects, 261
Lipid peroxidation, role in food quality, 247–248
Lung tumors
  benzo[a]pyrene-induced, inhibition by 2-N-butylthiophene and 2-N-heptylfuran, 283,285t
  4-(methylnitrosamino)-1-(3-pyridyl)-1-butanone-induced, inhibition by 2-N-butylthiophene and 2-N-heptylfuran, 286,287t
Lysinoalanine, mechanism of beneficial effects, 263,265–268

M

Maillard reaction
  formation of meaty aroma compounds during cooking, 181
  identification of antioxidants, 248
  model systems for sulfur-containing heterocyclic compounds with antioxidative activity, 247–256
  products, 51,248
  role in food aroma, 127
Maillard reaction extract antioxidative activity, sulfur-containing heterocyclic compounds with antioxidative activity in model systems, 250–252, 253f
Measured electrode potential, 139–140
Meat
  compounds identified, 50
  factors affecting flavor quality, 49–50
  formation of flavor compounds during analysis, 50
  sulfur-containing aroma volatiles, 180–186
Meaty aroma compounds formed during cooking, 181
Medicinal properties, sulfur compounds in foods, 4–5
Melons, aroma profiles, 36–37
  sulfur volatile content, 36–37
Methanethiol
  modulation of volatile sulfur compounds, 93,97f
  role in broccoli odor, 92
Methanethiol-related volatile sulfur compounds, formation, 90,91f
Methional, kinetics of formation via Maillard reaction, 127–136
Methionine, sulfur-containing flavor compound formation, 224–235
Methyl methanethiosulfinate, modulation of volatile sulfur compounds, 92–93, 94f

4-(Methylnitrosamino)-1-(3-pyridyl)-1-butanone-induced lung tumors, inhibition by 2-*N*-butylthiophene and 2-*N*-heptylfuran, 286,287*t*

Milk
  compounds responsible for sulfurous flavor, 24
  cooked, chemiluminescence detection of sulfur compounds, 22–33
  flavor importance, 22
  heat-induced generation, 23
  off-flavor categories, 22–33
  volatile formation, 23
Minimally processed vegetables, methods for control of physiological and microbiological changes, 92
Minimum detectable level, definition, 9
Modulation of volatile sulfur compounds in cruciferous vegetables, *See* Volatile sulfur compounds in cruciferous vegetables
Muscle foods, heat-induced changes of sulfhydryl groups, 177–179
Muskmelon aroma, sulfur volatile sensory evaluation by GC–olfactometry, 36–47

N

Natural flavors, definition, 63
Natural flavors of *Allium*, formation, 63
Natural sulfur-containing compounds, oxidative stress protection, 248
Nonenzymic reactions related to color, aroma, and taste, 80

O

Off-flavor, role of sulfur compounds, 2
Off-flavor formation in stored citrus products, L-cysteine and *N*-acetyl-L-cysteine, 80–88
Oilseeds of *Brassica*, thioglucosides, 106–122
Orange juice, stored, L-cysteine and *N*-acetyl-L-cysteine related to off-flavor formation, 80–88
Organic sulfur compounds, role in odor and flavor of foods, 8

P

Pentose–cysteine model systems, furfuryl mercaptan generation, 160–169
Photochemistry of proteins, and sulfur amino acids, 269,271–272
Protease inhibitors, mechanism of beneficial effects, 259–261,262*f*
Proteins, thermal denaturation, 177–179

R

Rapeseed, oil production, 107
Roasted coffee, furfuryl mercaptan generation, 160–169

S

Sauerkraut, suppression of undesirable flavors by caraway seed, 95–99
Scorodocarpus borneensis Beec., *See* Wood garlic
Sensory evaluation of sulfur volatiles in muskmelon aroma
  aroma compounds by odor units, 44,45*t*
  aroma compounds identified by aroma extraction dilution analysis, 38,43*t*,44
  blended and nonblended flesh analysis, 44,46*f*
  carbon of extract aroma extraction dilution, 38,39*f*
  experimental procedure, 37–38
  most dilute extract flame ionization detection and aroma group, 38,42*f*,44

Sensory evaluation of sulfur volatiles in muskmelon aroma—*Continued*
  nonblended and blended flesh aroma aromagrams, 44,47f
  previous studies, 37
  sulfur aroma extract, 38,39f
  volatile constituents, 38,40–41t
Sensory properties, sulfur compounds, 1–2
Sesquiterpene lactones, mechanism of beneficial effects, 263,265f
Sodium sulfite, use to minimize Maillard-type browning, 80
Stored citrus products, L-cysteine and N-acetyl-L-cysteine related to off-flavor formation, 80–88
Strecker aldehydes, kinetics of formation via Maillard reaction, 127–136
Strecker degradation, cysteine and methionine, 224
Strecker reaction, description, 127–128
Sugars, ^{13}C-labeled, sulfur-containing flavor compound formation, 224–235
Sulfhydryl groups
  cysteine and glutathione, 258,259
  heat-induced changes in muscle foods, 171–179
  role in functional and structural properties of heat-processed meat, 171–172
Sulfur amino acids, mechanisms of beneficial effects, 258–274
  aflatoxin B_1, 261–263,264f
  antioxidative effects, 272,274
  browning prevention, 266,269–271
  detoxification reactions, 272,273f
  lectins, 261
  lysinoalanine, 263,265–268
  photochemistry of proteins, 269, 271–272
  protease inhibitors, 259–262,262f
  sesquiterpene lactones, 263,265f
  urethane, 263
Sulfur chemiluminescence detector
  comparison to flame photometric and atomic emission detectors, 17,21

Sulfur chemiluminescence detector—*Continued*
  dynamic range, 15
  minimum detectable level, 13t,15
  principle, 13,15
  schematic representation, 13,14f
  selectivity, 15t
  sulfur compounds detection in cooked milk, 25–27
Sulfur compound(s)
  foods
    analysis, 2–3
    antioxidative properties, 4
    artifacts, 3
    formation, 3–4
    functional properties, 4–5
    identification, 1
    medicinal properties, 4–5
    occurrence, 8
    role in off-flavors, 2
    sensory properties, 1–2
    formulas and molecular weights, 9
  wood garlic
    antimicrobial activity, 244–245
    antimicrobial compound identification, 243–244,245f
    experimental materials, 240
    odor significance of volatile compounds, 242–243
    polysulfide formation, 243
    volatile flavor compound isolation and identification, 240–242
  yeast extracts, volatile, *See* Volatile sulfur compounds in yeast extracts
Sulfur compound detection in cooked milk
  chemiluminescence detection, 27–33
  description, 24–25
  flame photometric detector, 25
  sulfur chemiluminescence detector, 25–27
Sulfur-containing aroma volatiles in meat
  compound identified, 182–185
  experimental procedure, 181–182
  inosine 5′-monophosphate, 186

Sulfur-containing aroma volatiles in
  meat—*Continued*
  pH, 184*t*,185–186
Sulfur-containing flavor compounds
  beef
    categories, 50–51
    experimental procedure, 51–53*f*
    heating, 52,54
    injector port temperature, 56,58*f*
    purge temperature, 54–60
    storage, 54,55*f*
    formation from ^{13}C-labeled sugars,
      cysteine, and methionine
    degradation via 1-deoxydiketoses,
      231,233–235
    degradation via 3-deoxyaldoketose,
      227–232
    experimental description, 224–225
    [^{13}C]glucose–cysteine–methionine
      reactions, 225–227
  identification in meat, 138
Sulfur-containing heterocyclic
  compounds with antioxidative activity
  formed in Maillard reaction model
  systems
    antioxidative activity
      1,3-dithiolane, 252,254*f*
      Maillard reaction extracts,
        250–252,253*f*
      measurement, 248–249
      thiazolidine, 252,254*f*
      thiopene derivatives, 252,253–254*f*
    experimental description, 248
    Maillard reaction sample
      preparation, 249
    oxidation product identification
      procedure, 250
    reaction procedure, 250
    reactions with electrophilic oxidants,
      252,255*f*,256
Sulfur volatiles in muskmelon aroma,
  sensory evaluation by GC–
  olfactometry, 36–47
Sulfurous off-flavors, modulation in
  cruciferous vegetable foods, 95–102

T

Taste threshold, determination, 81
Thermal degradation of thiamin
  compounds
    reaction with cysteine, 211,214
    reaction with methionine, 211,217*f*
    experimental procedure, 200–201
    formation pathways
      compounds from reaction with
        cysteine, 211,215–216*f*
      compounds from reaction with
        methionine, 218,220*f*,222*f*
      new flavor compounds, 204,205–207*f*
      new flavor compounds, 201,203*f*,204
    previous studies, 200
    primary degradation compounds,
      201,202*f*
    sensory properties
      compounds from reaction with
        methionine, 218,221*f*
      new flavor compounds, 211,213*t*
    structure elucidation
      compounds from reaction with
        methionine, 211,218,219*f*
      new flavor compounds, 204,208–212
Thermal denaturation, proteins,
  177–179
Thermal generation of flavor, 199
Thermal interactions, volatile compounds
  generated from inosine
  5′-monophosphate and alliin or
  deoxyalliin, 188–197
Thermal treatment, kinetics of hydrogen
  sulfide release during thermal
  treatment, 138–145
Thiamin
  functions, 200
  role as flavor precursor, 199
  thermal degradation, 200–222
Thiazolidine antioxidative activity,
  sulfur-containing heterocyclic
  compounds with antioxidative activity
  formed in Maillard reaction model
  systems, 252,254*f*

Thioglucosides of *Brassica* oilseeds
  antinutritional effects, 107,111
  beneficial effects, 111
  chemical transformation, 116,120–124
  content of canola varieties,
    111,113,114–115f
  experimental description, 107
  toxicity, 111
Thiophene compounds in cooked meat
  factors affecting composition, 278
  structures, 278,279f
Thiophene derivative antioxidative
  activity, sulfur-containing heterocyclic compounds with antioxidative
  activity formed in Maillard reaction
  model systems, 252,253–254f

U

Urethane, mechanism of beneficial
  effects, 263

V

Vegetables, cruciferous, *See* Cruciferous
  vegetables
p-Vinylguaiacol
  L-cysteine and *N*-acetyl-L-cysteine
    effect on formation, 84,85f,86t
  generation, 81,82f
Vitamin B$_1$, *See* Thiamin
Volatile compounds generated from
  inosine 5′-monophosphate–alliin or
  deoxyalliin thermal interactions
  experimental procedure, 189–190
  GC profiles, 192,193f

Volatile compounds generated from
  inosine 5′-monophosphate–alliin or
  deoxyalliin thermal interactions—
  Continued
  identification, 192–196
  properties of model reaction systems,
    190–191
  yields vs. model system, 196–197
Volatile flavor compounds in foods,
  GC detector evaluation, 8–21
Volatile sulfur compound(s) in yeast
  extracts
  aldehyde reactions, 154,155f
  experimental procedure, 148,150
  formation routes, 150
  list, 148–150,152–153
  reducing sugar effect on formation,
    154,157t
  thiamin degradation product formation,
    154,156f
Volatile sulfur compound modulation in
  cruciferous vegetables
  objectionable flavor modulation in
    broccoli, 98,100–102t
  suppression of undesirable sauerkraut
    flavor by caraway seed, 95–99
  targets, 92–93,94f

W

Wood garlic, sulfur compounds, 240–246

Y

Yeast extracts
  factors affecting composition, 147
  flavor components, 147–148,150,151t
  preparation, 147
  volatile sulfur compounds, 147–157

Production: Susan Antigone
Indexing: Janet S. Dodd and Deborah H. Steiner
Acquisition: Rhonda Bitterli
Cover design: Neal Clodfelter

Printed and bound by Maple Press, York, PA

Bestsellers from ACS Books

The ACS Style Guide: A Manual for Authors and Editors
Edited by Janet S. Dodd
264 pp; clothbound ISBN 0–8412–0917–0; paperback ISBN 0–8412–0943–X

The Basics of Technical Communicating
By B. Edward Cain
ACS Professional Reference Book; 198 pp;
clothbound ISBN 0–8412–1451–4; paperback ISBN 0–8412–1452–2

Chemical Activities (student and teacher editions)
By Christie L. Borgford and Lee R. Summerlin
330 pp; spiralbound ISBN 0–8412–1417–4; teacher ed. ISBN 0–8412–1416–6

*Chemical Demonstrations: A Sourcebook for Teachers,
Volumes 1 and 2,* Second Edition
Volume 1 by Lee R. Summerlin and James L. Ealy, Jr.;
Vol. 1, 198 pp; spiralbound ISBN 0–8412–1481–6;
Volume 2 by Lee R. Summerlin, Christie L. Borgford, and Julie B. Ealy
Vol. 2, 234 pp; spiralbound ISBN 0–8412–1535–9

Chemistry and Crime: From Sherlock Holmes to Today's Courtroom
Edited by Samuel M. Gerber
135 pp; clothbound ISBN 0–8412–0784–4; paperback ISBN 0–8412–0785–2

Writing the Laboratory Notebook
By Howard M. Kanare
145 pp; clothbound ISBN 0–8412–0906–5; paperback ISBN 0–8412–0933–2

Developing a Chemical Hygiene Plan
By Jay A. Young, Warren K. Kingsley, and George H. Wahl, Jr.
paperback ISBN 0–8412–1876–5

Introduction to Microwave Sample Preparation: Theory and Practice
Edited by H. M. Kingston and Lois B. Jassie
263 pp; clothbound ISBN 0–8412–1450–6

Principles of Environmental Sampling
Edited by Lawrence H. Keith
ACS Professional Reference Book; 458 pp;
clothbound ISBN 0–8412–1173–6; paperback ISBN 0–8412–1437–9

Biotechnology and Materials Science: Chemistry for the Future
Edited by Mary L. Good (Jacqueline K. Barton, Associate Editor)
135 pp; clothbound ISBN 0–8412–1472–7; paperback ISBN 0–8412–1473–5

For further information and a free catalog of ACS books, contact:
American Chemical Society
Distribution Office, Department 225
1155 16th Street, NW, Washington, DC 20036
Telephone 800–227–5558